Instructor's Guide and Solutions

Modern Trigonometry

Revised Edition

Wooton
Beckenbach
Buchanan
Dolciani

HOUGHTON MIFFLIN COMPANY / BOSTON
Atlanta   Dallas   Geneva, Illinois
Hopewell, New Jersey   Palo Alto

CONTENTS

1976 Impression

Printed in the United States of America

ISBN: 0-395-21688-5

Features of the Textbook

MODERN TRIGONOMETRY, Revised Edition, develops trigonometry in the way best suited for modern-day applications. It is ideally suited for students who are interested in continuing their study of mathematics, since the material is presented in a way that reveals its natural role in present-day calculus and vector analysis courses. At the same time, the wealth of modern practical applications of the subject makes it a suitable course for those students who will not continue their study of mathematics, but who will need the concepts developed here in allied physics, engineering, or technical courses. The book is intended for students with a working knowledge of algebra and plane geometry.

The textbook includes the following features:

1. Emphasis is placed on the <u>logical development of circular functions as periodic functions of a real variable</u>. Circular functions are introduced and covered thoroughly before the trigonometric functions are studied. This order of presentation places the emphasis on functions of numbers rather than functions of angles and greatly increases the student's ability to use what he learns in this course in later courses in calculus and other areas of mathematics.

2. Students are introduced to <u>identities</u> very early in the book, and the presentation clearly emphasizes the importance of these identities in the study of periodic functions.

3. The study of systems of <u>vectors</u>, <u>complex numbers</u>, and <u>2 x 2 matrices</u> provides students with opportunities:

   (a) to learn how these three systems are closely related, by using an <u>ordered-pair approach</u>, and to learn how each of these systems can be used to help learn about the others.

   (b) to learn how trigonometry is involved in many areas of present-day mathematics.

   (c) to use the notion of an <u>isomorphism</u> in informal, natural ways, thus helping to build a sound intuitive foundation upon which a later course in modern algebra can be built.

4. Frequent use is made of the student's <u>intuition</u>, and great care is taken in the exposition to render the ideas clear and meaningful.

5. <u>Mathematical soundness without undue abstractness</u> characterizes the treatment of all topics. Although proofs are of fundamental concern, the formal definition-axiom-theorem format has been avoided.

6. Enrichment features include:

   (a) interesting and informative essays at the ends of chapters to illustrate vividly
       modern-day applications of trigonometry and to serve as powerful motivating
       factors; some of these essays discuss applications involving the use of computers
       for students who have access to an electronic computer that will accept BASIC.

   (b) a discussion of logarithms in Appendix B for those teachers who wish to give
       additional emphasis to computational aspects of trigonometry, or who wish to
       teach or review logarithms in general.

   (c) a discussion of spherical trigonometry in Appendix C.

7. A review of algebraic concepts which are especially pertinent to this course is
   provided in Appendix A.

8. Features of the exercises are the following:

   (a) Ample exercises provide for the development of skills and for a thorough under-
       standing of the applications of these skills in contemporary ways.

   (b) The grading of exercises in A, B, and C levels of difficulty facilitates adapting
       the text to classes of varying ability.

   (c) A number of exercises are of the discovery type.

   (d) Answers to the odd-numbered exercises are bound in the student's text.

9. The list of objectives preceding each chapter, together with chapter subheadings,
   provides an overview of the chapter.

10. Provisions for evaluation include:

    (a) Chapter summaries and chapter tests in each chapter.

    (b) Cumulative review tests, over Chapters 1-3, 4-6, and 7-9.

    (c) A comprehensive test (multiple-choice) over the entire book.

    (d) For ease of review, chapter tests are keyed to sections in the text; for flexibility
        in evaluation, cumulative review tests and the comprehensive test are arranged
        by chapters.

11. All tables needed for the course are included at the back of the book.

12. Supplementary materials include:

    (a) this Instructor's Guide, which contains solutions to all exercises in the text,
        including those that involve proofs.

    (b) Duplicating Masters: Progress Tests.

Basic Philosophy

MODERN TRIGONOMETRY, Revised Edition, helps the student to:

1. Understand the properties of periodic functions and their role in different areas of mathematics.

2. Recognize that the manipulative techniques involved in working with periodic functions depend upon fundamental properties of the functions.

3. Acquire facility in applying mathematical processes.

4. Appreciate the scope and depth of the study of functions.

5. Prepare for modern courses in calculus, vector analysis, complex variables, matrices, and abstract algebra.

6. Perceive the unity of mathematics.

Careful attention is given to the understanding of concepts and the development of skills, both of which are essential in mathematics. This textbook presents sound mathematics and furnishes the student with ample opportunity to strengthen his understanding of this area of mathematics and to develop his ability to apply it in a wide variety of situations.

It is important that students be encouraged to read this textbook carefully. Students tend to regard mathematics texts as collections of exercises. However, one cannot really learn mathematics by simply studying an illustrative example and then practicing on similar problems. The explanation of principles is important and the student should not rely exclusively on the teacher for this, but should try to learn by studying the text. By developing a student's ability and desire to read mathematical text, you equip him with a skill of inestimable value for his future development as an educated person in this age of science.

Also, students should realize that reading mathematics is a different technique than reading a magazine or newspaper. Mathematics deals with precise ideas and logical development. Therefore, the reader must be sure he understands exactly—not vaguely—what is being said and that he can actually fill in omitted details. Consequently, it is desirable to have all students read sections to be discussed at the next class before coming to class. The use of paper and pencil to fill in omitted details of computations and algebraic simplifications while reading the text should be encouraged.

Encourage questions and class discussion. Discourage students from making vague and imprecise statements or seeking meaningless prescriptions of "what do to." By emphasizing self-reliance, verbal precision, a readiness to look for general principles, and a questioning attitude, you will develop your students' mathematical maturity, while you teach the basic structure and skills of mathematics.

In preparing this text, the authors have utilized recent knowledge about the teaching and learning of mathematics. Consequently, they believe that their book is a mathematically sound and pedagogically feasible modern course in trigonometry.

Order of Development

MODERN TRIGONOMETRY, Revised Edition, is basically concerned with the study of periodic functions of real variables. Basic course content, therefore, centers around the circular functions, that is, functions defined by using a unit circle. Trigonometric functions of angles appear in Chapter 5, and, thereafter, both kinds of functions are studied concurrently.

Chapters 1 and 2 contain the heart of the course. Here, students are introduced to the periodic functions cosine and sine, and these are then used to define the remaining four circular functions: tangent, cotangent, secant, and cosecant. The sum and difference formulas and related reduction formulas are developed in these chapters, and students are provided with opportunities to discover identities and to explore the properties of the circular functions.

Graphs of the six circular functions are introduced early in these first two chapters since various properties of the circular functions are useful in drawing their graphs, and their graphs, in turn, assist us in recalling these properties. No angles are used in these two chapters except in one isolated case where the properties of $30°$-$60°$-$90°$ and $45°$-$45°$-$90°$ triangles are used to find values for cos x and sin x for x equal to $\frac{\pi}{6}$, $\frac{\pi}{4}$, and $\frac{\pi}{3}$. Even this exposure to angles can be avoided if desired by using the techniques set forth in this manual on pages 7-8. Thus, the functions studied are real functions of real variables, and they are the periodic functions of greatest importance in modern mathematics.

Chapter 3 deals with the effect of parameters (constants) on the graphs of the circular functions. Also, two sections are devoted to applications of circular functions to uniform circular motion and simple harmonic motion. These are clear applications of circular, as opposed to trigonometric, functions and offer the students an introduction to ideas which they will deal with in detail in physics courses.

Chapter 4 is concerned with inverse circular functions and the related matter of solving open sentences involving circular functions.

Chapter 5 is a comprehensive discussion of trigonometric functions and applications of these functions to solving right and oblique triangles.

Chapters 6, 7, and 8 contain a wealth of applications of circular and trigonometric functions to three very important branches of mathematics: vectors, complex numbers, and matrices. The discussion in Chapters 6 and 7 is organized around the ordered-pair concept; this presents a wonderful opportunity for students to see the power and usefulness of such an approach. They will also learn to appreciate the role the circular and trigonometric functions can play in studying the structure of systems whose elements are ordered pairs. Intuitive geometric notions are freely employed so that students always have before them a concrete model of the essentially abstract system they are exploring.

Chapter 9 exposes the student to the usefulness of infinite series, and, in particular, to the way in which power series can be used to represent transcendental functions.

Appendix A, Sets, Relations, and Functions, is a review of preliminary algebraic topics. This unit can be taught in its entirety early in the course (possibly prior to Chapter 1) or it can serve as a reference section. Included in this guide (pp. 46-49) are exercise sets which may be useful in teaching this material.

Planning Your Course

This book provides considerable flexibility in designing a one-semester course in trigonometry. Three particular courses are suggested. For each of these courses, the number of lessons per chapter is suggested in the table at the end of this section. A one-week period at the end of each course is included for review and testing. The time schedules should be modified as necessary according to the mathematical background, ability, and interests of the students in the class.

I. Basic Course. Includes the nine chapters, and excludes the appendices.

II. Course for exceptionally well prepared students. Includes the nine chapters and spherical trigonometry.

III. Course for less well prepared students. Includes a review (Appendix A) at the beginning of the course, Chapters 1-7, and the first three sections of Chapter 8, Matrices. Excludes series, logarithms, and spherical trigonometry.

Numerous variations of these courses are possible. For example:

In Course I, substitute Appendix B (logarithms) for Chapter 9 (series).

In Courses I or II, include two or three days on Appendix A early in the course.

In Course III, substitute Appendix B (logarithms) for Chapter 8 (matrices).

Suggested 90-Day Time Schedule (Including Testing)

Chapter / Course	App. A	1	2	3	4	5	6	7	8	9	App. B	App. C	Review	Total
I	0	13	8	8	8	13	11	8	8	8	0	0	5	90
II	0	13	7	7	8	12	11	8	7	8	0	4	5	90
III	5	15	8	8	10	14	11	10	4	0	0	0	5	90

Teaching Suggestions

CHAPTER 1

Originally trigonometry was developed as a set of relationships and procedures specifically designed for solving problems concerning sides and angles of triangles.

The functions that arose from this development are called <u>trigonometric</u> functions. Studying
trigonometry strictly along the lines of its historical development, however, promotes initial
concepts that are too limited for present-day needs; therefore, this chapter approaches the
subject in a generalized way, defining the circular functions cosine and sine in terms of a unit
circle in a coordinate plane. The process of winding a real number line around the unit circle
assigns a real number x to each point P on the unit circle. Each point P has two rectangular
coordinates, u and v. It is this point, P(u, v), which is used to define the circular functions
cos and sin.

    <u>1-1 (p. 2, par. 1)</u>   The periodic function f defined here can be obtained by partitioning
the integers into residue classes mod 5. Dividing each integer x by 5 produces a quotient q
and a remainder r; that is, x = 5q + r. It is the remainder, or residue, r, which is
associated with an integer x, forming the set of ordered pairs $\{(x, r)\}$. It may be of interest
to note that this function, or, indeed, any periodic function, can be interpreted schematically
using a circle, with the circumference corresponding to the fundamental period of the function.
As shown in the figure at the right, the function f described above
can be illustrated by regarding the real number line as being
wound around a circle of circumference 5.

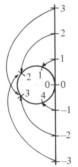

    <u>(p. 4)</u>   Even though they are C exercises, a discussion of
the example on page 4 and possibly one of the exercises 9-12
would be helpful to the entire class.   Such a discussion can
also help make Exercises 7 and 8 more meaningful.

    <u>1-2</u>   The variables u and v are deliberately used in
discussing the unit circle so that the variable x used to denote arc length will carry over
naturally into the xy-plane when graphs of circular functions are introduced.   The variables
u and v are replaced immediately with y so that the equations y = cos x and y = sin x become
familiar to the students early in the course.   One of the important objectives of this development is to suggest to the student that the sine and cosine functions are real-valued functions
of a real variable.   Therefore, it is helpful if students can think of cos x and sin x (rather
than x and y) as the abscissa and ordinate, respectively, of a point on the unit circle in the
Cartesian coordinate plane.

    Students can examine the winding process described in this section by drawing a unit
circle with a one-inch radius and wrapping a strip of paper around it.   If a scale graduated
in inches is drawn on the strip, the winding process produces a set of ordered pairs (x, P),
where x is a real number corresponding to a point on the strip and P is a point on the unit
circle.   The circle and paper strip provide a physical model of Figure 1-3, page 45 with the
paper strip representing the tangent line.

If desired, the process of winding the x number line onto the unit circle with center at the origin of the Cartesian coordinates system and then deriving the cosine and sine functions from this can be viewed as a composition of functions. In the first pairing, each real number x is associated with the coordinates (u, v) of a point P on the unit circle. Thus, the domain of the function, call it g, is $\mathcal{R}$, and the range is the set of all coordinates of points P on the unit circle. The function

$$g = \{(x, (u, v))\}$$

is periodic, for each real number x is mapped into the same point as $x + 2k\pi$ for every $k \in J$. In the second pairing, call if f_1 or f_2, the coordinates (u, v) of each point P on the unit circle are paired with either the first (in f_1) or second (in f_2) component of the ordered pair (u, v). Then the function cosine is just the composite function $f_1 \circ g$, and the function sine is the composite function $f_2 \circ g$. Note that in this composition f_1 and f_2 are not periodic; the periodicity of the circular functions is inherent in the winding function g.

(p. 7, par. 1) Emphasize that the symbols cos and sin denote the functions cosine and sine. The symbols cos x and sin x are then special cases of the more general notation f(x), except that in the case of cos x and sin x, the parentheses are omitted. Just as f(x) denotes a value of a function f, so do cos x and sin x denote values of the functions cos and sin.

(p. 7, par. 5) The coordinate axes separate the Cartesian plane into five subsets: the sets of points in the four quadrants and the set of points lying on the axes. Values of cos x or sin x, such as 0, 1, -1, that are associated with points belonging to the intersection of the axes and the unit circle [i.e., (1, 0), (0, 1), (-1, 0), (0, -1)] are called quadrantal values of the function cos or the function sin. Values of x such that cos x or sin x are quadrantal values of cos or sin are called quadrantal values of the variable x over the real numbers. It should always be clear from context whether "quadrantal values" refers to elements in the domain or elements in the range of a function.

1-3 (pp. 10-11) The method used to determine the quadrantal values illustrates the fact that the definitions of sin and cos are independent of the more restrictive right-triangle concept for defining these functions. This point is worth making, particularly since many students have seen these functions defined by means of right triangles in earlier courses. The use of degree measure and a right triangle for finding the circular function value for $x = \frac{\pi}{4}$ and then again for $x = \frac{\pi}{6}$ and $x = \frac{\pi}{3}$ is not a departure from the coordinate-point development, but rather an expedient device for establishing the coordinates of B as $(\frac{\sqrt{2}}{2}, \frac{\sqrt{2}}{2})$ (Figure 1-8) and then later as $(\frac{\sqrt{3}}{2}, \frac{1}{2})$ and C as $(\frac{1}{2}, \frac{\sqrt{3}}{2})$ (Figure 1-9). Students are familiar with the right triangles 45°-45° and 30°-60°, and these lend themselves conveniently to the discussion here. You can suggest that these triangles are just as useful when referred to as the $\frac{\pi}{4} - \frac{\pi}{4} - \frac{\pi}{2}$ and $\frac{\pi}{6} - \frac{\pi}{3} - \frac{\pi}{2}$ triangles.

There is an alternative method of finding function values for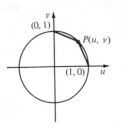
$\frac{\pi}{4}$, $\frac{\pi}{6}$, and $\frac{\pi}{3}$ which is independent of triangle considerations and
which depends on the distance formula.   For example, to find
$\cos \frac{\pi}{4}$ and $\sin \frac{\pi}{4}$, you can observe that the point P(u, v)
associated with $\frac{\pi}{4}$ is equidistant from the points (0, 1) and
(1, 0).   Hence, $\sqrt{(u - 1)^2 + (v - 0)^2} = \sqrt{(u - 0)^2 + (v - 1)^2}$,
and, after squaring and simplifying both members, you have u = v.   Then, since $u^2 + v^2 = 1$,
you can substitute u = v in $u^2 + v^2 = 1$ to obtain $u^2 + u^2 = 1$, or $2u^2 = 1$, or $u = \pm\frac{1}{\sqrt{2}} = v$.
Thus, since u > 0, $\cos \frac{\pi}{4} = \frac{1}{\sqrt{2}}$ and $\sin \frac{\pi}{4} = \frac{1}{\sqrt{2}}$.

To find values for $\cos \frac{\pi}{6}$ and $\sin \frac{\pi}{6}$, you can observe that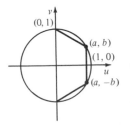
the distance from the point (a, b) associated with $x = \frac{\pi}{6}$ to the
point with coordinates (a, -b) is the same as the distance from
(a, b) to (0, 1).   Then
$$\sqrt{(a - 0)^2 + (b - 1)^2} = \sqrt{(a - a)^2 + (b + b)^2},$$
or $a^2 + b^2 - 2b + 1 = 4b^2$.
Since $a^2 + b^2 = 1$, you have $2b^2 + b - 1 = 0$, $(2b - 1)(b + 1) = 0$.   Then either $b = \frac{1}{2}$ or
b = -1, and since b > 0, you choose $\frac{1}{2}$.   From $a^2 + b^2 = 1$, you find that $a = \frac{\sqrt{3}}{2}$, so that
$$\cos \frac{\pi}{6} = \frac{\sqrt{3}}{2} \text{ and } \sin \frac{\pi}{6} = \frac{1}{2}.$$

Values for $\cos \frac{\pi}{3}$ and $\sin \frac{\pi}{3}$ can be obtained by using the symmetry of the circle about the
line with equation y = x.

(pp. 11-13)   Note how the example and accompanying diagram on page 13 tie in with
paragraph 2 on page 11; these symmetries lead to the table of circular-function values for
the particularly useful numbers 0, $\frac{\pi}{6}$, $\frac{\pi}{4}$, $\frac{\pi}{3}$, $\frac{\pi}{2}$, and so forth, as displayed in Figure 1-10
and on page 12.   Emphasize the necessity and the advantages of having these values well in
mind.

1-4 (p. 15, par. 1)   Most students will have had a considerable amount of experience
with algebraic computations involving real numbers, and they will have found the distributive
property very effective.   It is not surprising, then, that some students will tend to think
that $\cos (x_2 - x_1)$ is equal to $\cos x_2 - \cos x_1$.   This incorrect assumption is due to confusing
the symbol cos x with a product.   Strong emphasis on the fact that cos x and sin x denote
elements in the range of the functions cosine and sine will help to minimize this confusion.

The development of the highly useful $\cos (x_2 - x_1)$ identity is dependent upon the
properties of real numbers.   Emphasizing that the winding process (Figures 1-11 and 1-12)
associates points A', B', C', and so forth, with real numbers will enable the student to justify
each step of the derivation as being a real-number manipulation justified by the properties of
the real-number field.

(pp. 17-19)   The identities (2) through (6) are all easily developed from the difference identity for cos $(x_2 - x_1)$, i.e., identity (1), and consequently are all corollaries of (1).   Also, Figures 1-13, 1-14, and 1-15 help the students keep an intuitive concept of the analytical proofs in the foreground by allowing them to picture the geometric properties of the unit circle.

1-5   Now that our study of trigonometry has had a brief beginning (Sections 1-1 through 1-4), it is timely for us to turn our attention to some of the algebraic concepts which are so deeply embedded in trigonometry.   In this section, we therefore examine the general concept of an identity, with examples from algebra.

It is recommended that references to appropriate algebraic techniques and identities be made throughout the course.   Notice especially Exercise 7 on page 26.

(p. 23, par. 1)   Here we have a definition of the term "identity."   This may be an appropriate time to refer your students to the discussion of open sentences, in general, in Appendix A, Section A-2, page 359.

(pp. 24-25)   A distinction is made between deriving identities (middle of page 24) and verifying identities (top of page 25).   Examples and suggestions illustrate these two processes.

This section should be viewed in relation to several other closely related sections of the textbook: Additional basic identities are developed in Sections 1-6, 1-8, and 2-4; then Section 2-5 consists of a follow up on identities by offering further suggestions for deriving and verifying identities.

1-6   Once more the power of the sum and difference identities is shown by their use as the basis for the derivation of the reduction identities for cos and sin.   Figures 1-16 and 1-17 permit the student to examine the geometric properties along with the manipulative derivation of the reduction identities.   In this way he can recognize that for any nonquadrantal value x, there is a related Quadrant I value such that the circular-function value at x is equal to, or the negative of, a corresponding circular-function value in Quadrant I.

1-7   One way to develop the graphs of cos and sin is to plot the points corresponding to special values of x in the interval $0 \le x \le 2\pi$, to connect the points with a smooth curve, and then to extend the graphs to domain R using the property of periodicity.   This method is related to the method of "unwinding" described on page 35, which illustrates the geometric significance of replacing the variables u and v, respectively, with y to define cos and sin, as discussed in Section 1-2.

However, the method which this book emphasizes (pp. 32-34) for developing the graphs of cos and sin is to plot the points corresponding to special values of x in the interval $0 \le x \le \frac{\pi}{2}$, to connect these points with a smooth curve, and then to extend the graphs as follows: (1) to the domain $0 \le x \le \pi$, using the properties cos $(\pi - a)$ = -cos a and sin $(\pi - a)$ = sin a;   (2) to the domain $0 \le x \le 2\pi$, using the properties

cos (a + π) = -cos a and sin (a + π) = -sin a; (3) to the domain R, using the property
of periodicity.

This development utilizes the properties stated in (1), (2), and (3), above, and, as
should be strongly emphasized, the graphs of cos and sin, in turn, assist us in recalling these,
and other, properties of cos and sin.   Note that Exercises 1-7 on page 36 are designed to help
the student relate various properties of cos and sin to their graphs.

When graphing cos and sin over an interval whose left-hand endpoint is a negative number,
the identities cos (-x) = cos x and sin (-x) = -sin x are useful.   Emphasize that cos is an even
function (its graph is symmetric with respect to the y-axis) and that sin is an odd function (its
graph is symmetric with respect to the origin).

(p. 32, par. 3)   In the absence of limit processes, it is necessary to depend on the
students' intuition for justifying the continuity of the cosine and sine functions.   For a
discussion of this continuity, see Dolciani, Beckenbach, Donnelly, Jurgensen, Wooton,
Modern Introductory Analysis (Boston, Houghton Mifflin Co., 1970), pp. 445-449.

(p. 33, par. 3, and p. 34)   In Section 1-1, the student learned that a periodic function
has the property of repeating a pattern.   This discussion should make it clear that when
graphing a periodic function, it need be done over only one period to ascertain the pattern.
If an extension of the graph is required, the student has only to duplicate his graph over the
first period in either direction.   The sequence of rectangles suggested in Figure 1-23 is a
useful device for obtaining a mental picture of periodicity.

(p. 35, Figures 1-27 and 1-28)   The student should be aware of the fact that when the
cosine and sine functions were defined by means of the unit circle, the domains of these
functions were the set of lengths along arcs of the circle and the ranges were the set of distances from the points on the circle to the vertical axis for cos and the set of distances from
the same points to the horizontal axis for sin.   Applying this type of geometrical interpretation
to the Cartesian graphs of cos and sin, as in Figures 1-27 and 1-28, we observe that the
domain of cos is the set of linear distances from the points on the graph to the vertical axis and
the range is the set of linear distances from the points on the graph to the horizontal axis.
Similarly for sin.   The scales on the vertical and horizontal axes are uniform to preserve the
shape of the graph.   If students use different scales, they should understand that the graphs
they produce will be distortions of the uniform-scale graphs.

(p. 35, pars. 4-7)   The terms in boldface type here are elaborated upon further in
Chapter 3.   Note that the terms "sine waves," "cycle," and "amplitude," may be applied
either to the function or to its graph.

(p. 37, Exercises 7-12)   Exercises of this sort point out a valuable way in which the
graphs of cos and sin can be used, that is, as an aid in solving equations of the form

cos x = c or sin x = c, -1 ≤ c ≤ 1.   Students should learn to produce sketches of cos and sin, together with horizontal lines, as in these exercises.   You may find numerous uses of a similar nature for quick sketches of the graphs of the circular functions throughout the remainder of the course.

1-8   The identities derived here are again corollaries of the sum and difference identities.   This suggests that a good student could, by means of deductive reasoning, derive all of the identities mentioned thus far.   These derivations, re-enforced with exercises, usually provide enough familiarization so that a student will be able to recall most of them as needed. The cos 2x and sin 2x identities are commonly referred to as the double-angle identities, the sin $\frac{x}{2}$ and cos $\frac{x}{2}$ identities are usually called the half-angle identities, as is pointed out in Chapter 5, page 152.   Notice that the half-angle identities (7) and (8) are listed with alternate signs.   They have been listed separately to indicate to the student that the proper sign is to be determined by the quadrantal position of $\frac{x}{2}$.

1-9   This section introduces students to tables of values for the circular functions. Since no immediate use will be made of the tables, the primary purpose of the section is to acquaint the students with the general nature of tables of function values.

(p. 41, par. 1)   Stress the fact that the identities

$$\sin\left(\tfrac{\pi}{2} - x\right) = \cos x \quad \text{and} \quad \cos\left(\tfrac{\pi}{2} - x\right) = \sin x$$

permit using just one column for both sin x and cos x, one reading from top to bottom, and the other from bottom to top.

(p. 41, par. 5)   Make sure students understand why the table shown here can be "folded." Because the entries for values of sine and cosine are symmetric with respect to $\frac{\pi}{4}$, one need only list function values over the interval $0 \le x \le \frac{\pi}{4}$.   Not all tables of circular functions are made in this way.   For example, Table 3 on page 397 which lists radian measures at intervals of 0.01^R, can not be "folded" because there is no entry in the table for $\frac{\pi}{4}$, and hence no columnar symmetry for the function values.

(p. 41, par. 6)   Point out to students that function values listed in Table 1 are only approximations to the true values.   This is because almost all of the true values are irrational numbers for which finite decimal numerals do not exist.   This also accounts for the frequent use of the symbol $\doteq$ in the text and examples here and later in the book—the purpose being to keep the students constantly aware of when they are working with approximations and when they are not.

(p. 46) Computer Investigations   This is the first of five sections on computer investigations for those students and teachers who have access to a computer that will accept BASIC programs.

Because many of the programs were designed to print out graphs of various trigonometric functions discussed in the text, a few general comments about these graphs might prove helpful to the user.

It should be emphasized that these graphs are only approximations to the actual graphs. Certain limitations in terminal operation make exact reproduction of the graphs impossible.

Functions such as TAB, which are responsible for the spacing involved in printing values on the graph, must round off calculations. This is the result of the fact that the terminal must print characters in fixed spaces across the printout line. In addition, computers have limitations on the number of decimal places which may be used in calculations; therefore, round-offs are frequently necessary.

You will notice that the x-axis is printed vertically and the y-axis horizontally in these programs. This is necessary in order that x be used as the independent variable. Because the computer prints information one horizontal line at a time, given a particular value of y which is being printed, all values x associated with that value of y would have to be known before printing. This would require a far more complex program in each instance.

(p. 47)   Notice the flattening of the curve in some areas. As noted earlier, this is the result of the TAB function rounding off results. The terminal cannot print a character in a fraction of a space.

The range of values for x was chosen so that two "cycles" of the function would appear: -6.6 is approximately equal to -2π; 6.6 is approximately equal to 2π. However, because most computers work all computations in binary form (powers of 2), using the values -6.6 and 6.6 and "steps" of .2 results in slight computational errors. Using statement 60 as written (steps of 2) prevents these errors. Values for x are then multiplied by .1 in statement 70.

By changing values in statement 60, the range of values may be varied. This might be helpful when terminal usage is heavy because the programs, as they appear, take a considerable amount of terminal time to execute.

At the top of the next page is the output for the program when the indicated changes are made.

CHAPTER 2

The four remaining circular functions are introduced in this chapter; they are defined in terms of the cosine and sine functions. The fact that the six circular functions are actually three pairs of reciprocal functions is emphasized and utilized to advantage in studying the properties of the four new functions. The graphs of sec, csc, and cot are derived with ease by relating the graph of each one to the graph of its corresponding reciprocal function.

The remainder of the chapter is concerned with further development of techniques for deriving and verifying identities.

```
10 PRINT "Y = COS(X+P1)"
60 FOR X1=-32 TO 32 STEP 2
75 LET P1=3.14159
80 LET Y=COS(X+P1)
RUN
```

Y = COS(X+P1)

X	Y
-3.2	.998295
-3	.989193
-2.8	.942223
-2.6	.85689
-2.4	.737395
-2.2	.588503
-2	.416149
-1.8	.227205
-1.6	2.92018E-02
-1.4	-.169965
-1.2	-.362356
-1	-.540301
-.8	-.696706
-.6	-.825334
-.4	-.92106
-.2	-.980066
0	-1.
.2	-.980067
.4	-.921062
.6	-.825337
.8	-.696708
1	-.540305
1.2	-.362361
1.4	-.16997
1.6	2.91969E-02
1.8	.2272
2	.416145
2.2	.588498
2.4	.737393
2.6	.856888
2.8	.942221
3	.989992
3.2	.998295

---Y

X

END

2-1   The tangent function is defined in terms of the sine and the cosine functions; tan x is the ratio of sin x to cos x.   The two basic properties of tan are developed, and a comparison with corresponding properties of cos and sin can be helpful.   For example, tan and sin are both odd functions, and tan is periodic, although its fundamental period is π rather than 2π.

To develop the graph of tan, we begin somewhat as we did with the graphs of cos and sin; that is, we first graph y = tan x in the interval $0 \le x < \frac{\pi}{2}$.   Then we use the identity tan (-x) = -tan x to extend the domain to $-\frac{\pi}{2} < x < \frac{\pi}{2}$.   At this point, we note that tan has fundamental period π (see page 52), so that we can extend its graph by reproducing in either direction along the x-axis the portion of the graph over the interval $-\frac{\pi}{2} < x < \frac{\pi}{2}$.   Thus the graph of tan displays the two most basic properties of the tan function: that it is an odd function, and that

it is periodic, with fundamental period π.   Keep in mind that these two properties of tan
suffice as reduction formulas (see p. 52, last paragraph, and p. 53).   Moreover, sketches of
the graph of tan are important for illustrating geometrically other identities (see Exercises
13-16, p. 54).

(p. 51)   You may wish to discuss the concept of an asymptote in more detail, especially
if your students have not encountered this topic before.

(p. 55, Exercises 18-20)   See the remarks in this manual regarding Exercises 7-12
on page 37.

2-2   This section introduces the three remaining circular functions secant, cosecant,
and cotangent, and discusses the domain, range, and properties of even-odd and periodicity
for these functions.   (Accordingly, in Exercise 12, page 59, the student is asked to compare
the domain and range of all six circular functions.)   Since these three new functions, as well
as tan, are defined in terms of cos and sin, the student need only remember the zeros of cos
and sin to determine restrictions on their domains.

2-3   In this section, we graph sec, csc, and cot by making use of the fact that sec and
cos, csc and sin, and cot and tan are each pairs of reciprocal functions.

We note that all three of these pairs of functions have certain characteristics in common.
We list these characteristics below in terms of a general definition, which you may wish to
discuss with the class:

Definition: Two functions f and g are said to be a pair of <u>reciprocal functions</u> if the following
conditions are met:

1. If $x \in Dom_f \cap Dom_g$, then $f(x) g(x) = 1$
2. If $x \in Dom_f$ and $x \notin Dom_g$, then $f(x) = 0$, and
   if $x \in Dom_g$ and $x \notin Dom_f$, then $g(x) = 0$.

It can be helpful to graph various pairs of algebraic functions, for example,

$f(x) = x^2$, $x \in R$, and $g(x) = \dfrac{1}{x^2}$, $x \in R$, $x \neq 0$;

$f(x) = x - 1$, $x \in R$, and $g(x) = \dfrac{1}{x - 1}$, $x \in R$, $x \neq 1$;

$f(x) = 2^x$, $x \in R$, and $g(x) = 2^{-x}$, $x \in R$.

The functions sec and csc are defined directly as reciprocals of cos and sin, respectively.   But cot is not defined by $\cot x = \dfrac{1}{\tan x}$, for if it were, then the domain of cot would
<u>exclude</u> all numbers of the form $\dfrac{\pi}{2} + k\pi$, $k \in J$ (that is, the numbers for which tan is
undefined).   We could, of course, define cot by

$$\cot x = \begin{cases} \dfrac{1}{\tan x} & \text{if } x \neq \dfrac{\pi k}{2} \\[2ex] 0 & \text{if } x = \dfrac{\pi}{2} + k\pi \end{cases} \quad k \in J$$

but it is simpler to define cot by $\cot x = \frac{\cos x}{\sin x}$. It follows that the question "For what real numbers x is $\cot x = \frac{1}{\tan x}$?" may lead to a very interesting class discussion. The answer is $\{x : x \in R, \ x \neq \frac{\pi}{2}k\}$

2-4   Notice that identities (2) - (5) are all special cases of identity (1).

You will probably find this to be an excellent time to review very carefully the collection of identities listed on page 66. This activity can include recalling the proofs of some of the identities, relating various identities to each other, relating appropriate identities to the unit circle, and recalling how certain identities can be remembered quite easily by means of the graph of cos, sin, or tan (as, for example, Exercises 1-6 on page 36 and Exercises 13-16 on page 54). It is helpful to keep in mind which identities are special cases of other identities in the list and precisely how they are special cases.

2-5   This section continues the study of identities. Students should be alerted to the fact that there is no unique method for proving an identity, but that some methods are more direct than others.

(p. 68, par. 2)   Emphasize this paragraph. Some students are prone to "symbol pushing" when proving identities and often show little or no regard for the validity of whatever operations or substitutions they may be performing. By insisting that students specifically state restrictions on the domains of the variables involved in identities, you will enable them to make more meaningful applications of identities in future work.

(p. 70, bottom of page)   These five suggestions for proving identities can be very useful. You may wish to refer to them frequently in class.

(p. 72, Exercises 24-29)   The results stated in Exercises 24-29 should be noted by all students, even though the difficulty of their proofs results in their "C" rating. The significance of these identities is that they permit the transformation of sums and differences (of circular functions) into products. This property is important in certain topics in calculus.

(p. 74)   This is the output for the program after the suggested changes have been made.

```
10 PRINT "Y = COT X"
60 FOR X1=6 TO 26 STEP 2
80 LET Y=1/TAN(X)
RUN
```

Y = COT X

```
    X       Y
   .6     1.4617                              !              *
   .8      .971215                            !         *
   1       .642093                            !      *
   1.2     .388779                            !   *
   1.4     .172477                            !*
   1.6    -.029212                          * !
   1.8    -.233303                       *    !
   2      -.457658                    *       !
   2.2    -.727896              *             !
   2.4    -1.09169          *                 !
   2.6    -1.66224      *                     !
 END                                          X
```

(p. 75)   Notice that the order from left to right of the functions to be printed must be
considered in advance so that the correct PRINT statements can be included.   In the case of
cos x and sec x, this order depends on the choice of values for x.   In the first example of
output, the values of cos x always appear to the left (less, actually) of the values for sec x,
while the reverse is true in the second example.

CHAPTER 3

This chapter investigates the roles of the parameters, or constants, a, b, c, and d,
in the equation

$$y = a \sin b(x - c) + d,$$

using the ideas of cycle, period, amplitude, and phase shift.   Rather than confronting the
student all at once with the entire family of functions defined by $y = a \sin b(x - c) + d$, the
investigation is conducted in stages, in which each constant is explored individually.

The ideas of graphing which are developed for the sine function can be applied similarly
to the five other circular functions.   Inherent in this development are applications of the
concept of a translation.   Graphing by addition of ordinates follows.

Having gained skills in making rapid, meaningful curve sketches, the student is ready
to investigate examples of uniform circular motion and simple harmonic motion—two important
modern applications of circular functions in the physical sciences.

3-1   For continuity of presentation, the examples and principal discussions in Sections
3-1, 3-2, and part of 3-3 are concerned with only one circular function (and accompanying
families of functions), namely sine.   There exists, therefore, the need for class discussions
of graphs involving the other circular functions, both before and after assignment of the
exercises in each section.

(p. 80, pars. 1-2)   The objectives for Sections 3-1, 3-2, and 3-3 are presented.

(p. 80, pars. 3-5, and p. 81, par. 1)   The emphasis is on the fact that the equation
$y = a \sin x$ describes a family—that is, a set—of functions.   The symbols x and y are
variables, whereas the symbol a plays a role different from that played by x and y.   This
role may be better expressed by referring to a as a parameter rather than as a constant.
The important point is that for every real number a, except 0, we obtain a particular member
of the family $y = a \sin x$.

Keep in mind that the preceding remarks apply also to discussions of other families of
functions in Sections 3-1, 3-2, and 3-3.

You may find it interesting, as a side adventure, to regard the family of functions
$y = a \sin x$, $a \in R$, together with the operation of addition, as a mathematical system.   The
sum of $y = a_1 \sin x$ and $y = a_2 \sin x$ is defined to be $y = (a_1 + a_2) \sin x$.   The properties of

closure, associativity, and commutativity are easily proved. In order for the system to be a commutative group, it is necessary to include y = 0 · sin x, or simply y = 0, which becomes the additive identity element. Then the additive inverse of any element y = a sin x exists and is y = (-a) sin x. A similar discussion involving the family y = a sin x + d can be held.

(p. 81, bottom half of page)   Emphasize that the equation y = sin x + d defines a family of functions. The notion of a translation is inherent in graphing members of this family.

(p. 82)   Each of the families y = a sin x and y = sin x + d may be referred to as a "subfamily" of the family y = a sin x + d.

3-2   Keep in mind that many of the remarks in this guide for Section 3-1 apply also to Sections 3-2 and 3-3.

A function defined by either y = sin bx, b ≠ 0, or y = sin (x - c) can be thought of as the composition of two functions. For example, if h(x) = sin 3x, then h(x) = f[g(x)], where g(x) = 3x and f(x) = sin x; also, if j(x) = sin (x - 1), then j(x) = f[t(x)], where t(x) = x - 1 and f(x) = sin x.

As is true of Section 3-1, this section offers an opportune time for the students to discover many ideas about graphing. The three main generalizations of this section are in the following locations:

1. bottom of page 84 and top of page 85;

2. middle of page 86;

3. top of page 87.

(p. 83, par. 2)   A clear understanding of this example, and similar examples, can help immeasurably toward grasping the generalization stated at the bottom of page 84. Choose at least one example with a negative value for b.

(p. 85, pars. 2-4, and p. 86)   Students often find translations especially exciting. Spending a liberal amount of time on this topic is time well spent as preparation for, or as a follow-up to, applications of translations in other areas of mathematics.

The graph of each member of the family y = sin b(x - c) can be made to coincide with a corresponding member of the family y = cos b(x - c) by means of a translation. For example, if the graph of y = sin x is translated to the left $\frac{\pi}{2}$ units, it coincides with the graph of y = cos x; thus y = cos x and y = sin (x + $\frac{\pi}{2}$) have the same graph. This means, in terms of sets of ordered pairs, that {(x, y): y = cos x} = {(x, y)· y = sin (x + $\frac{\pi}{2}$)}; it may be helpful to illustrate this result by listing several ordered pairs in each set.

(p. 87, par. 1)   To graph an equation of the form y = sin (bx - c), it is usually advantageous first to factor out the b, obtaining y = sin b(x - $\frac{c}{b}$), so that the graph is simply a translation of the graph of y = sin bx; the translation is $\frac{c}{b}$ units to the right if $\frac{c}{b} > 0$ and $|\frac{c}{b}|$ units to the left if $\frac{c}{b} < 0$.

3-3   Keep in mind that many of the remarks in this guide for Sections 3-1 and 3-2 apply to some extent to Section 3-3.

As illustrated by Examples 1 and 2, it is recommended that students show preliminary graphs prior to the final graph.

Since a prime purpose of the graph of a function is simply to present a geometric picture of the function, it is not very often that one needs the precise and accurate graphing of a function obtained by plotting many points.   The chief exceptions are those cases in which values must be read from the graph, as, for example, in vacuum tube charts of various kinds.

As a class activity, you may want to have the students make graphs of assorted variations of cos and sin, similar to the example shown here, by first sketching the curve;

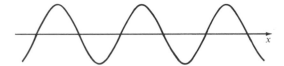

then locating the y-axis;

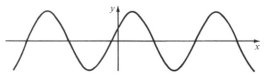

and then establishing a scale.

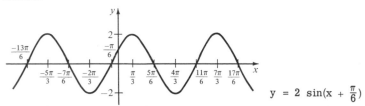

$y = 2 \sin(x + \frac{\pi}{6})$

Graphing a selected group of functions in this manner will reinforce the students' understanding of the role of the parameters a, b, c, and d in the equation y = a sin b(x - c) + d, so that they should be able to make sketches of these graphs with reasonable ease and rapidity.

3-4   The graph of the sum of the cos and sin functions is shown in Example 1.   The student should be able to extend the method developed in this section to include the sums or differences of nonfundamental functions, such as that defined by y = 3 cos x + 2 sin 3x, and the combination of a circular function with a nonperiodic function, such as that defined by y = e^x + sin x, even though these are not considered here.

3-5   It is interesting to note that the ideas expressed in this section involve the use of circular functions and not trigonometric functions; nowhere in the discussion is the notion of an angle used.

(p. 95, par. 1)   Uniform circular motion is discussed here as an example of one of the applications of circular functions to the solution of problems that are quite basic in the fields of physics and electrical engineering.   Note that the term "speed" is used here to denote a quantity having magnitude only.   The term "velocity" as used in "rotational velocity" has the same general meaning as speed but it is more restrictive in that it involves, in addition to the numerical value of speed, the direction of the motion.

(p. 97, par. 2)   Since many problems to which these ideas are applied cannot be conveniently stated in terms of a unit circle, it is important for the student to understand that while $d(\overline{MP_1})$ = sin ωt for a unit circle such as the one in Figure 3-11, $d(\overline{NP_2})$ = a sin ωt for a circle of radius a.

3-6   Here, again, the functions involved are circular and not trigonometric functions.

(p. 105)   Below is an example of partial output for y = sin 2x + cos x. The remaining values repeat the curve. The suggested changes to obtain greater printout width have also been made.

```
10 PRINT "Y = SIN 2X + COS X"
80 LET Y=SIN(2*X) + COS(X)
RUN

Y = SIN 2X + COS X

X
 -6.6                                       !   *
 -6.4                                       !        *
 -6.2                                       !              *
 -6                                         !                 *
 -5.8                                       !                      *
 -5.6                                       !                      *
 -5.4                                       !                      *
 -5.2                                       !                *
 -5                                         !          *
 -4.8                                       ! *
 -4.6                                 *     !
 -4.4                           *           !
 -4.2                     *                 !
 -4                   *                     !
 -3.8               *                       !
 -3.6               *                       !
 -3.4                 *                     !
 -3.2                     *                 !
 -3                           *             !
 -2.8                           *   !
 -2.6                               !*
 -2.4                               ! *
 -2.2                               !  *
 -2                                 !   *
 -1.8                               ! *
 -1.6                               !*
 -1.4                          *    !
 -1.2                     *          !
 -1                     *           !
 -.8                    *           !
 -.6                       *  !
 -.4                               ! *
 -.2                               !     *
  0                                         *                 ---Y
END                                 X
```

CHAPTER 4

Given a function f, sometimes we are concerned with the number f(x) in the range for a particular value of x in the domain, and sometimes we are concerned with the number, or numbers, x in the domain which correspond to a particular value of f(x) in the range.   This chapter deals with the second of these two concerns, for the six circular functions, and does so in the two ways:

1. (Sections 4-1, 4-2, 4-3)   We restrict the domains of cos, sin, tan, sec, csc, and cot to suitable domains to obtain the one-to-one functions Cos, Sin, Tan, Sec, Csc, and Cot.   Since these functions are one-to-one, their inverses are also functions, which we denote by Cos^{-1}, Sin^{-1}, Tan^{-1}, Sec^{-1}, Csc^{-1}, and Cot^{-1}.   The properties and graphs of these principal inverse circular functions are investigated.

2. (Sections 4-4 and 4-5)   Procedures are developed for solving open sentences involving circular functions and their inverses.

4-1 (pp. 109-110)   At this time you may wish to review the general notions of inverse relations and functions and to discuss some familiar pairs of inverse algebraic functions, such as:

$f(x) = x^2$, $x \geq 0$ and $f^{-1}(x) = \sqrt{x}$, $x \geq 0$;

$g(x) = \log_{10} x$, $x > 0$, and $g^{-1}(x) = 10^x$, $x \in \mathcal{R}$.

The discussion of the relation $\cos^{-1}$ (note that the c is not capitalized) should bring out the facts that its graph is a reflection of the graph of cos in the line $y = x$ and that it is not a function, since cos is not one-to-one.

(p. 110, last par., and p. 111)   It is important to realize that Cos and cos are different functions, since they have different domains.   This is an appropriate time to point out that two functions f and g are equal, provided: (i) Dom $_f$ = Dom$_g$, and (ii) for every $x \in$ Dom$_f$, $f(x) = g(x)$.

Moreover, the notation Cos^{-1} does not mean $\frac{1}{Cos}$; that is, the -1 is not an exponent.

(p. 112, Example 1)   In each case, we are given a number in the domain of Cos^{-1} and are asked to find the corresponding number in the range of Cos^{-1}.   This is equivalent to being given a number in the range of Cos and being asked to find the corresponding number in the domain of Cos.   It is a good idea to refer to a sketch of $y = Cos^{-1} x$, and sometimes to a sketch of $y = Cos x$ simultaneously, when working problems such as Example 1.

(p. 114, Example 2)   Remarks similar to those stated for Example 1, p. 112, apply here also.

4-2   Here we use the same notion of restricting the domain of a circular function in order to obtain a one-to-one function as was discussed in Section 4-1.

In equation (1), the <u>domain</u> of Tan, $-\frac{\pi}{2} < x < \frac{\pi}{2}$, is stated inside the set-builder braces, whereas in Equation (5) the <u>range</u> of Tan^{-1}, $-\frac{\pi}{2} < y < \frac{\pi}{2}$, is stated. Displaying several ordered pairs of each of these two functions, in conjunction with a close examination of the graph in Figure 4-6 on page 118, can help to make this notation meaningful. Similar remarks apply to each of the following pairs of equations: (2) and (6), (3) and (7), (4) and (8).

Notice that the domain of each of the six principal circular functions, Cos, Sin, Tan, Sec, Csc, and Cot, includes the interval $0 < x < \frac{\pi}{2}$ and one of the intervals $-\frac{\pi}{2} < x < 0$ or $\frac{\pi}{2} < x < \pi$, with variations in the inclusion of endpoints of these intervals.

(p. 120) The generalization which contains Equations (16) and (17) is of paramount importance in the concept of inverse functions. Indeed, this generalization can be used as a definition. The concept of the composition of inverse functions f and g is involved in this generalization, together with the fact that $f \circ g = h$, where h is the <u>identity function</u> which is defined by $h(x) = x$ and which has domain $= \text{Dom}_g$. When applying Equations (16) and (17), it is of crucial importance to state the set of values of x for which each statement is true.

Equations (16) and (17) can be very useful in evaluating expressions involving inverse circular functions. For example, students should evaluate an expression such as Cos $(\text{Cos}^{-1}\ 0.8090)$ using the principle stated here rather than trying to "grind out" the answer by determining that $\text{Cos}^{-1}\ 0.8090 = 0.6283$ and that $\text{Cos}\ 0.6283 = 0.8090$. Students should also be encouraged to understand the meaning of the symbols involved. Thus, they should observe that Cos $(\text{Cos}^{-1}\ 0.8090)$ asks for the Cosine of the number whose Cosine is 0.8090.

4-3 (p. 122, par. 1) These geometrical interpretations give both reason and opportunity for reviewing the definitions of cos and sin (Section 1-2) in terms of the winding process, and the further the students' understanding of the Cos^{-1} and Sin^{-1} functions.

(p. 123, par. 1) The "arc" notation is important; it is encountered in many textbooks and is in general use by many mathematicians.

(p. 123, par. 3—p. 125, Equations (1) and (2)) This discussion may be regarded as an extension of the discussion on page 119 (last paragraph) and page 120. Notice that compositions of circular functions and their principal inverses often yield algebraic expressions, as, for example, in Equations (12) - (14) on pages 119 and 120 and in Equations (1) and (2) on page 125. In evaluating cos $(\text{Sin}^{-1} x)$, for example, do not let the essential emphasis on the proper interval and sign blind the student from the fact that cos $(\text{Sin}^{-1} x)$ asks for the cosine of <u>the</u> number whose Sine is x, and, of course, this is $\sqrt{1 - x^2}$.

4-4 Various common techniques and procedures which are associated with solving open sentences in algebra extend readily to solving open sentences involving circular functions. It might be helpful to refer to Section A-2 of Appendix A at this time.

Since we have restricted the examples in this section to those involving only one circular function, we can substitute a single variable, such as z, for the circular function value to obtain a corresponding algebraic equation.   In Example 1(a), we have z = cos x; in 1(b), z = sin 3x; in Example 2(a), z = cos x, and in 2(b), z = tan x.

Emphasize the paragraph following the solution on page 129.

4-5   This section extends the ideas of the preceding section to solving equations involving two circular functions.

CHAPTER 5

This chapter introduces trigonometric functions, that is, functions having angles rather than real numbers as elements of the domain.   The definitions of the six trigonometric functions relate them to the circular functions by means of radian measure of angles.   The relationship between the circular functions and their trigonometric counterparts guarantees the validity for trigonometric functions of all the identities established thus far for circular functions.   This chapter includes most of the trigonometry of right and oblique triangles.

5-1 (pp. 143-145)   The material on these three pages concerns two principal topics: definitions of an angle and angle measure.

Regarding definitions of an angle, first (p. 143, par. 2) we recall the usual geometrical definition that an angle is the union of two (noncollinear) rays having a common endpoint.   Next (p. 144, par. 1, and p. 145), we state a new definition for an angle, one which includes the undefined (but intuitively clear) notion of a rotation.   This latter definition is the one which is used throughout the remainder of the book.

Regarding angle measure, we define both degree measure and radian measure and relate the two units by the equation $\dfrac{m°(\alpha)}{m^R(\alpha)} = \dfrac{180}{\pi}$, which is often restated simply as $180° = \pi^R$.

We now have the two measure functions, $m°$ and m^R, each having as domain the set of all angles (as defined in terms of a rotation).   The range of each of these functions is $\mathcal{R}$. Note that we write $\alpha = 30°$ to mean $m°(\alpha) = 30$, or $\alpha = \pi^R$ to mean $m^R(\alpha) = \pi$.   It is a matter of notation that we use the symbols α, $m°(\alpha)$, and $m^R(\alpha)$ interchangeably to refer to members of the range of $m°$ or m^R when the intent is clear from the context.

Incidentally, it should be noted that an angle θ such that $m°(\theta) = m^R(\theta) = 0$ is the union of two coincident rays with zero rotation.

5-2 (p. 148, pars. 1-3)   There are various ways in which the trigonometric functions might be defined.   One way would be to choose any point (x, y) other than the origin on the terminal side of an angle α and define the trigonometric functions using ratios, in the manner used in Section 5-5.   Rather than this, we have elected to define the trigonometric functions in terms of the circular functions (with which the students have been working for four chapters) and then (in Sections 5-5 and 5-7) to obtain the familiar ratios as a consequence of the definitions.

(pp. 149-151)   The notion of a reference angle as introduced here is very useful, and students should be encouraged to make quick sketches showing reference angles when working with angles in quadrants other than the first.   The idea of a reference angle α is a simple extension of the idea of a reference arc x, as shown in Figures 1-16 and 1-17, pp. 28, 29.

5-3   This section contains little that is new and is chiefly for the purpose of providing some additional experience in proving identities.   Because of the way in which the trigonometric functions were defined, all previously proven circular-function identities become immediately valid for trigonometric functions, and students should be encouraged to use them as needed when proving the identities in the exercises for this section.

5-4   This section introduces the conventional tables for trigonometric functions. Although Table 1, page 391, is useful for demonstrating why certain tables of circular and trigonometric functions can be "folded" and presented in a form that is readable from bottom to top as well as from top to bottom, it is not widely encountered.   The introduction of the conventional tables was postponed to this point in the book because they were not needed until now.

(p. 154, Example 1)   Stress to students that the "approximately equal to" symbol ($\doteq$) used in Example 1 is necessary because entries in tables are, in general, only rational-number approximations to irrational numbers.   Students should be alerted to the fact that although the symbol $\doteq$ is used consistently in this book whenever such approximations are employed, the equals symbol = is frequently encountered in similar contexts in other books.

(p. 154, par. 1)   By having students examine the differences of a few successive entries in Table 2, you can easily convince them that over any short interval the trigonometric function is nearly linear.   For example, the differences between $\sin 27°00'$, $\sin 27°10'$, $\sin 27°20'$, and $\sin 27°30'$ are 0.0026, 0.0026, and 0.0025, which suggests that for $27°00' \le \theta \le 27°30'$, sine is approximately linear, making the method of linear interpolation feasible.   You may want to discuss Exercises 39-41, page 158, in class; these exercises should be particularly helpful in developing in the student a clearer understanding of linear interpolation and an ability to determine when linear interpolation will provide a good approximation and when it will not.

5-5   Formulas (1) and (2) are fundamental to the application of trigonometric functions. Indeed, it is possible to define the trigonometric functions cosine and sine by using Formulas (1) and (2) and then showing that these definitions are consistent with those made in Section 5-2, which, in turn, relate trigonometric and circular functions.

5-6   This section offers another opportunity for students to become accustomed to working with inverses of periodic functions and to increase their skill in using notation for

such functions.   Both the notation involving the superscript -1 and that involving the prefix
Arc are employed in the exercises so that students can become familiar with both.   Use of
angle terminology in discussing inverses and triangles, such as in the solution to Example 1,
can enhance a student's understanding of some of the ideas presented in Chapter 4.   You may
therefore wish to refer back to portions of Chapter 4 at this time.

5-7   It is likely that most students will have encountered the ideas in this section in
earlier mathematics courses.   In solving problems, primary emphasis should be placed on
finding right triangle relationships and on choosing the most convenient trigonometric function
for the purpose.

(p. 168)   Emphasize to students the convention for approximating solutions stated here:
lengths to the nearest hundredth and angle measures to the nearest minute.

5-8   Students sometimes think that arguments of the kind used to derive the Law of
Cosines in this section are not general enough because the triangle is situated in a particular
position relative to the coordinate axes.   Emphasize that the geometric properties used in the
derivation are independent of the location of the origin and axes of the coordinate system.   It
is wise before going through the actual derivation to outline for students the general method
of attack.

(p. 170, last par.)   To verify that students understand the derivation of the Law of
Cosines, you may wish to have them derive independently formula (2) or formula (3), or both,
on this page.

(p. 171, par. 1, below solution)   Emphasize this paragraph.   Students should not only
be able to apply the Law of Cosines when necessary, but should also know when it can be applied.

5-9   This section includes a standard development of the formula for the area of a
triangle in terms of two sides and the included angle, followed by the derivation of the Law of
Sines.   If you desire students to use logarithms in the exercise set in this section, a brief
discussion of the use of logarithms in computations is found in Appendix B, beginning on page
369, and may be covered before beginning Section 5-9.

5-10   Figures 5-19 and 5-20 should be emphasized when discussing this section, since
they help students visualize the possible kinds of triangles with given lengths of sides and
angle measures.

(pp. 181-183)   In these programs, restrictions that can easily be checked mentally have
been omitted.

In Program (1), ASA (p. 181), for instance, the measures of the angles must be positive,
and the sum of the measures of angles A and B must be less than 180.

Below is a solution of Exercise 1 on page 183, giving the output when the values for angle A and side C are fixed at 30 and 5, respectively. In addition, slight changes have been made in the program to enhance the readability of the output and a FOR-NEXT loop has been inserted to test various values from 10° to (170-A)°. The original program could simply have been rerun for each of these values instead.

```
5   PRINT "ANGLE A=30 AND SIDE C=5"
6   PRINT
7   PRINT "ANGLE B","ANGLE C","SIDE A","SIDE B"
10  LET A1=30
30  LET C=5
40  LET P1=3.14159
44  FOR B1=10 TO 170-A1 STEP 10
45  LET C1=180-A1-B1
50  LET A2=A1*P1/180
55  LET B2=B1*P1/180
60  LET C2=C1*P1/180
65  LET A=C*SIN(A2)/SIN(C2)
70  LET B=C*SIN(B2)/SIN(C2)
75  DEF FNR(X)=INT(100*X+.5)/100
80  PRINT FNR(B1),FNR(C1),FNR(A),FNR(B)
86  NEXT B1
90  END


RUN


ANGLE A=30 AND SIDE C=5
```

ANGLE B	ANGLE C	SIDE A	SIDE B
10	140	3.89	1.35
20	130	3.26	2.23
30	120	2.89	2.89
40	110	2.66	3.42
50	100	2.54	3.89
60	90	2.5	4.33
70	80	2.54	4.77
80	70	2.66	5.24
90	60	2.89	5.77
100	50	3.26	6.43
110	40	3.89	7.31
120	30	5	8.66
130	20	7.31	11.2
140	10	14.4	18.51

```
END
```

In Program (2), SAS (p. 182), the Law of Cosines is used in Statements 50 and 75 to calculate side a. The procedure should be clear from reading these two statements.

The calculations for tan B are somewhat more difficult. Since in most BASIC systems the only inverse circular function is the arctangent, it is necessary to find tan B in order to find B. The following statements justify statement 55.

$$\tan B = \frac{\sin B}{\cos B}$$

$$\sin B = \frac{\sin A \cdot B}{A} \qquad \text{(Law of Sines)}$$

$$\cos B = \frac{A^2 + C^2 - B^2}{2AC} \qquad \text{(Law of Cosines)}$$

$$\therefore \tan B = \dfrac{\dfrac{\sin A \cdot B}{A}}{\dfrac{A^2 + C^2 - B^2}{2AC}} = \dfrac{2 \sin A}{A^2 + C^2 - B^2}$$

Since statement 50 assigns S the value of A^2 (again, from the Law of Cosines), the justification for statement 55 is now complete.

Notice the restriction that $B \le C$ in line 10.   When this program is used to solve exercises in Section 5-9, it may be necessary to reletter the figure before solving.

Below is an example of output for Exercise 3, Section 5-9.   Following it is an example using data where the restriction that $B \le C$ has been violated.   Notice the incorrect results.

```
RUN


TWO  SIDES  AND  THE  INCLUDED  ANGLE

SIDE B (CHOOSE B<=C):?18
SIDE C:?40
ANGLE A (DEGREES):?28
SIDE A: 25.55
ANGLE B: 19.32 DEGREES    ANGLE C: 132.68 DEGREES

END

RUN


TWO  SIDES  AND  THE  INCLUDED  ANGLE

SIDE B (CHOOSE B<=C):?40
SIDE C:?18
ANGLE A (DEGREES):?28
SIDE A: 25.55
ANGLE B:-47.32 DEGREES    ANGLE C: 199.32 DEGREES

END
```

At the right is a solution of Exercise 2 on page 183, giving the output for program 2 when values for sides b and c are fixed at 18 and 40, respectively. A FOR-NEXT loop has been included to test values from 10° to 170° for angle A.

```
RUN


SIDE B=18 AND SIDE C=40
```

ANGLE A	SIDE A	ANGLE B	ANGLE C
10	22.49	7.99	162.01
20	23.89	14.93	145.07
30	26.02	20.24	129.76
40	28.65	23.82	116.18
50	31.6	25.87	104.13
60	34.7	26.7	93.3
70	37.84	26.56	83.44
80	40.91	25.67	74.33
90	43.86	24.23	65.77
100	46.63	22.34	57.66
110	49.16	20.13	49.87
120	51.42	17.65	42.35
130	53.38	14.97	35.03
140	55.02	12.14	27.86
150	56.31	9.2	20.8
160	57.25	6.17	13.83
170	57.81	3.1	6.9

```
END
```

In Program (3), SSS (p. 182), the formula for tan $\frac{A}{2}$ and the corresponding one for tan $\frac{B}{2}$ are used.

Below is the output of Program 3 for exercise 7, Section 5-8.

RUN

THREE SIDES

SIDE A:?5
SIDE B:?6
SIDE C:?8
ANGLE A: 38.62 DEGREES   ANGLE B: 48.51 DEGREES   ANGLE C: 92.87 DEGREES

END

Below is a solution of Exercise 3 on page 183, giving the output for Program 3 when values for sides a and b are fixed at 5 and 6, respectively. A FOR-NEXT loop has been included to test values from 1 to (a + b - 1) for side c.

RUN

SIDE A=5 AND SIDE B=6

SIDE C	ANGLE A	ANGLE B	ANGLE C
1	****** NO TRIANGLE ******		
2	51.32	110.49	18.19
3	56.25	93.82	29.93
4	55.77	82.82	41.41
5	53.13	73.74	53.13
6	49.25	65.38	65.38
7	44.42	57.12	78.46
8	38.62	48.51	92.87
9	31.59	38.94	109.47
10	22.33	27.13	130.54

END

In Program (4), SSA (p. 183), Statements 70 and 75 make use of the fact that

$$\tan B2 = \frac{\sin B2}{\sqrt{1 - \sin^2 B2}}$$ and the arctangent function to determine angle B2.

Notice that this program gives no test for the situation when angle A is obtuse and a < b. There is clearly no triangle in this case. If run for this case, the program prints negative values.

At the top of next page is a solution of Exercise 4 on page 183, giving the output for Program 4 when values for sides a and b are fixed at 8 and 4, respectively. A FOR-NEXT loop has been included to test values from 10° to 170° for Angle A. If you wish, you may ask students to find out what happens when a = b and when a < b.

If your students have had programming experience, they may wish to revise the programs to include tests for all possibilities so that if values for A, B, C, a, b, and c are put in at random, the output will be appropriate.

```
RUN

SIDE A=8 AND SIDE B=4
```

ANGLE A	ANGLE B	ANGLE C	SIDE C
10	4.98	165.02	11.91
20	9.85	150.15	11.64
30	14.48	135.52	11.21
40	18.75	121.25	10.64
50	22.52	107.48	9.96
60	25.66	94.34	9.21
70	28.02	81.98	8.43
80	29.5	70.5	7.66
90	30	60	6.93
100	29.5	50.5	6.27
110	28.02	41.98	5.69
120	25.66	34.34	5.21
130	22.52	27.48	4.82
140	18.75	21.25	4.51
150	14.48	15.52	4.28
160	9.85	10.15	4.12
170	4.98	5.02	4.03

```
END
```

CHAPTER 6

One very practical application of trigonometric functions is to the study of vectors. This chapter discusses vectors both from an abstract viewpoint (as ordered pairs of real numbers) and from a concrete viewpoint (as directed line segments in the plane). Following a development of the properties of vector operations, applications of vectors to navigational problems, to force problems, and to a polar coordinate system are studied in detail.

6-1   A distinction is made in this section between a vector, which is an ordered pair of real numbers, and a geometric vector, which is a directed line segment. The main reason for introducing the notion of a geometric vector at this time is to provide a concrete way of visualizing the properties of vectors. All mathematical properties of vectors are developed using ordered pairs, not their geometric representations. While it is possible to discuss vectors and vector operations solely in terms of geometric vectors, to do so is to limit greatly the applications that can be made of vectors in areas where geometric ideas are absent, as, for example, in factor analysis and other statistical problems. Moreover, students whose only experience with vectors is geometric have difficulty later when they encounter vector spaces of more than three dimensions.

(p. 186, par. 1)   An alternative means of denoting a vector in handwritten equations is to use the wavy underscore, $\underset{\sim}{v}$ or $\underset{\sim}{a}$, which, in printer's symbols, indicates boldface type.

(p. 186, par. 3)   With the introduction of an arrow, or directed line segment, as a geometric picture of an ordered pair (x, y), the student now has two ways to visualize such an ordered pair, the other way being simply as a point. In some mathematical contexts, as, for example, in graphing solution sets, the point picture is more useful. In other situations, as,

for example, in discussing forces, the arrow picture is preferable.   Students should understand that in using either the point or the arrow representation, the mathematical ideas
involved do not depend on the pictures.   Any geometric interpretation made of an algebra of
ordered pairs is a matter of convenience and not of necessity.   The more one tends to confine
a concept such as that of an ordered pair to one pictorial representation, the more restricted
one's thinking becomes, which in turn reduces the usefulness of the abstract concept.

(p. 186, pars. 4, 5)   Point out to students that although the norm of a vector can be
thought of in terms of the length of a line segment, a line segment is not used in the definition
itself.   The assignment of a norm to a vector is another example of a function, with domain
the set of all vectors (x, y) and range the set of nonnegative real numbers.   The function is
not one-to-one, of course, because infinitely many vectors have any given positive real
number for a norm.   The symbol for the norm of a vector $\vec{v}$, $\|\vec{v}\|$, is similar to that of the
absolute value of a real number x, $|x|$, and it might be worthwhile observing that for the
subset $\{(x, 0)\} \subset V$, we have $\|(x, 0)\| = |x|$, and for the subset $\{(0, y)\} \subset V$, we have
$\|(0, y)\| = |y|$.   In fact, in geometric context, it is customary to refer to the norm as the
length of $\vec{v}$ and to denote this length by the symbol $|\vec{v}|$.

(p. 187, last par.)   Make sure students clearly distinguish between the real number 0
and the zero vector $\vec{0} = (0, 0)$.

(p. 188, par. 1)   You may wish to point out to students that these are definitions of
operations with vectors.   Because vectors have been defined in terms of real numbers, the
basic properties of vectors depend upon properties of real numbers.   Thus, these properties
are theorems, whereas the corresponding properties of real numbers are axioms.   The same
is also true for the properties of complex numbers discussed in the next chapter.   Therefore,
the fact that the set C of complex numbers is a field is a matter of proof rather than of
assumption.

6-2   The notion of a basis for a vector space as introduced in this section extends to
vector spaces of any finite number of dimensions.   Although we are interested only in a two-
dimensional space in this course, students should be reminded of the fact that the ideas being
discussed are not necessarily restricted to such a space and that if they pursue further courses
in mathematics they will find that the concepts developed here extend quite easily and directly
to spaces of higher dimension.

(p. 192—p. 193, par. 2)   Students may understand the general result discussed at the
bottom of page 192 and the top of page 193, and summarized by Equations (1), (3), and (4),
more easily if you first discuss the example given in paragraph 2, page 192.   The general
result on page 193 then leads to the definitions of "basis" and "basis vectors."   It should be
emphasized that given any two noncollinear vectors $\vec{u}$ and $\vec{v}$, any other vector $\vec{a}$ can be written
as a linear combination of $\vec{u}$ and $\vec{v}$.

It may be helpful at this point to discuss the requirement that $\vec{u}$ and $\vec{v}$ be noncollinear. To see why this requirement is necessary, suppose (using indirect reasoning) that $\vec{u}$ and $\vec{v}$ are collinear, with $\vec{v} \neq \vec{0}$. Then there exists a real number r_1 for which $\vec{u} = r_1 \vec{v}$. If $\vec{a}$ is any vector, then, for r, $s \in R$, $\vec{a} = r\vec{u} + s\vec{v} = r(r_1\vec{v}) + s\vec{v} = (rr_1 + s)\vec{v}$; or, letting $rr_1 + s = r_2$, $\vec{a} = r_2\vec{v}$. But (since $\vec{a}$ is any vector) this implies that every vector is collinear with $\vec{v}$, which is clearly not true.

(p. 193, par. 3)   Although the term "orthogonal" is defined here in terms of the directions of two given vectors, it is quite possible to define orthogonal vectors without recourse to the notion of direction.   However, because the concept of geometric perpendicularity is so strongly ingrained in the students' mathematical background, it seems preferable for our use to give a definition of orthogonality that lends itself readily to this geometric interpretation.

(p. 193, par. 4)   The symbols $\vec{i}$ and $\vec{j}$ used here for unit orthogonal basis vectors are those commonly employed in discussing such vectors.   There is little danger of students confusing the vector $\vec{i}$ with the complex number i so long as the different symbols are used consistently within their respective contexts.

(p. 194, par. after Solution)   Students must distinguish between __vector__ components and __scalar__ components.   As used here, a vector component can be regarded as the product of a vector and a scalar—hence a vector, whereas a scalar component is simply a scalar—a number. Geometrically, a scalar component can be thought of as the directed length of a vector component.

6-3   The inner-product concept introduced here is fundamental to a detailed study of vector algebra, and students completing the material here will have an advantage in their later study of mathematics over those who encounter the idea for the first time in a college course.

(p. 198, par. 1)   It may be observed by some students that Equation (1) defines a function of two variables, with domain the set of all ordered pairs of real numbers and range R.

(p. 199)   Equation (3) can serve as motivation for the definition of inner product;   its development can be accompanied by an examination of one or more particular pairs of vectors. This is an excellent opportunity to apply both the Law of Cosines and the distance formula.

(p. 200, par. 2)   Students may enjoy exhibiting pairs of orthogonal vectors.

(p. 202)   Note that when $\vec{u}$ and $\vec{v}$ are unit orthogonal vectors, Equation (14) reduces to

$$\vec{a} = (\vec{u} \cdot \vec{a})\vec{u} + (\vec{v} \cdot \vec{a})\vec{v}.$$

You could illustrate this fact with actual examples.

6-4   This section broadens the discussion of vectors to include the notion of free vectors which is so useful in the physical sciences.   No attempt is made here to place the discussion on a rigorous axiomatic foundation, but rather the students' geometric intuition is exploited as much as possible.

(p. 204, pars. 1-3)   Be sure students understand that "equivalent" does not mean "equal" in this usage.   A concept closer to "equivalence" as the term is employed in this context is that of "congruency," although in order for two geometric vectors to be equivalent they must not only be congruent in the usual sense but must also have the same direction.

It might prove helpful to students if several pictures showing how given vectors can be translated to the origin are drawn on the chalkboard.   For example, the following diagram shows two such pictures.   Notice that only two rigid "slidings" of each vector are needed to place it in standard position.

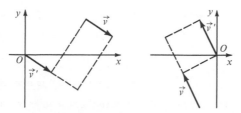

(p. 204, par. 4)   The idea of defining a "vector" as an equivalence class of free vectors is analogous to what is sometimes done in defining a rational number.   In particular, it is possible to build a consistent system of rational numbers by defining an equivalence class of ordered pairs, for example,

$$\{(3, \ 5), \ (6, \ 10), \ (9, \ 15), \ \ldots\},$$

to be a rational number.   Similarly, if $\vec{v}_i$ denotes one of the set of all geometric free vectors equivalent to a given free vector, say $\vec{v}_1$, then the set

$$\{\vec{v}_1, \ \vec{v}_2, \ \vec{v}_3, \ \ldots, \ \vec{v}_i, \ \ldots\}$$

can be defined to be a "vector."   Clearly, the complexity and abstractness of such a viewpoint renders it something less than satisfactory as a first introduction to the subject.

(p. 204, par. 5 and p. 205, par. 1)   Students should see that the two different geometric models for the sum of two vectors actually involve the same parallelogram.   Thus, the two models simply show two different locations for $\vec{v}$.

6-5   This section continues the discussion of physical applications of vectors; the ideas considered are from the elementary physics of forces.

6-6   Polar coordinates offer still another widely used application of vector concepts. The most difficult idea for students to grasp is that each point in the geometric plane is paired with infinitely many pairs of polar coordinates, since ρ can be positive or negative and θ can be measured in either a clockwise or a counterclockwise direction. A good way to clarify this matter in the students' minds is to require them from time to time to give alternative coordinates for a given point.

(pp. 220-222)   Because students often tend to transform a Cartesian equation into a polar equation by rote substitution, it is worthwhile spending class time on Example 3, particularly parts (b) and (c). In (b), the equation $\rho (\cos \theta - \sin \theta) = 0$ is equivalent to $\cos \theta - \sin \theta = 0$; that is, the factor ρ can properly be discarded. However, careless discarding of a factor can lead to an erroneous conclusion. For example, discarding the factor ρ in $\rho (\rho - 2) = 0$ removes the origin from the graph, although the origin is clearly part of the graph of the corresponding Cartesian equation, namely $x^2 + y^2 = \pm 2\sqrt{x^2 + y^2}$. Examples 3(b) and 3(c) illustrate that a given Cartesian equation may have more than one equivalent polar equation.

6-7   In plotting graphs of polar equations, as in plotting graphs of Cartesian equations, students should focus on determining general properties of the graphs rather than spending time on laborious point-plotting. A student's progress in a calculus course may be slowed by an inability to sketch the graphs of a variety of equations quickly and accurately, and help given in prior courses to improve this technique will be extremely valuable later.

(p. 225, Example 1)   A useful aid in sketching graphs such as the one in this example (as well as graphs of other polar equations) is the concept of an auxiliary equation. In this example, if one first sketches the graph of $y = 2 \cos \theta$ over one period, it is evident by inspection that the value of $2 \cos \theta$ decreases steadily for $0 < \theta < \pi$, increases steadily for $\pi < \theta < 2\pi$, and is 0 when $\theta = \frac{\pi}{2}$ or $\theta = \frac{3\pi}{2}$. Since, in the given

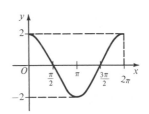

equation $\rho = 2 \cos \theta$, the value of ρ is just the value of $2 \cos \theta$, the graph of $\rho = 2 \cos \theta$ can quickly be sketched from the information obtained from the auxiliary graph. In effect, such an auxiliary graph offers the same information presented by the table shown in this example, but does so pictorially.

It is worthwhile to discuss the effect of the constant "2" in Example 1 and of the constant "2" preceding the expression "cos 2θ" in Example 2, page 226. For example, the question can be posed, "How will the graph be affected if the '2' is changed to '3' or to '-2'?"

CHAPTER 7

This chapter offers students another example of an algebra of ordered pairs in which circular functions play a role. First, we list the axioms for a field. Then we define the complex numbers in terms of ordered pairs and we prove (with some of the proofs left as exercises) that the system of complex numbers satisfies the field axioms (Sections 7-1 and 7-2).

Then the standard form for complex numbers is introduced and complex numbers are graphed on the Argand plane. This is followed by a discussion of the polar form for complex numbers, and this form is then used to determine products and quotients of complex numbers. The chapter concludes with De Moivre's Theorem and its extensions, which are used to obtain roots of complex numbers.

7-1   By defining complex numbers as ordered pairs of real numbers, we gain several advantages over the alternative procedure of adjoining i to the set of real numbers. For one thing, there is the immediately apparent suggestion that complex numbers can be represented either by points in the plane or by vectors in the plane. Also, the way in which sums and products of complex numbers are defined makes it comparatively easy to show that the set C of complex numbers is a field with respect to these operations, and hence we inherit a wealth of facts about C from what we know about R.

(p. 240, par. 5)   Students will very likely wonder why the product of complex numbers is defined as it is. You can explain that we want a definition that will:

1. Preserve the correspondence $(a, 0) \leftrightarrow a$ under the operation of multiplication (and the operation of addition). This will be evident in Section 7-3; see in particular p. 247, par. 1.

2. Yield a number system satisfying the field properties.

3. Preserve the equality $|a \cdot b| = |a| \cdot |b|$

Very able students may be referred to D. J. Hansen, "A Note on the Operation of Multiplication in the Complex Plane," American Mathematical Monthly, Vol. 71 (February, 1964), pp. 185-186. This brief article shows that the definition of multiplication given here is the only one that can be used to extend the algebra of vectors described in Chapter 6 to a system fulfilling the three foregoing requirements.

(p. 241, par. 1)   Make sure students understand that the zero complex number is the ordered pair (0, 0). In Section 7-3, after we have established a correspondence between the subset $\{(x, 0): x \in R\}$ of C and R, 0 may be used in appropriate context for the zero complex number.

7-2   Here we complete the discussion of the complex numbers as a field, considering the operations of subtraction and division separately from the two basic operations of addition and multiplication. We relate subtraction to addition by the additive inverse and division to multiplication by the multiplicative inverse.

7-3 (pp. 245-246)   In more formal language, you can say that there is a one-to-one correspondence between the set $\{(a, 0): a \in R\} \subset C$ and the set $\{a: a \in R\} = R$ which pairs (a, 0) with a and which has the property that if (a, 0) ↔ a and (b, 0) ↔ b, then

(a, 0) + (b, 0) ↔ a + b and (a, 0)(b, 0) ↔ ab.   Such correspondences are called <u>isomorphisms.</u>

In addition to referring to the members of the set {yi, y ≠ 0} as <u>pure imaginary numbers,</u> the following terminology is also commonly used·

1. {x + 0i} = {x}   Real numbers

2. {x + yi} with y ≠ 0   Imaginary numbers

3. {x + yi}   Complex numbers

The difference between the set of imaginary numbers and the set of complex numbers is that each imaginary number must have a nonzero imaginary part.   Thus {x + yi, y ≠ 0} ⊂ {x + yi}.

(p. 247, par. 1)   Perhaps the most enlightening application of this paragraph is a justification of the definition of multiplication of complex numbers given in Section 7-1.

7-4 (p. 249, par. 1)   Students may sometimes see the axes of the Argand plane labeled "real axis" and "imaginary axis."   We have used x and y for this purpose in keeping with the basic definition of a complex number as an ordered pair of real numbers.

(p. 249, par. 2)   In Section 7-1, it was shown that the definitions for addition and equality of two vectors are the same as for two complex numbers.   If you make a further comparison of these two systems, you will find that under the operation of addition, the system of vectors and the system of complex numbers each constitutes a commutative group.   The fact that these two groups are isormorphic provides the basis for the correspondence suggested between two-dimensional vectors and complex numbers.

(p. 249, par. 3)   You may wish to point out to students that the definition of absolute value, or modulus, of a complex number stated here is consistent with the use of the term "absolute value" for real numbers.   If $z = x + 0i$, then $|z| = \sqrt{x^2 + (0)^2} = \sqrt{x^2} = |x|$.

It should be pointed out that whereas R is an <u>ordered</u> field, C <u>is not!</u>

(p. 250, par. 1)   As a further comparison of vectors and complex numbers, it is helpful to note that the amplitude of any complex number z is the same as the direction angle of the corresponding geometric vector (x, y).   (See page 186.)

(p. 250, par. 4 and p. 251, pars. 1, 2)   The material here is based on the concept of the inner product of two vectors.   If desired, another argument for orthogonality of the vectors corresponding to the nonzero complex numbers (x_1, y_1) and $(-cy_1, cx_1)$ can be made using the slope formula or the distance formula.   If students have learned that line segments with slopes m_1 and m_2 are perpendicular if and only if $m_1 m_2 = -1$, then by showing that the slopes of the directed line segments from the origin to

(x_1, y_1) and $(-cy_1, cx_1)$ are $\dfrac{y_1}{x_1}$ and $\dfrac{cx_1}{-cy_1}$, respectively, you have $\dfrac{y_1}{x_1} \cdot \dfrac{cx_1}{-cy_1} = -1$, and the line segments are perpendicular. Alternatively, an application of the distance formula will show that

$$x_1^2 + y_1^2 + c^2y_1^2 + c^2x_1^2 = (x_1 + cy_1)^2 + (y_1 - cx_1)^2$$

is an identity in x_1 and y_1, and hence, by the converse of the Pythagorean theorem, the line segments corresponding to (x_1, y_1) and $(-cy_1, cx_1)$ are perpendicular. Notice, however, that both of these arguments are essentially geometric, and hence, in our context, heuristic.

(p. 251, par. 3)   You may wish to point out to students that

$$f(n) = i^n, \ n \in \{\text{nonnegative integers}\}$$

defines a periodic function with domain {nonnegative integers} and range {1, i, -1, -i}.

7-5 (p. 255, bottom of page)   You may wish to make a summarizing statement to emphasize the convenience of the polar form for obtaining the product of two complex numbers: the product $z_1 z_2 = \rho_1 \rho_2 [\cos (\theta_1 + \theta_2) + i \sin (\theta_1 + \theta_2)]$ can be represented by a bound vector whose magnitude is the product of the respective moduli and whose direction is the sum of the respective arguments. Similarly, the quotient $\dfrac{z_1}{z_2} = \dfrac{\rho_1}{\rho_2} [\cos (\theta_1 - \theta_2) + i \sin (\theta_1 - \theta_2)]$ of two complex numbers in polar form can be represented by a bound vector whose magnitude is the quotient of the respective moduli and whose direction is the difference of the respective arguments.

(p. 257, top of page)   From the form of the quotient

$$\frac{z_1}{z_2} = \frac{\rho_1}{\rho_2} [\cos (\theta_1 - \theta_2) + i \sin (\theta_1 - \theta_2)],$$

students can easily determine whether two given complex numbers have bound-vector geometric representations that are collinear, orthogonal, or oblique. If $\theta_1 - \theta_2 = k\pi$, $k \in J$, then the quotient must be a real number, $\dfrac{\rho_1}{\rho_2}$ or $-\dfrac{\rho_1}{\rho_2}$; if $\theta_1 - \theta_2 = \dfrac{(2k + 1)\pi}{2}$, $k \in J$, then the quotient must be a pure imaginary number, $\dfrac{\rho_1}{\rho_2}i$ or $-\dfrac{\rho_1}{\rho_2}i$; if $\theta_1 - \theta_2 \neq \dfrac{k\pi}{2}$, $k \in J$, then the quotient must be an imaginary number with a nonzero real part. The converse of each of these assertions is clearly true, so that an inspection of the difference of the arguments of two complex numbers z_1 and z_2 suffices to determine the orthogonality or collinearity (if such exists) of their vector representations.

7-6   This section contains a presentation of De Moivre's Theorem and an extension of the theorem to the case of all integral exponents, with proofs left as "C" exercises.

A proof of De Moivre's Theorem is as follows: Since, by definition, $z^1 = \rho^1 (\cos 1 \cdot \theta + i \sin 1 \cdot \theta)$, the statement is true for $n = 1$.

Assume that for any positive integer k,

$$z^k = \rho^k (\cos k\, \theta + i \sin k\, \theta) \text{ is true.}$$

Then, by Equation (3), page 255,

$$z^{k+1} = z^k z^1 = \rho^k \rho^1 [\cos (k\theta + \theta) + i \sin (k\theta + \theta)]$$
$$= \rho^{k+1} [\cos (k+1)\theta + i \sin (k+1)\theta],$$

so that the statement of the theorem is true for $k + 1$ whenever it is true for k.   Hence, by the Axiom of Induction, the statement is true for all natural numbers.

To prove De Moivre's Theorem extended to negative integers, for $z \neq 0$: From Exercise 27, page 258,

$$z^{-n} = (\frac{1}{\rho}[\cos (-\theta) + i \sin (-\theta)])^n.$$

This is equal to $(\frac{1}{\rho})^n [\cos n(-\theta) + i \sin n(-\theta)]$

Thus $z^{-n} = \rho^{-n}[\cos (-n\theta) + i \sin (-n\theta)]$

<u>7-7 (p. 264, par. 1)</u>   For a demonstration of the fact that the sum of the q distinct qth roots of a complex number is the zero complex number, see S. L. Green, The Theory and Use of the Complex Variable (London, Pitman, 1939), pp. 24-25.

CHAPTER 8

This chapter develops the properties of 2×2 matrices and explores a few ways in which they can be applied to the study of vectors and complex numbers.   The material included gives students experience with a significant algebraic system different from the real or complex number system, particularly in the area of commutativity with respect to multiplication.   While the discussion in this chapter pertains primarily to 2×2 matrices, most of the results obtained apply to $n \times n$ matrices as well.

<u>8-1 (p. 272, par. 1-2)</u>   Although in this book entries for matrices are restricted to real numbers, the theory of matrices does not depend on this fact.   Indeed, all that is required is that the entries be from sets in which a sum and product of elements is defined.   Students should be reminded that the symbol $S_{2\times2}$ denotes a set of matrices rather than a single matrix.

<u>(p. 272, par. 5)</u>   Make sure students distinguish the symbol $\mathcal{O}$ used here for the zero matrix from the symbol 0.

<u>8-2</u>   Because students may find the "row-by-column" product of matrices strange at first, you may wish to go through several examples in class while they state individual entries for the product.   To further clarify the process, you can have them extend the definition and find several products of 3×3 matrices with 3×1, 3×2, and 3×3 matrices.   Yet another clarifying procedure is to lead them to discover that if two matrices A and B have order $a \times b$ and $c \times d$, respectively, then A and B have a product if and only if $b = c$.   Moreover, they can determine that the product must be of order $a \times d$.

(p. 279, Example 2)   This example provides a concrete illustration of how the absence of a commutative law for multiplication in a system affects the performance of seemingly routine processes in the system.   Long familiarity with the polynomial identity

$$(x + y)^2 = x^2 + 2xy + y^2$$

tends to obscure the role of the commutative property of multiplication, and students are often genuinely surprised to find that in working with matrices it does not always hold.   Many other things they have taken for granted about algebraic manipulations are no longer to be trusted blindly.   For example, another very common relationship that fails when noncommutative products are involved is

$$(x + y)(x - y) = x^2 - y^2.$$

Also, since it is possible for two nonzero matrices to have the zero matrix for a product, it is not true that for all matrices A, B, and C (C $\neq$ 0), A · C = B · C implies A = B.

8-3 (p. 282, par. 1)   Before deriving the inverse of the general 2 × 2 matrix $\begin{bmatrix} a & b \\ c & d \end{bmatrix}$, you may wish to consider one or two specific matrices with the students.   For example, a step-by-step derivation of the inverse of the matrix A = $\begin{bmatrix} 1 & 1 \\ 1 & 2 \end{bmatrix}$ might be made using

$$A^{-1}A = \begin{bmatrix} w & x \\ y & z \end{bmatrix}\begin{bmatrix} 1 & 1 \\ 1 & 2 \end{bmatrix}$$

or

$$AA^{-1} = \begin{bmatrix} 1 & 1 \\ 1 & 2 \end{bmatrix}\begin{bmatrix} w & x \\ y & z \end{bmatrix}.$$

Since the determinant of the matrix A is 1, you will obtain integral entries for the inverse, and thus avoid having to use fractions in the discussion.

Point out to students that all that is necessary in finding an inverse for a given 2 × 2 matrix A is to find one such inverse, either a left-inverse or a right-inverse, because if it is true that for the 2 × 2 matrix B one of the equations BA = I or AB = I is satisfied, then both must be.   This can be shown by solving

$$\begin{bmatrix} w & x \\ y & z \end{bmatrix}\begin{bmatrix} a & b \\ c & d \end{bmatrix} = \begin{bmatrix} 1 & 0 \\ 0 & 1 \end{bmatrix}$$

and

$$\begin{bmatrix} a & b \\ c & d \end{bmatrix}\begin{bmatrix} w & x \\ y & z \end{bmatrix} = \begin{bmatrix} 1 & 0 \\ 0 & 1 \end{bmatrix}$$

separately for w, x, y, and z and comparing the results.

To show in another way that the inverse A^{-1} of an invertible matrix A is unique, you can proceed as follows:

If A^{-1} and A* are matrices such that

$$AA^{-1} = I \quad \text{and} \quad AA* = I,$$

then

$$AA^{-1} = AA*.$$

But it then follows that

$$A^{-1}(AA^{-1}) = A^{-1}(AA^*),$$
$$(A^{-1}A)A^{-1} = (A^{-1}A)A^*,$$
$$IA^{-1} = IA^*,$$

from which you obtain $\qquad A^{-1} = A^*.$

(p. 283, par. 1)   If you wish, you can point out to students that the assignment of a determinant to each 2×2 matrix constitutes a function δ.   The domain of δ is the set $S_{2 \times 2}$, and the range is R.

You can also emphasize to students that once they have determined that for

$A = \begin{bmatrix} a & b \\ c & d \end{bmatrix}$, $A^{-1} = \dfrac{1}{\delta(A)} \begin{bmatrix} d & -b \\ -c & a \end{bmatrix}$, they can write the inverse of any 2×2 nonsingular

matrix A immediately by interchanging the entries d and a, replacing b and c with their negatives, and multiplying the resulting matrix by $\dfrac{1}{\delta(A)}$.

8-4   This section is intended to show students yet another example of an interesting and effective isomorphism, although the term "isomorphism" does not occur in the text.

(p. 287, bottom of page)   You will want the students to note that, as defined in the text, $S_R \subset S_{2 \times 2}$.

In order to reinforce the concept of a field, you may wish to have students prove that the set

$$S_R = \left\{ \begin{bmatrix} a & 0 \\ 0 & a \end{bmatrix} : a \in R \right\}$$

is a field under the matric operations of addition and multiplication.   The proof is computational, and consists of showing directly that the matrices in the set obey the field axioms. For example, to prove closure for addition, you observe that if

$$\begin{bmatrix} a & 0 \\ 0 & a \end{bmatrix} \in S_R \text{ and } \begin{bmatrix} b & 0 \\ 0 & b \end{bmatrix} \in S_R,$$

then

$$\begin{bmatrix} a & 0 \\ 0 & a \end{bmatrix} + \begin{bmatrix} b & 0 \\ 0 & b \end{bmatrix} = \begin{bmatrix} a+b & 0 \\ 0 & a+b \end{bmatrix}.$$

Since $a + b \in R$, $\begin{bmatrix} a+b & 0 \\ 0 & a+b \end{bmatrix} \in S_R$.

The other demonstrations are similar.

8-5   One very important application of matrices lies in the study of transformations of the plane, and these transformations are perhaps best discussed in terms of vectors.   By identifying the matrix $\begin{bmatrix} x \\ y \end{bmatrix}$ or the matrix $\begin{bmatrix} x & y \end{bmatrix}$ with the ordered pair (x, y), we find another powerful isomorphism, one that lets us study vectors by studying the corresponding matrices, or vice versa.

(p. 293, top of page)   You may wish to point out here that if a linear transformation of the plane is into, but not onto, the plane, the transformation matrix must be singular, in which case all the (bound) vectors in the plane are mapped into a set of collinear vectors or else into the zero vector.

(p. 294, Example 2)   The use of the square with vertices (0, 0), (1, 0), (1, 1), and (0, 1) to visualize the results of a linear transformation of the plane is an effective pedagogical device.   Your students will find it rewarding if you, with the class helping, go over a sample of each kind of transformation described in this section using this square to illustrate the geometric effects of the transformation.   However, it is wise to guard against students' obtaining the impression that it is only squares, or even primarily squares, which are being transformed.   For this reason, and for variety, examples involving other geometrical figures, such as triangles, are important.

8-6   This section covers the application of matrices and trigonometric functions to those linear transformations known as rotations of the plane or rotations of the axes.   For a more detailed treatment, see School Mathematics Study Group, Introduction to Matrix Algebra (New Haven, Conn., Yale University Press, 1962).

(p. 304)   The output for exercise 8, Section 8-3 is shown below.

                                    RUN

                              1 - 1

                               2   1

                               1

                              1 1

                               4

                               3

                                    END

It should be emphasized to students that once a program has been completed to the extent that it solves a particular problem correctly, modifications in the format of the output can and frequently should be made.   Such changes should improve the readability of the output so that people with little or no actual programming experience will able to understand the meaning of the output.   Notice the difference in the output if the program is changed to read as follows:

```
10  DIM A[2,2],B[2],V[2,2],X[2]
20  MAT  READ A,B
30  PRINT A[1,1];"X + (";A[1,2];")Y = ";B[1]
35  PRINT A[2,1];"X + (";A[2,2];")Y = ";B[2]
40  PRINT
50  MAT V=INV(A)
60  MAT X=V*B
70  PRINT "THE SOLUTION IS (";X[1];",";X[2];") ."
80  DATA 1,-1,2,1,1,11
90  END
```

```
RUN
```

```
1X + (-1)Y =   1
2X + ( 1)Y =  11
```

```
THE SOLUTION IS ( 4, 3) .
```

```
END
```

Students should be encouraged to make similar changes in the format of program output.

(p. 305)   The program to find the coordinates of points in the plane resulting from a rotation of the plane may be re-written as follows:

```
10   DIM A[2],M[2,2],P[2]
20   PRINT "WHAT IS ANGLE OF ROTATION (DEGREES)?"
30   INPUT D
40   PRINT
50   LET D1=D*3.14159/180
60   LET M[1,1]=COS(D1)
65   LET M[2,2]=COS(D1)
70   LET M[1,2]=-SIN(D1)
80   LET M[2,1]=SIN(D1)
90   FOR N=1 TO 4
100  MAT  READ A
110  MAT P=M*A
120  MAT  PRINT P
130  PRINT
140  NEXT N
150  DATA 0,0,0,1,1,1,1,0
160  END
```

This version uses the matrix $\begin{bmatrix} \cos\theta & -\sin\theta \\ \sin\theta & \cos\theta \end{bmatrix}$

directly, and though it is, perhaps, simpler, may not provide as much exposure to the use of matrices.   The output, however, is the same for both programs.

CHAPTER 9

The principal aim of this chapter is to introduce students to another means of studying trigonometric and circular functions: infinite series, and particularly, power series.   In earlier chapters, we have seen how the symmetry of the unit circle and the properties of special right triangles can be used to determine values of cos x and sin x for certain values of x, and how reduction formulas, double-angle formulas, half-angle formulas, and other

identities enable us to use these known values to find additional values.   In many situations,
however, such methods are extremely laborious.   In this chapter, students learn about
various relationships between circular and trigonometric functions and infinite series and see
how these series can be used to find entries for trigonometric tables.   In the process, a
startling connection between circular and trigonometric functions and exponential functions is
discovered (p. 329, Equation (3)).

   9-1   This section is a rather compact introduction to the basic ideas of sequences.
For students with limited prior experiences with these ideas, it is highly recommended that
this section be supplemented with additional examples and discussions.   An appropriate
reference is Buchanan, Limits: A Transition to Calculus (Boston, Houghton Mifflin Co., 1970).

   (p. 307 and p. 308, par. 1)   Stress that a sequence is actually a function and that it is
the members of the range of the function in which we are interested.   A given member of the
range may appear repeatedly, as, for example, in a constant sequence, in which every term
is the same number, or in the sequence $\{(-1)^n\}$, whose range consists only of the two numbers
-1 and 1.   There is always an infinite number of terms in any sequence, even when the range
is finite.

   Graphs of a sequence function may be meaningful and may be drawn in either or both of
two ways: (1) graphing the set of ordered pairs $\{(n, a_n)\}$ in a Cartesian plane, (2) graphing
only the terms a_n on a single number line as shown in the next drawing.
   (p. 308, par. 2, and Example 1)   The definition of the limit of a sequence is extremely
important in mathematics.   Although the remainder of the chapter does not assume a working
knowledge of this definition, some discussion of it at this time will be extremely valuable to
students in their later study of calculus.

   The number line can be used as a schematic aid in visualizing the idea that the terms of
a sequence "concentrate" about a fixed number.   For example, graphing the first few
terms of $\{\frac{n+1}{n}\}$ on the number line produces the diagram shown below.

The "crowding-in" process is clearly evident, since the closer one approaches the graph of 1
from the right, the more graphs of terms in the sequence one encounters.   This schematic
device can be improved somewhat by suggesting to students that if they were to draw circles
with center at the point corresponding to 1, then no matter how small the radius of the given
circle, there will always be infinitely many terms of the sequence inside the circle and there
will always be at most a finite number of terms of the sequence outside the circle.

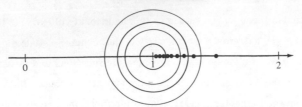

This second diagram suggests the topological "neighborhood" concept of a limit.

(p. 309, par. 1)   The important thing at this time is that students understand what each
of the nine assertions means.   One way to help students do this is to provide a specific
example of each in class.   One such example (there are many good ones) might be:

For 1: $\frac{2}{1}$, $\frac{2}{2}$, $\frac{2}{3}$, $\frac{2}{4}$, $\frac{2}{5}$, $\cdots$, $\frac{2}{n}$, $\cdots$

For 2: $\frac{3}{4}$, $\frac{9}{16}$, $\frac{27}{64}$, $\frac{81}{256}$, $\cdots$, $(\frac{3}{4})^n$, $\cdots$

For 3: 1, 1, 1, 1, $\ldots$

For 4: $(\frac{2}{1} + \frac{1}{2})$, $(\frac{3}{2} + \frac{2}{3})$, $(\frac{4}{3} + \frac{3}{4})$, $(\frac{5}{4} + \frac{4}{5})$, $\ldots$, $(\frac{n+1}{n} + \frac{n}{n+1})$, $\cdots$

For 5: $(1 - \frac{1}{2})$, $(1 - \frac{1}{4})$, $(1 - \frac{1}{8})$, $(1 - \frac{1}{16})$, $\ldots$, $(1 - \frac{1}{2^n})$, $\cdots$

For 6: $(\frac{1}{1})(\frac{2}{1})$, $(\frac{3}{2})(\frac{5}{2})$, $(\frac{5}{3})(\frac{8}{3})$, $(\frac{7}{4})(\frac{11}{4})$, $\ldots$, $(\frac{2n-1}{n})(\frac{3n-1}{n})$, $\cdots$

For 7: $\dfrac{\frac{2}{1}}{\frac{2}{2}}$, $\dfrac{\frac{3}{2}}{\frac{5}{4}}$, $\dfrac{\frac{4}{3}}{\frac{8}{6}}$, $\dfrac{\frac{5}{4}}{\frac{11}{8}}$, $\ldots$, $\dfrac{\frac{n+1}{n}}{\frac{3n-1}{2n}}$, $\cdots$

For 8: $(0)^3$, $(\frac{1}{2})^3$, $(\frac{2}{3})^3$, $(\frac{3}{4})^3$, $\ldots$, $(\frac{n-1}{n})^3$, $\cdots$

For 9: $3(\frac{3}{2})$, $3(\frac{7}{4})$, $3(\frac{11}{6})$, $3(\frac{15}{8})$, $\ldots$, $3(\frac{4n-1}{2n})$, $\cdots$

9-2 (p. 311 and p. 312, par. 1)   One good way to begin the teaching of this section is to
discuss "sequences of partial sums" in some detail before using the phrase "infinite series."
Students usually need several concrete examples of sequences of partial sums.   It is important not to confuse the terms of a sequence of partial sums $\{A_n\}$, where $A_n =$
$a_1 + a_2 + \ldots + a_n$, with the terms of the corresponding sequence $\{a_n\}$.   Moreover,
convergence of $\{a_n\}$ does not imply convergence of $\{A_n\}$.

(pp. 312-313)   Because the summation notation $\sum\limits_{i=1}^{\infty} a_i$ will be new to many students,
and because it is used frequently throughout the chapter, you will find it advantageous to go
over several different examples of it in the classroom while the students supply information.
Some of the properties of the notation are developed in Exercises 19-21, on page 315.   It is
important that students understand the following properties:

1. $\sum\limits_{i=1}^{n} a = na$, for each constant a.

2. $\displaystyle\sum_{i=1}^{n} as_i = a \sum_{i=1}^{n} s$

3. $\displaystyle\sum_{i=1}^{n} s_i + \sum_{i=1}^{n} r_i = \sum_{i=1}^{n} [s_i + r_i]$

4. $\displaystyle\sum_{i=1}^{k} s_i + \sum_{i=k+1}^{n} s_i = \sum_{i=1}^{n} s_i$, where $k \in N$ and $1 < k < n$.

(p. 313, last par., and p. 314)   The meaning of convergence and divergence of infinite series is stated, in terms of limits of sequences, but procedures for determining convergence or divergence are not discussed in this section.   Since it is seldom possible to find a general term to describe a sequence of partial sums, the theorems for sequences on page 309 are rarely useful directly for showing convergence or divergence of infinite series.   Stress the last paragraph on page 314.

· 9-3   This section provides some relatively easy applications and reinforcement of basic ideas of the two preceding sections.   This section also leads directly to power series (Section 9-4); you may wish to notice at this time that in Example 1 of Section 9-4, the power series $\displaystyle\sum_{i=1}^{\infty} x^{i-1}$ is a geometric series.

Equation (3), page 316, is often easier to apply when written in the form $S_n = a\left(\dfrac{1 - r^n}{1 - r}\right)$.

9-4 (p. 320, pars. 2 and 3)   Although the harmonic series and the alternating harmonic series are not power series, they are introduced here because they appear in Example 2 on page 321 in connection with the endpoints of the interval of convergence of $\displaystyle\sum_{i=1}^{\infty} \dfrac{x^{i-1}}{i}$.   For a proof of the convergence of the alternating harmonic series, see any good calculus textbook.   One proof is based on the property:

An alternating series is convergent if each term is less in absolute value than its predecessor and if the sequence of terms in the series approaches 0 as a limit.

From this, it follows immediately that the alternating harmonic series converges, because $\left|\dfrac{(-1)^n}{n+1}\right| < \left|\dfrac{(-1)^{n-1}}{n}\right|$ is equivalent to $\dfrac{1}{n+1} < \dfrac{1}{n}$, which is true for all natural numbers n, and $\displaystyle\lim_{n \to \infty} \dfrac{1}{n} = 0$.

(p. 320, last par.)   A proof of the ratio test can be found in many calculus books. See, for example, Alfred B. Willcox, et al., Introduction to Calculus 1 and 2, Boston, Houghton Mifflin Co., 1971), pp. 440-41.   The proof is within the range of many high school students but is beyond the scope of this text.

(p. 322, middle of page)   Motivation for the formula given here for the product of two power series can be provided by showing the application of the familiar multiplication algorithm to a few terms in two such series.   For example,

$$a_0 + a_1 x \quad + a_2 x^2 \quad + a_3 x^3 \quad + \cdots$$
$$b_0 + b_1 x \quad + b_2 x^2 \quad + b_3 x^3 \quad + \cdots$$

$$a_0 b_0 + a_1 b_0 x + a_2 b_0 x^2 + a_3 b_0 x^3 + \cdots$$
$$a_0 b_1 x + a_1 b_1 x^2 + a_2 b_1 x^3 + a_3 b_1 x^4 + \cdots$$
$$a_0 b_2 x^2 + a_1 b_2 x^3 + a_2 b_2 x^4 + a_3 b_2 x^5 + \cdots$$
$$a_0 b_3 x^3 + a_1 b_3 x^4 + a_2 b_3 x^5 + a_3 b_3 x^6 + \cdots$$
$$\vdots$$

$$a_0 b_0 + (a_1 b_0 + a_0 b_1)x + (a_2 b_0 + a_1 b_1 + a_0 b_2)x^2 + \cdots$$

9-5 (p. 323, par. 1)   For a formal proof of the Binomial Theorem, see Dolciani, Beckenbach, Donnelly, Jurgensen, Wooton, Modern Introductory Analysis (Boston, Houghton Mifflin Co., 1970), pp. 91-92.

(p. 324, top of page)   Students need considerable practice with factorial notation if they are to be proficient in using it.   Exercises 1-6 on page 327 provide some practice with typical problems, and you may wish to supplement these six exercises with additional ones.

9-6   A nonrigorous derivation of the infinite power series expansion for sin x and cos x is developed in this section, using the expansion for $(1 + \frac{ix}{n})^n$ as a vehicle.

(p. 330, Example 2)   Although the method used to solve this example is typical of the procedure used to find values for cos x and sin x when x is small (in particular, when x < 1), it is not typical of the procedure involved in finding cos x and sin x for relatively large x.   The series (4) and (5) on this page do not converge very rapidly for larger values of x.   Digital computers can handle evaluation of these and similar series for reasonably small values of x, and can use them to print out tables for cosine and sine with great rapidity.   Other series are available for computing sin x and cos x for larger values of $|x|$.

9-7 (p. 332, par. 2)   Make sure students understand that in using the expression $\frac{e^x - e^{-x}}{2}$ in defining sinh, the factor i in the right-hand member of (6) is not included.

(p. 335, top of page)   For a discussion of the relationship of cosh x and sinh x to the equilateral hyperbola, see George B. Thomas, Jr., Calculus and Analytic Geometry (Reading, Mass., Addison-Wesley Publishing Co., Inc., 1968) pp. 271-272.

(p. 343)   Both versions of the program will produce the same output.   At the top of the next page is the case where A = 1 and R = .5.

```
RUN

INPUT A AND R? 1,.5

NUMBER              TERM              SUM
1                   1                 1
2                   .5                1.5
3                   .25               1.75
4                   .125              1.875
5                   .0625             1.9375
6                   .03125            1.96875
7                   .015625           1.98437
8                   7.81250E-03       1.99219
9                   3.90625E-03       1.99609
10                  1.95312E-03       1.99805

END
```

(p. 344)   The output for the revised program to find an approximation for e is shown

below.   Nine terms were necessary in order for T to be less than .000005.

```
RUN

NUMBER              TERM              SUM
1                   1                 1
2                   1                 2
3                   .5                2.5
4                   .166667           2.66667
5                   4.16667E-02       2.70833
6                   8.33333E-03       2.71667
7                   1.38889E-03       2.71806
8                   1.98413E-04       2.71825
9                   2.48016E-05       2.71828

END
```

Using the fact from Section 9-6 that:

$$\sin x = x - \frac{x^3}{3!} + \frac{x^5}{5!} - \frac{x^7}{7!} + \cdots + (-1)^{r+1} \frac{x^{2n-1}}{(2n-1)!} + \cdots$$

the following changes can be made in the program for cos x:

```
50    PRINT "NUMBER", "TERM", "SIN"; D
60    LET T=D1
105   IF N>2 THEN 110
106   LET T=D1
110   LET T=T*(-1)*D1↑2/((2*N-2)*(2*N-1))
150   PRINT "SIN"; D; " ="; SIN(D1)
```

Statement 60 sets the first term equal to x (the angle measure in radians).   Statements 105

and 106 are included to establish the initial value of the exponent of x as 3 in the second term.

Statement 110 then increases this exponent by 2 in each subsequent term.   Below is an

example of output for this program.

```
RUN

INPUT ANGLE IN DEGREES ?80

NUMBER          TERM            SIN 80
   1          1•39626         1•39626
   2          -•45368          •942582
   3          4•42236E-02      •986806
   4         -2•05276E-03      •984753
   5          5•55827E-05      •984808
   6         -9•85101E-07      •984807

SIN 80 = •984807

END
```

APPENDIX A (Solutions to these exercises start on page 231 of this manual)

The material in this appendix can be used, as needed, to review selected algebraic concepts which are especially pertinent to the contents of this textbook. For a class whose background warrants it, a concentrated study of this appendix could be conducted at the beginning of the course. The following exercise sets, provided here for each of the sections of this appendix, may be useful in teaching this material.

A-1

In Exercises 1-6, let A = {1, 2, 3, 4}, B = {3, 4, 5, 6} and C = {1, 4, 5, 7}, and specify each set by listing its members.

1. A ∪ B   4. A ∩ (B ∩ C)

2. A ∩ C   5. C ∪ (A ∩ B)

3. (A ∪ B) ∪ C   6. (A ∪ B) ∩ (A ∪ C)

In Exercises 7-11, state which axiom, listed on page 358, justifies the statement. Assume that all variables denote real numbers.

7. $2 + (8 + x) = (2 + 8) + x$ 8. $10(4 + \frac{1}{10}) = 40 + 1$

9. $3 + (7 + 8) = (7 + 8) + 3$

10. $(x - y)(x + y) = (x - y)(x) + (x - y)(y)$

11. $5(7x) = (5 \cdot 7)x$

In Exercises 12-21, state whether the statement is true or whether it is false.

12. N ∩ J = N   15. J ∪ Q = $\mathcal{R}$

13. J ∩ Q = ∅   16. $\mathcal{R}$ ∩ J = $\mathcal{R}$

14. N ⊂ Q   17. Q ∪ Q' = $\mathcal{R}$

18. Subtraction in $\mathcal{R}$ is commutative.   20. Division in $\mathcal{R}$ is commutative.

19. Subtraction in $\mathcal{R}$ is associative.   21. Division in $\mathcal{R}$ is associative.

22. What is the subset of $A = \{\sqrt[3]{-8},\ -\sqrt{3},\ 10,\ 3.14,\ 7^2,\ 2\frac{1}{4},\ \frac{16}{2}\}$ that consists of:

(a) all natural numbers in A?

(b) all integers in A?

(c) all rational numbers in A?

(d) all irrational numbers in A?

23. Answer the questions in Exercise 22 for $B = \{1.1,\ -3^2,\ \sqrt{25},\ \sqrt{7},\ \sqrt[3]{27},\ \frac{5}{3}\}$

A-2

Find the solution set of each of the following open sentences.

1. $12x - 3 = 9$

2. $5 - \frac{1}{2}x = 7$

3. $3(x + 3) = 7(x - 1)$

4. $11 - 2(x + 4) = 14(2 - x)$

5. $\dfrac{5x - 2}{3} + \dfrac{2x + 4}{4} = 2x + 1$

6. $x(x + 3) + 2 = (x + 1)(x + 2)$

7. $x^2 = x$

8. $x(x + 4) = 0$

9. $10x^2 = 8x$

10. $x^2 - 2x = 15$

11. $2x^2 - 4x - 30 = 0$

12. $5 - x^2 = 0$

13. $x^2 + 8x + 1 = 0$

14. $\dfrac{1}{x - 3} - \dfrac{1}{x + 3} = \dfrac{6}{x^2 - 9}$

15. $(x - 2)^2 = 49$

16. $(x - 5)^2 = 10$

A-3

In Exercises 1-5, find each solution set over R.

1. $-4x \le -12$

2. $3(x - 2) \le x + 8$

3. $6(x - 3) < -2(x + 5)$

4. $2x(x - 4) \ge 2x^2 - 5x + 1$

5. $\dfrac{x + 5}{2} > 7$

Graph each interval on a real number line. In each case, assume $x \in R$.

6. $\{x: 1 \le x \le 4\}$

7. $\{x: -3 < x < 1\}$

8. $\{x: |x| \le 4\}$

9. $\{x: |x - 1| \le 4\}$

10. $\{x: |x + 3| < \frac{4}{3}\}$

11. $\{x: x \le 2\} \cap \{x: x \ge -2\}$

12. $\{x: x < 0\} \cup \{x: x > 4\}$

The solution set of each of the following open sentences is one of the sets: R, $R^+ = \{x: x \ge 0\}$, $R^- = \{x: x \le 0\}$, and $\emptyset$. Find each solution set.

13. $|x| = |-x|$

14. $|x| = x$

15. $\sqrt{x^2} = x$

16. $x^3 \le 0$

17. $x^2 \ge 0$

18. $(x - 3)^2 < 0$

19. $|x|^2 = x^2$

20. $\sqrt{x^2} = |x|$

A-4

In Exercises 1-8, graph the relation and state whether or not it is a function.

1. $\{(x, y): y = x\}$

2. $\{(x, y): x = 3\}$

3. $\{(x, y): y = 4\}$

4. $\{(x, y): y = x + 4\}$

5. $\{(x, y): x^2 + y^2 = 1\}$

6. $\{(x, y): x^2 + y^2 \leq 1\}$

7. $\{(x, y): x^2 = y^2\}$

8. $\{(x, y): |y| = 3\}$

9-16. Describe the range for each of the functions whose graph is shown in Figures A-4— A-11 on page 365.

17. On the same coordinate plane with a scale of approximately 4 inches = 1 unit, sketch the graph of each of the following equations for $0 \leq x \leq 1$. $y = x$, $y = x^2$, $x = x^3$, $y = \sqrt{x}$, $y = \sqrt[3]{x}$. Carefully plot at least six points on each graph.

In Exercises 18-24, describe the range of each function, either in set notation or as a real number line.

18. $\{(x, y): y = 2x, -3 \leq x \leq 3\}$

19. $\{(x, y): y = x^2, -2 \leq x \leq 6\}$

20. $\{(x, y): y = x^3, -2 \leq x \leq 2\}$

21. $\{(x, y): y = \sqrt{x}, 4 \leq x \leq 16\}$

22. $\{(x, y): y = |x|, -2 \leq x \leq 2\}$

23. $\{(x, y): y = \frac{1}{x}, x \geq 2\}$

24. $\{(x, y): y = -x, -4 \leq x \leq 1\}$

In Exercises 25-27, graph each function and describe its range.

25. $f(x) = \begin{cases} x^2 & \text{for } -2 \leq x \leq 1 \\ 1 & \text{for } 1 < x \leq 2 \end{cases}$

26. $g(x) = \begin{cases} x & \text{for } -4 \leq x \leq 0 \\ \sqrt{x} & \text{for } 0 < x \leq 4 \end{cases}$

27. $h(x) = \begin{cases} x^3 & \text{for } -2 \leq x \leq -1 \\ x & \text{for } -1 < x < 1 \\ x^2 & \text{for } 1 \leq x \leq 3 \end{cases}$

A-5

In Exercises 1 and 2, list the ordered pairs in each of the functions $f + g$, $f \cdot g$, and $\frac{f}{g}$. List the elements in the domain and the range of f, g, $f + g$, $f \cdot g$, and $\frac{f}{g}$.

1. $f = \{(1, 4), (2, 3), (3, 2), (4, 1)\}$

   $g = \{(1, 0), (2, 2), (3, 4), (4, 6)\}$

2. $f = \{(-2, 4), (-1, 2), (0, 0), (1, 2), (2, 4)\}$

   $g = \{(-2, 2), (-1, 1), (0, 0), (1, 1), (2, 2)\}$

In Exercises 3-5, find the rules of correspondence for $f + g$, $f \cdot g$, $\frac{f}{g}$, and $f \circ g$.

<u>3.</u> $f(x) = 6x + 1$, $g(x) = 3x$

<u>4.</u> $f(x) = x^2 - 1$, $g(x) = (x - 1)^2$

<u>5.</u> $f(x) = \frac{x + 4}{x - 2}$, $g(x) = x - 2$

In each of Exercises 6-9, graph the functions f, g, and h on the same coordinate plane over the same domain D, as given. State the range for f, g, and h.

<u>6.</u> $D = \{x: -2 \le x \le 2\}$ $\qquad$ $f(x) = x^3$, $g(x) = f(x) + 1$, $h(x) = 2f(x)$

<u>7.</u> $D = \{x: -3 \le x \le 3\}$ $\qquad$ $f(x) = |x|$, $g(x) = f(x) - 4$, $h(x) = 3f(x)$

<u>8.</u> $D = \{x: 1 \le x \le 4\}$ $\qquad$ $f(x) = \frac{1}{x}$, $g(x) = f(x) + 2$, $h(x) = -2g(x)$

<u>9.</u> $D = \{x: 1 \le x \le 4\}$ $\qquad$ $f(x) = \frac{1}{x}$, $g(x) = -f(x)$, $h(x) = -f(x) + 1$

In Exercises 10 and 11, find the domain and the rule of correspondence for (a) $f \circ g$ and (b) $g \circ f$.

<u>10.</u> $f(x) = \sqrt{x}$, $g(x) = 5 - x$, $\text{Dom}_f = \{x: x \ge 0\}$, $\text{Dom}_g = \mathcal{R}$

<u>11.</u> $f(x) = -x$, $g(x) = x^2$, $\text{Dom}_f = \text{Dom}_g = \mathcal{R}$

<u>12.</u> Let $f(x) = x + 1$ and $f[g(x)] = 4x - 2$. Find the rule of correspondence for g.

SOLUTIONS

CHAPTER 1. The Cosine and Sine Functions

Exercises 1-1 · Pages 3-5

A   1. a. The values repeat in the sequence 3, 4, 5, 6.   Hence a = 4.

b. f(20) = f(0 + 20) = f(0) = 3   c. f(-10) = f(2 + (-12)) = f(2) = 5

2. a. The values repeat in the sequence 1, 3, 5, 7, 9.   Hence a = 5.

b. f(20) = f(0 + 20) = f(0) = 1   c. f(-10) = f(0 + (-10)) = f(0) = 1

3. a. The values repeat in the sequence 1, 0, -1, 0.   Hence a = 4.

b. f(20) = f(0 + 20) = f(0) = 1   c. f(-10) = f(2 + (-12)) = f(2) = -1

4. a. The values repeat in the sequence 0, 1.   Hence a = 2.

b. f(20) = f(0 + 20) = f(0) = 0   c. f(-10) = f(0 + (-10)) = f(0) = 0

5. a. The values repeat in the sequence -2, -1, 0, 1, 2, 1, 0, -1.   Hence a = 8.

b. f(20) = f(4 + 16) = f(4) = 2   c. f(-10) = f(6 + (-16)) = f(6) = 0

6. a. The values repeat in the sequence -2, -1, 3, 1.   Hence a = 4.

b. f(20) = f(0 + 20) = f(0) = -2   c. f(-10) = f(2 + (-12)) = f(2) = 3

B   7. a. Yes, f(2p) = f(p + p) = f(p)

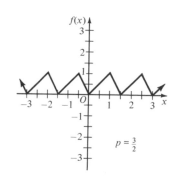

b. Yes, f(3p) = f(2p + p) = f(p) by part (a).

c. No, in periodic function f (at the right),

where $p = \frac{3}{2}$, $f(\frac{3}{2}) \neq f(\frac{2}{3})$

d. Yes, f(0) = f(0 + p) = f(p)

e. No, in the function f shown in (c),

$f(\frac{3}{4}) \neq f(\frac{3}{2})$

f. Yes, f(-p) = f(-p + p) = f(0) = f(p)

by part d

g. Yes, f(-2p) = f(-2p + p) = f(-p) = f(p) by part f

h. No, in the function f shown in (c), $f(\frac{9}{4}) \neq f(\frac{3}{2})$

8. a. f(x + p) = f(x) by def.

b. No. In 7(c), in the periodic function illustrated, $f(p) = f(\frac{3}{2}) \neq f(x)$ for x = 1.

c. f(x - p) = f[(x - p) + p] = f(x) by def.

d. No. In the periodic function illustrated in 7(c), f(0) ≠ f(x) for x = 1.

e. No. In the periodic function illustrated in 7(c), $f(-p) = f(-\frac{3}{2}) \neq f(x)$ for x = 1.

$\underline{f}$. No. In the periodic function illustrated in 7(c), $f(2x) \neq f(x)$ for $x = 1$.

$\underline{g}$. No. In the periodic function illustrated in 7(c), $f(x^2) \neq f(x)$ for $x = -1$.

$\underline{C}$ $\underline{9.}$

$$f(x) = \begin{cases} -x & \text{if } -1 \le x < 0 \text{ (for } k = 0) \\ 1 - x & \text{if } 0 \le x < 1 \text{ (for } k = 1) \\ 2 - x & \text{if } 1 \le x < 2 \text{ (for } k = 2) \\ 3 - x & \text{if } 2 \le x < 3 \text{ (for } k = 3) \end{cases}$$

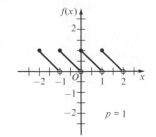

$\underline{10.}$

$$g(x) = \begin{cases} x + 6 & \text{if } -6 \le x < -3 \text{ (for } k = -2) \\ x + 3 & \text{if } -3 \le x < 0 \text{ (for } k = -1) \\ x & \text{if } 0 \le x < 3 \text{ (for } k = 0) \\ x - 3 & \text{if } 3 \le x < 6 \text{ (for } k = 1) \\ x - 6 & \text{if } 6 \le x < 9 \text{ (for } k = 2) \end{cases}$$

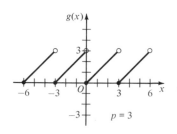

$\underline{11.}$

$$h(x) = \begin{cases} x + 4 & \text{if } -5 \le x < -3 \text{ (for } k = -1) \\ x + 2 & \text{if } -3 \le x < -1 \\ x & \text{if } -1 \le x < 1 \text{ (for } k = 0) \\ x - 2 & \text{if } 1 \le x < 3 \\ x - 4 & \text{if } 3 \le x < 5 \text{ (for } k = 1) \\ x - 6 & \text{if } 5 \le x < 7 \\ x - 8 & \text{if } 7 \le x < 9 \text{ (for } k = 2) \end{cases}$$

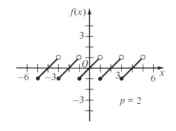

$\underline{12.}$

$$g(x) = \begin{cases} x^2 & \text{if } 0 \le x < 1 \text{ (for } k = 0) \\ 2 - x & \text{if } 1 \le x < 2 \\ (x - 2)^2 & \text{if } 2 \le x < 3 \text{ (for } k = 2) \\ 4 - x & \text{if } 3 \le x < 4 \\ (x - 4)^2 & \text{if } 4 \le x < 5 \text{ (for } k = 4) \\ 6 - x & \text{if } 5 \le x < 6 \end{cases}$$

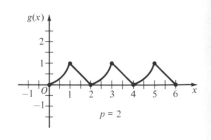

Exercises 1-2 · Pages 8-9

$\underline{A}$ $\underline{1.}$ $\frac{5\pi}{2} - 2\pi = \frac{\pi}{2}$ and $0 \le \frac{\pi}{2} < 2\pi$, so $\frac{5\pi}{2} = \frac{\pi}{2} + 2\pi$ and $k = 1$

$\underline{2.}$ $\frac{17\pi}{3} - 2\pi = \frac{11\pi}{3}$, but $\frac{11\pi}{3} > 2\pi$; $\frac{11\pi}{3} - 2\pi = \frac{5\pi}{3}$ and $0 \le \frac{5\pi}{3} < 2\pi$, so $\frac{17\pi}{3} = \frac{5\pi}{3} + 4\pi$

and $k = 2$

3. $-\frac{5\pi}{4} + 2\pi = \frac{3\pi}{4}$ and $0 \le \frac{3\pi}{4} < 2\pi$, so $-\frac{5\pi}{4} = \frac{3\pi}{4} + (-2\pi)$ and k = -1

4. $\frac{9\pi}{4} - 2\pi = \frac{\pi}{4}$ and $0 \le \frac{\pi}{4} < 2\pi$, so $\frac{9\pi}{4} = \frac{\pi}{4} + 2\pi$ and k = 1

5. $-\frac{9\pi}{4} + 2\pi = -\frac{\pi}{4}$, but $-\frac{\pi}{4} < 0$; $-\frac{\pi}{4} + 2\pi = \frac{7\pi}{4}$ and $0 \le \frac{7\pi}{4} < 2\pi$, so $-\frac{9\pi}{4} = \frac{7\pi}{4} + (-4\pi)$

   and k = -2

6. $-\frac{17\pi}{3} + 2\pi = -\frac{11\pi}{3}$, but $-\frac{11\pi}{3} < 0$; $-\frac{11\pi}{3} + 2\pi = -\frac{5\pi}{3}$, but $-\frac{5\pi}{3} < 0$; $-\frac{5\pi}{3} + 2\pi = \frac{\pi}{3}$

   and $0 \le \frac{\pi}{3} < 2\pi$, so $-\frac{17\pi}{3} = \frac{\pi}{3} + (-6\pi)$ and k = -3

7. $5\pi - 2\pi = 3\pi$, but $3\pi > 2\pi$; $3\pi - 2\pi = \pi$ and $0 \le \pi < 2\pi$, so $5\pi = \pi + 4\pi$ and k = 2

8. $-3\pi + 2\pi = -\pi$, but $-\pi < 0$; $-\pi + 2\pi = \pi$ and $0 \le \pi < 2\pi$, so $-3\pi = \pi + (-4\pi)$

   and k = -2

9. By repeated subtractions, $2107\pi - 2106\pi = \pi$, so $2107\pi = \pi + 2106\pi$ and k = 1053

10. By repeated additions, $-145\pi + 146\pi = \pi$, so $-145\pi = \pi + (-146\pi)$ and k = -73

11. Since $\frac{13\pi}{2} = \frac{\pi}{2} + 6\pi$, $\sin \frac{13\pi}{2} = \sin (\frac{\pi}{2} + 6\pi) = \sin \frac{\pi}{2}$

12. Since $\frac{18\pi}{5} = \frac{8\pi}{5} + 2\pi$, $\cos \frac{18\pi}{5} = \cos (\frac{8\pi}{5} + 2\pi) = \cos \frac{8\pi}{5}$

13. Since $-\frac{\pi}{2} = \frac{3\pi}{2} + (-2)\pi$, $\sin (-\frac{\pi}{2}) = \sin (\frac{3\pi}{2} + (-2)\pi) = \sin \frac{3\pi}{2}$

14. Since $-\pi = \pi + (-2)\pi$, $\cos (-\pi) = \cos (\pi + (-2)\pi) = \cos \pi$

15. Since $-\frac{13\pi}{3} = \frac{5\pi}{3} + (-6)\pi$, $\sin (-\frac{13\pi}{3}) = \sin (\frac{5\pi}{3} + (-6)\pi) = \sin \frac{5\pi}{3}$

16. Since $18\pi = 0 + 18\pi$, $\sin 18\pi = \sin (0 + 18\pi) = \sin 0$

17. Since $26\pi = 0 + 26\pi$, $\cos 26\pi = \cos (0 + 26\pi) = \cos 0$

18. Since $13\pi = \pi + 12\pi$, $\cos 13\pi = \cos (\pi + 12\pi) = \cos \pi$

19. Since $4.125\pi = 0.125\pi + 4\pi$, $\cos 4.125\pi = \cos (0.125\pi + 4\pi) = \cos 0.125\pi$

20. Since $3.215\pi = 1.215\pi + 2\pi$, $\sin 3.215\pi = \sin (1.215\pi + 2\pi) = \sin 1.215\pi$

21. Since $2315\pi = \pi + 2314\pi$, $\sin 2315\pi = \sin (\pi + 2314\pi) = \sin \pi$

22. Since $-473\pi = \pi + (-474)\pi$, $\cos (-473\pi) = \cos (\pi + (-474)\pi) = \cos \pi$

23. $(\frac{\sqrt{2}}{2})^2 + \sin^2 x = 1$, so $\sin^2 x = 1 - \frac{2}{4} = \frac{2}{4}$ and $\sin x$ is either $\frac{\sqrt{2}}{2}$ or $-\frac{\sqrt{2}}{2}$.

    Since P is in the first quadrant, $\sin x = \frac{\sqrt{2}}{2}$.

24. $(\frac{\sqrt{2}}{2})^2 + \sin^2 x = 1$, so $\sin^2 x = 1 - \frac{2}{4} = \frac{2}{4}$, and $\sin x$ is either $\frac{\sqrt{2}}{2}$ or $-\frac{\sqrt{2}}{2}$.

    Since P is in the fourth quadrant, $\sin x = -\frac{\sqrt{2}}{2}$.

25. $(-\frac{\sqrt{3}}{2})^2 + \sin^2 x = 1$, so $\sin^2 x = 1 - \frac{3}{4} = \frac{1}{4}$ and sin x is either $\frac{1}{2}$ or $-\frac{1}{2}$.
   Since P is in the second quadrant, $\sin x = \frac{1}{2}$.

26. $(-\frac{\sqrt{3}}{2})^2 + \sin^2 x = 1$, so $\sin^2 x = 1 - \frac{3}{4} = \frac{1}{4}$ and sin x is either $\frac{1}{2}$ or $-\frac{1}{2}$.
   Since P is in the third quadrant, $\sin x = -\frac{1}{2}$.

27. $(\frac{1}{2})^2 + \sin^2 x = 1$, so $\sin^2 x = 1 - \frac{1}{4} = \frac{3}{4}$ and sin x is either $\frac{\sqrt{3}}{2}$ or $-\frac{\sqrt{3}}{2}$.
   Since P is in the fourth quadrant, $\sin x = -\frac{\sqrt{3}}{2}$.

28. $(-\frac{1}{2})^2 + \sin^2 x = 1$, so $\sin^2 x = 1 - \frac{1}{4} = \frac{3}{4}$ and sin x is either $\frac{\sqrt{3}}{2}$ or $-\frac{\sqrt{3}}{2}$.
   Since P is in the second quadrant, $\sin x = \frac{\sqrt{3}}{2}$.

B 29. If x = 0, P(cos x, sin x) = P(cos 0, sin 0) = (1, 0). Hence cos 0 = 1 and sin 0 = 0.

   If $x = \frac{\pi}{2}$, P(cos x, sin x) = P(cos $\frac{\pi}{2}$, sin $\frac{\pi}{2}$) = (0, 1). Hence cos $\frac{\pi}{2}$ = 0 and

   sin $\frac{\pi}{2}$ = 1. If x = π, P(cos x, sin x) = P(cos π, sin π) = (-1, 0). Hence

   cos π = -1 and sin π = 0. If $x = \frac{3\pi}{2}$, P(cos x, sin x) = P(cos $\frac{3\pi}{2}$, sin $\frac{3\pi}{2}$) =

   (0, -1). Hence cos $\frac{3\pi}{2}$ = 0 and sin $\frac{3\pi}{2}$ = -1.

C 30. Suppose p is a period of f. Then f(x + p) = f(x). Since f(x) = 4π + sin x and

   f(x + p) = 4π + sin(x + p), sin(x + p) = sin x. Thus p is a period of sin x. Now

   suppose p is a period of sin x. Then sin (x + p) = sin x. Thus 4π + sin x =

   sin (x + p) and f(x) = f(x + p). Hence p is a period of f. The fund. period of f is

   therefore the fund. period of sin x, or 2π.

31. The fund. period of sin x + cos x is 2π. By examining the unit circle, we see the

   sum u + v will be the same after a period of 2π.

32. The fund. period of sin x cos x is π. For every point $P_1(u_1, v_1)$ in the first

   quadrant, there are corresponding points $P_2(u_2, v_2)$, $P_3(u_3, v_3)$, and $P_4(u_4, v_4)$

   in the other quadrants such that $u_2 = -u_1$, $u_3 = -u_1$, $u_4 = u_1$, $v_2 = v_1$, $v_3 = -v_1$,

   and $v_4 = -v_1$. Hence $u_2 v_2 = -u_1 v_1$, $u_3 v_3 = (-u_1)(-v_1) = u_1 v_1$, and $u_4 v_4 = -u_1 v_1$

   and the product $u_1 v_1$ = cos x sin x is cyclic with period π.

33. The fund. period of $\sin^2 x$ is π. By examining the unit circle, we see the values of

   v = sin x go from 0 to 1 to 0 and then from 0 to -1 to 0. In squaring the values of v

   in Quadrants III and IV, we repeat the values of v^2 in Quadrants I and II.

Exercises 1-3 · Pages 13-14

A **1.**

x	cos x	sin x
0	1	0
$\frac{\pi}{6}$	$\frac{\sqrt{3}}{2} \doteq 0.87$	$\frac{1}{2} = 0.5$
$\frac{\pi}{4}$	$\frac{\sqrt{2}}{2} \doteq 0.71$	$\frac{\sqrt{2}}{2} \doteq 0.71$
$\frac{\pi}{3}$	$\frac{1}{2} = 0.5$	$\frac{\sqrt{3}}{2} \doteq 0.87$
$\frac{\pi}{2}$	0	1
π	-1	0
$\frac{3\pi}{2}$	0	-1

2.

x	cos x	sin x
$\frac{\pi}{6}$	$\frac{\sqrt{3}}{2} \doteq 0.87$	$\frac{1}{2} = 0.5$
$\frac{5\pi}{6}$	$-\frac{\sqrt{3}}{2} \doteq -0.87$	$\frac{1}{2} = 0.5$
$\frac{7\pi}{6}$	$-\frac{\sqrt{3}}{2} \doteq -0.87$	$-\frac{1}{2} = -0.5$
$\frac{11\pi}{6}$	$\frac{\sqrt{3}}{2} \doteq 0.87$	$-\frac{1}{2} = -0.5$

3.

x	cos x	sin x
$\frac{\pi}{4}$	$\frac{\sqrt{2}}{2} \doteq 0.71$	$\frac{\sqrt{2}}{2} \doteq 0.71$
$\frac{3\pi}{4}$	$-\frac{\sqrt{2}}{2} \doteq -0.71$	$\frac{\sqrt{2}}{2} \doteq 0.71$
$\frac{5\pi}{4}$	$-\frac{\sqrt{2}}{2} \doteq -0.71$	$-\frac{\sqrt{2}}{2} \doteq -0.71$
$\frac{7\pi}{4}$	$\frac{\sqrt{2}}{2} \doteq 0.71$	$-\frac{\sqrt{2}}{2} \doteq -0.71$

4.

x	cos x	sin x
$\frac{\pi}{3}$	$\frac{1}{2} = 0.5$	$\frac{\sqrt{3}}{2} \doteq 0.87$
$\frac{2\pi}{3}$	$-\frac{1}{2} = -0.5$	$\frac{\sqrt{3}}{2} \doteq 0.87$
$\frac{4\pi}{3}$	$-\frac{1}{2} = -0.5$	$-\frac{\sqrt{3}}{2} \doteq -0.87$
$\frac{5\pi}{3}$	$\frac{1}{2} = 0.5$	$-\frac{\sqrt{3}}{2} \doteq -0.87$

5.

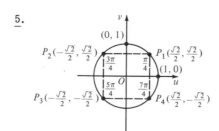

6.

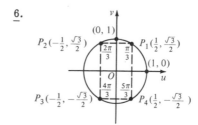

7.

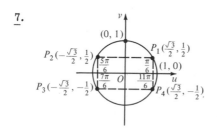

8.

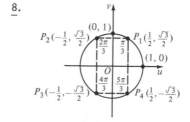

9. $\sin 3\pi = \sin (\pi + 2\pi) = \sin \pi = 0$

10. $\cos 7\pi = \cos (\pi + 6\pi) = \cos \pi = -1$

11. $\cos \frac{9\pi}{4} = \cos (\frac{\pi}{4} + 2\pi) = \cos \frac{\pi}{4} = \frac{\sqrt{2}}{2}$

12. $\sin \frac{71\pi}{6} = \sin\left(\frac{11\pi}{6} + 10\pi\right) = \sin \frac{11\pi}{6} = -\frac{1}{2}$

13. $\sin(-8\pi) = \sin(0 + (-8)\pi) = \sin 0 = 0$ 14. $\cos(-11\pi) = \cos(\pi + (-12)\pi) = \cos \pi = -1$

15. $\cos\left(-\frac{11\pi}{2}\right) = \cos\left(\frac{\pi}{2} + (-6)\pi\right) = \cos \frac{\pi}{2} = 0$

16. $\sin\left(-\frac{21\pi}{4}\right) = \sin\left(\frac{3\pi}{4} + (-6)\pi\right) = \sin \frac{3\pi}{4} = \frac{\sqrt{2}}{2}$

17. $x = 0, \pi, 2\pi$ 18. $x = \frac{\pi}{3}, \frac{5\pi}{3}$

19. $x = \frac{5\pi}{6}, \frac{7\pi}{6}$ 20. $x = \frac{5\pi}{4}, \frac{7\pi}{4}$

21. $x = \frac{3\pi}{2}$ 22. $x = \frac{\pi}{2}, \frac{3\pi}{2}$

B   23. $x = \frac{\pi}{4}$ 24. $x = \frac{3\pi}{4}$ 25. $x = 0, \pi$

26. $0 \le x \le \pi$ 27. $0 \le x \le \pi$ 28. $0 \le x \le \pi$

29.    30.

31.   32.

C   33. a. $x = \frac{\pi}{4}, \frac{3\pi}{4}, \frac{5\pi}{4}, \frac{7\pi}{4}$ b. $x = \frac{\pi}{4}, \frac{3\pi}{4}, \frac{5\pi}{4}, \frac{7\pi}{4}$ c. $x = \frac{\pi}{6}, \frac{5\pi}{6}, \frac{7\pi}{6}, \frac{11\pi}{6}$

34. a. 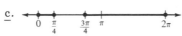   b.

    c.

Exercises 1-4 · Pages 20-23

A   1. $\cos\left(\frac{\pi}{4} + \frac{\pi}{3}\right) = \cos \frac{\pi}{4} \cos \frac{\pi}{3} - \sin \frac{\pi}{4} \sin \frac{\pi}{3}$

$$= \frac{\sqrt{2}}{2} \cdot \frac{1}{2} - \frac{\sqrt{2}}{2} \cdot \frac{\sqrt{3}}{2}$$

$$\doteq \frac{1.414 - 2.449}{4} \doteq -0.26$$

Thus $\cos \frac{7\pi}{12} \doteq -0.26$

2. $\cos\left(\frac{\pi}{4} + \frac{\pi}{2}\right) = \cos \frac{\pi}{4} \cos \frac{\pi}{2} - \sin \frac{\pi}{4} \sin \frac{\pi}{2}$

$$= \frac{\sqrt{2}}{2} \cdot 0 - \frac{\sqrt{2}}{2} \cdot 1 \doteq -0.71$$

Thus $\cos \frac{3\pi}{4} \doteq -0.71$

<u>3</u>. $\cos\left(\frac{\pi}{4} + \frac{2\pi}{3}\right) = \cos\frac{\pi}{4}\cos\frac{2\pi}{3} - \sin\frac{\pi}{4}\sin\frac{2\pi}{3}$

$$= \frac{\sqrt{2}}{2}\cdot\left(-\frac{1}{2}\right) - \frac{\sqrt{2}}{2}\cdot\frac{\sqrt{3}}{2}$$

$$\doteq \frac{-1.414 - 2.449}{4} \doteq -0.97$$

Thus $\cos\frac{11\pi}{12} \doteq -0.97$

<u>4</u>. $\cos\left(\frac{\pi}{4} + \frac{5\pi}{6}\right) = \cos\frac{\pi}{4}\cos\frac{5\pi}{6} - \sin\frac{\pi}{4}\sin\frac{5\pi}{6}$

$$= \frac{\sqrt{2}}{2}\cdot\left(-\frac{\sqrt{3}}{2}\right) - \frac{\sqrt{2}}{2}\cdot\frac{1}{2}$$

$$= \frac{-2.449 - 1.414}{4} \doteq -0.97$$

Thus $\cos\frac{13\pi}{12} \doteq -0.97$

<u>5</u>. $\cos\left(\frac{\pi}{3} - \frac{\pi}{6}\right) = \cos\frac{\pi}{3}\cos\frac{\pi}{6} + \sin\frac{\pi}{3}\sin\frac{\pi}{6}$

$$= \frac{1}{2}\cdot\frac{\sqrt{3}}{2} + \frac{\sqrt{3}}{2}\cdot\frac{1}{2}$$

$$\doteq \frac{1.732 + 1.732}{4} \doteq 0.87$$

Thus $\cos\frac{\pi}{6} \doteq 0.87$

<u>6</u>. $\cos\left(\frac{\pi}{3} - \frac{\pi}{4}\right) = \cos\frac{\pi}{3}\cos\frac{\pi}{4} + \sin\frac{\pi}{3}\sin\frac{\pi}{4}$

$$= \frac{1}{2}\cdot\frac{\sqrt{2}}{2} + \frac{\sqrt{3}}{2}\cdot\frac{\sqrt{2}}{2}$$

$$\doteq \frac{1.414 + 2.449}{4} \doteq 0.97$$

Thus $\cos\frac{\pi}{12} \doteq 0.97$

<u>7</u>. $\cos\left(\frac{\pi}{3} - \frac{\pi}{2}\right) = \cos\frac{\pi}{3}\cos\frac{\pi}{2} + \sin\frac{\pi}{3}\sin\frac{\pi}{2}$

$$= \frac{1}{2}\cdot 0 + \frac{\sqrt{3}}{2}\cdot 1$$

$$\doteq \frac{1.732}{2} \doteq 0.87$$

Thus $\cos\left(-\frac{\pi}{6}\right) \doteq 0.87$

<u>8</u>. $\cos\left(\frac{\pi}{3} - \frac{2\pi}{3}\right) = \cos\frac{\pi}{3}\cos\frac{2\pi}{3} + \sin\frac{\pi}{3}\sin\frac{2\pi}{3}$

$$= \frac{1}{2}\cdot\left(-\frac{1}{2}\right) + \frac{\sqrt{3}}{2}\cdot\frac{\sqrt{3}}{2}$$

$$= \frac{-1 + 3}{4} = \frac{1}{2} = 0.5$$

Thus $\cos\left(-\frac{\pi}{3}\right) = 0.5$

9. $\cos \left(\frac{\pi}{3} - \frac{5\pi}{6}\right) = \cos \frac{\pi}{3} \cos \frac{5\pi}{6} + \sin \frac{\pi}{3} \cos \frac{5\pi}{6}$

$$= \frac{1}{2} \cdot \left(-\frac{\sqrt{3}}{2}\right) + \frac{\sqrt{3}}{2} \cdot \frac{1}{2}$$

$$\doteq \frac{-1.732 + 1.732}{4} = 0$$

Thus $\cos \left(-\frac{\pi}{2}\right) = 0$

10. $\cos \left(\frac{\pi}{3} - \frac{3\pi}{4}\right) = \cos \frac{\pi}{3} \cos \frac{3\pi}{4} + \sin \frac{\pi}{3} \sin \frac{3\pi}{4}$

$$= \frac{1}{2} \left(-\frac{\sqrt{2}}{2}\right) + \frac{\sqrt{3}}{2} \left(\frac{\sqrt{2}}{2}\right)$$

$$\doteq \frac{-1.414 + 2.449}{4} \doteq 0.26$$

Thus $\cos \left(-\frac{5\pi}{12}\right) \doteq 0.26$

In Ex. 11-18, $\left(\frac{5}{13}\right)^2 + \sin^2 x_1 = 1$ (since $\cos x_1 = \frac{5}{13}$) and $\left(\frac{3}{5}\right)^2 + \sin^2 x_2 = 1$ (since

$\cos x_2 = \frac{3}{5}$). Hence $\sin x_1 = \pm\frac{12}{13}$ and $\sin x_2 = \pm\frac{4}{5}$, depending on the specified values

of x_1 and x_2.

11. Since $0 < x_2 < \frac{\pi}{2}$ and $0 < x_1 < \frac{\pi}{2}$,

$\sin x_2$ and $\sin x_1$ are both positive.

Thus, $\sin x_2 = \frac{4}{5}$ and $\sin x_1 = \frac{12}{13}$.

$\cos (x_1 - x_2) = \cos x_1 \cos x_2 +$

$\sin x_1 \sin x_2 = \frac{5}{13} \cdot \frac{3}{5} + \frac{12}{13} \cdot \frac{4}{5} =$

$\frac{15}{65} + \frac{48}{65} = \frac{63}{65}$.

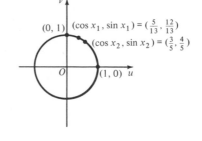

12. Since $\frac{3\pi}{2} < x_2 < 2\pi$ and $0 < x_1 < \frac{\pi}{2}$,

$\sin x_2 < 0$ and $\sin x_1 > 0$.

Thus $\sin x_2 = -\frac{4}{5}$ and $\sin x_1 = \frac{12}{13}$.

$\cos (x_2 - x_1) = \cos x_2 \cos x_1 +$

$\sin x_2 \sin x_1 = \frac{3}{5} \cdot \frac{5}{13} + \left(-\frac{4}{5}\right) \cdot \frac{12}{13} =$

$\frac{15}{65} - \frac{48}{65} = -\frac{33}{65}$.

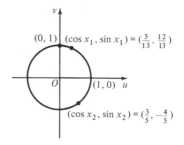

<u>13.</u> Since $0 < x_2 < \frac{\pi}{2}$, $\sin x_2 > 0$.

Hence $\sin x_2 = \frac{4}{5}$.

$\cos\left(\frac{\pi}{4} + x_2\right) = \cos\frac{\pi}{4}\cos x_2 -$

$\sin\frac{\pi}{4}\sin x_2 = \frac{\sqrt{2}}{2}\cdot\frac{3}{5} - \frac{\sqrt{2}}{2}\cdot\frac{4}{5} =$

$\frac{3\sqrt{2}}{10} - \frac{4\sqrt{2}}{10} = -\frac{\sqrt{2}}{10}$

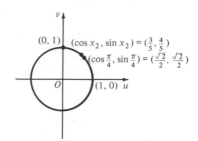

<u>14.</u> Since $\frac{3\pi}{2} < x_2 < 2\pi$ and $\frac{3\pi}{2} < x_1 < 2\pi$,

$\sin x_2$ and $\sin x_1$ are both negative.

Thus, $\sin x_2 = -\frac{4}{5}$ and $\sin x_1 = -\frac{12}{13}$.

$\cos(x_2 + x_1) = \cos x_2 \cos x_1 -$

$\sin x_2 \sin x_1 = \frac{3}{5}\cdot\frac{5}{13} - \left(-\frac{4}{5}\right)\left(-\frac{12}{13}\right) =$

$\frac{15}{65} - \frac{48}{65} = -\frac{33}{65}$.

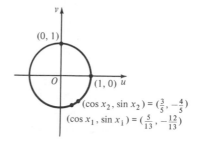

<u>15.</u> Since $\frac{3\pi}{2} < x_2 < 2\pi$, $\sin x_2 < 0$.

Hence $\sin x_2 = -\frac{4}{5}$.

$\cos\left(x_2 - \frac{\pi}{6}\right) = \cos x_2 \cos\frac{\pi}{6} +$

$\sin x_2 \sin\frac{\pi}{6} = \frac{3}{5}\cdot\frac{\sqrt{3}}{2} + \left(-\frac{4}{5}\right)\frac{1}{2} =$

$\frac{3\sqrt{3}}{10} - \frac{4}{10} = \frac{3\sqrt{3} - 4}{10} \doteq 0.12$.

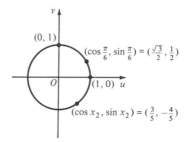

<u>16.</u> Since $0 < x_1 < \frac{\pi}{2}$, $\sin x_1 > 0$.

Hence $\sin x_1 = \frac{12}{13}$.

$\cos\left(\frac{\pi}{3} - x_1\right) = \cos\frac{\pi}{3}\cos x_1 +$

$\sin\frac{\pi}{3}\sin x_1 = \frac{1}{2}\cdot\frac{5}{13} + \frac{\sqrt{3}}{2}\cdot\frac{12}{13} =$

$\frac{5}{26} + \frac{12\sqrt{3}}{26} \doteq 0.80$.

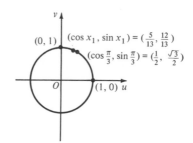

<u>17.</u> Since $0 < x_1 \le \frac{\pi}{2}$, $\sin x_1 > 0$.

Hence $\sin x_1 = \frac{12}{13}$.

$\cos\left(x_1 - \frac{\pi}{3}\right) = \cos x_1 \cos\frac{\pi}{3} +$

$\sin x_1 \sin\frac{\pi}{3} = \frac{5}{13}\cdot\frac{1}{2} + \frac{12}{13}\cdot\frac{\sqrt{3}}{2} =$

$\frac{5}{26} + \frac{12\sqrt{3}}{26} \doteq 0.99$.

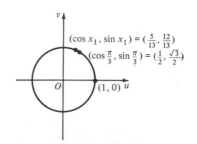

<u>18.</u> Since $\frac{3}{2}\pi < x_2 < 2\pi$, $\sin x_2 < 0$.

Hence $\sin x_2 = -\frac{4}{5}$.

$\cos (x_2 - \frac{\pi}{4}) = \cos x_2 \cos \frac{\pi}{4} +$

$\sin x_2 \sin \frac{\pi}{4} = \frac{3}{5} \cdot \frac{\sqrt{2}}{2} + (-\frac{4}{5}) \cdot \frac{\sqrt{2}}{2} =$

$\frac{3\sqrt{2} - 4\sqrt{2}}{10} = \frac{-\sqrt{2}}{10} \doteq 0.14$

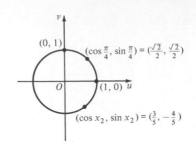

<u>19.</u>

x	cos x	sin x
0	1	0
$\frac{\pi}{6}$	$\frac{\sqrt{3}}{2} \doteq 0.87$	$\frac{1}{2} = 0.50$
$\frac{\pi}{4}$	$\frac{\sqrt{2}}{2} \doteq 0.71$	$\frac{\sqrt{2}}{2} \doteq 0.71$
$\frac{\pi}{3}$	$\frac{1}{2} = 0.5$	$\frac{\sqrt{3}}{2} \doteq 0.87$
$\frac{\pi}{2}$	0	1
$-\frac{\pi}{6}$	$\frac{\sqrt{3}}{2} \doteq 0.87$	$-\frac{1}{2} = -0.50$
$-\frac{\pi}{4}$	$\frac{\sqrt{2}}{2} \doteq 0.71$	$-\frac{\sqrt{2}}{2} \doteq -0.71$
$-\frac{\pi}{2}$	0	-1

<u>20.</u> <u>a.</u> (1)   <u>b.</u> (5)   <u>c.</u> (4)   <u>d.</u> (10)   <u>e.</u> (7)   <u>f.</u> (8)   <u>g.</u> (6)   <u>h.</u> (7)

<u>B</u>   <u>21.</u> $\cos \frac{\pi}{4} \sin x - \sin \frac{\pi}{4} \cos x = 0$.   Substituting values for $\cos \frac{\pi}{4}$ and $\sin \frac{\pi}{4}$, we have

$\frac{\sqrt{2}}{2} \sin x - \frac{\sqrt{2}}{2} \cos x = 0$.   Divide all members by $\frac{\sqrt{2}}{2}$, yielding $\sin x - \cos x = 0$.

Hence $\sin x = \cos x$ and $x = \frac{\pi}{4}$ or $\frac{5\pi}{4}$.

<u>22.</u> $\cos (x - \frac{\pi}{3}) = \frac{1}{2}$.   Since $\cos (-\frac{\pi}{3}) = \frac{1}{2}$, $\cos \frac{\pi}{3} = \frac{1}{2}$, and $\cos \frac{5\pi}{3} = \frac{1}{2}$, $x - \frac{\pi}{3} = -\frac{\pi}{3}$ or

$x - \frac{\pi}{3} = \frac{\pi}{3}$ or $x - \frac{\pi}{3} = \frac{5\pi}{3}$.   Hence $x = 0$, $x = \frac{2\pi}{3}$, or $x = 2\pi$.

<u>23.</u> $\cos (\frac{\pi}{2} - x) + \sin (\frac{\pi}{2} - x) = 0$.   Using the identities for $\cos (\frac{\pi}{2} - x)$ and $\sin (\frac{\pi}{2} - x)$

we have $\sin x + \cos x = 0$.   Thus $\sin x = - \cos x$ and $x = \frac{3\pi}{4}$ or $\frac{7\pi}{4}$.

<u>C</u>   <u>24.</u> $\cos (x + \frac{\pi}{3}) \cos \frac{\pi}{6} - \sin (x + \frac{\pi}{3}) \sin \frac{\pi}{6} = \cos [(x + \frac{\pi}{3}) + \frac{\pi}{6}] = \cos (x + \frac{\pi}{2}) =$

$\cos (\frac{\pi}{2} + x) = -\sin x$.

$\underline{25.}$ $\cos (x + \frac{\pi}{3}) \cos x + \sin (x + \frac{\pi}{3}) \sin x = \cos [(x + \frac{\pi}{3}) - x] = \cos \frac{\pi}{3} = \frac{1}{2}$

$\underline{26.}$ $\cos (x - y) \cos y - \sin (x - y) \sin y = \cos [(x - y) + y] = \cos x.$

Exercises 1-5 · Pages 26-27

$\underline{A}$ $\underline{1.}$ Since $\cos (x_2 - x_1) = \cos x_2 \cos x_1 + \sin x_2 \sin x_1,$

$\cos (x - \frac{\pi}{2}) = \cos x \cos \frac{\pi}{2} + \sin x \sin \frac{\pi}{2}$

$= \cos x \cdot 0 + \sin x \cdot 1$

$= 0 + \sin x = \sin x$

$\underline{2.}$ Since $\cos (x_2 + x_1) = \cos x_2 \cos x_1 - \sin x_2 \sin x_1,$

$\cos (\pi + x) = \cos \pi \cos x - \sin \pi \sin x$

$= -1 \cdot \cos x - 0 \cdot \sin x$

$= -\cos x - 0 = -\cos x$

$\underline{3.}$ Since $\cos (x_2 - x_1) = \cos x_2 \cos x_1 + \sin x_2 \sin x_1,$

$\cos (x - \pi) = \cos x \cos \pi + \sin x \sin \pi$

$= \cos x (-1) + \sin x \cdot 0$

$= -\cos x + 0 = -\cos x$

$\underline{4.}$ Since $\cos (x_2 - x_1) = \cos x_2 \cos x_1 + \sin x_2 \sin x_1,$

$\cos (x - \frac{\pi}{3}) = \cos x \cos \frac{\pi}{3} + \sin x \sin \frac{\pi}{3}$

$= \cos x \cdot \frac{1}{2} + \sin x \cdot \frac{\sqrt{3}}{2}$

$= \dfrac{\cos x + \sqrt{3} \sin x}{2}$

$\underline{5.}$ Since $\cos (x_2 - x_1) = \cos x_2 \cos x_1 + \sin x_2 \sin x_1,$

$\cos (2\pi - x) = \cos 2\pi \cos x + \sin 2\pi \sin x$

$= 1 \cdot \cos x + 0 \cdot \sin x$

$= \cos x + 0 = \cos x$

$\underline{6.}$ Since $\cos (x_2 + x_1) = \cos x_2 \cos x_1 - \sin x_2 \sin x_1,$

$\cos (\frac{\pi}{4} + x) = \cos \frac{\pi}{4} \cos x - \sin \frac{\pi}{4} \sin x$

$= \frac{\sqrt{2}}{2} \cos x - \frac{\sqrt{2}}{2} \sin x$

$= \dfrac{\sqrt{2} (\cos x - \sin x)}{2}$

7. a. (2) by prop. (i)   b. (12) by prop. (ii)   c. (5) by prop. (iii)

   d. (6) by prop. (iii)   e. (9) by prop. (ii)   f. (1) by prop. (i)

   g. (10) by prop. (i)   h. (3) by prop. (i)

8. $\sin^2 x = 1 - \cos^2 x$.   By adding $\cos^2 x$ to both sides, we have the known identity
   $\sin^2 x + \cos^2 x = 1$.

9. $2 \cos^2 x - 1 = \cos^2 x - \sin^2 x$.   By subtracting $\cos^2 x$ from each side, we have
   $\cos^2 x - 1 = -\sin^2 x$.   Adding $\sin^2 x$ to each member gives $\cos^2 x + \sin^2 x - 1 = 0$.
   Adding 1 to each side gives the known identity, $\cos^2 x + \sin^2 x = 1$.

10. Begin with the left-hand member $\sin x \cos^2 x + \sin^3 x$. $\sin x \cos^2 x + \sin^3 x =$
    $\sin x (\cos^2 x + \sin^2 x) = \sin x \cdot 1 = \sin x$.   Thus we have the equivalent sentence
    $\sin x = \sin x$ and the given sentence is an identity.

11. $1 - 2 \cos^2 x = 2 \sin^2 x - 1$.   Adding $1 + 2 \cos^2 x$ to each member gives
    $2 = 2 \sin^2 x + 2 \cos^2 x$.   Dividing each member by 2 yields the known identity
    $1 = \sin^2 x + \cos^2 x$.

12. $\cos (-x) \cos x = \cos x \cos x = \cos^2 x$ since $\cos (-x) = \cos x$.   Thus we have
    the equivalent sentence $\cos^2 x = 1 - \sin^2 x$.   Adding $\sin^2 x$ to each member gives
    the known identity $\cos^2 x + \sin^2 x = 1$.

13. $\sin (-x) \sin x = -\sin x \sin x = -\sin^2 x$, since $\sin (-x) = -\sin x$.   Thus we have
    the equivalent sentence $-\sin^2 x = \cos^2 x - 1$.   Adding $\sin^2 x$ to each member and
    then adding 1 to each member yields the known identity $1 = \sin^2 x + \cos^2 x$.

14. Begin with $(1 + \sin x)(1 - \sin x)$, which equals $1 - \sin^2 x$.   Thus we have the
    equivalent sentence $1 - \sin^2 x = \cos^2 x$.   Adding $\sin^2 x$ to each member gives the
    known identity $1 = \sin^2 x + \cos^2 x$.

15. Begin with the left-hand member. $\dfrac{(\sin x + \cos x)^2 - 1}{\sin x \cos x} =$
    $\dfrac{\sin^2 x + 2 \sin x \cos x + \cos^2 x - 1}{\sin x \cos x} = \dfrac{2 \sin x \cos x + 1 - 1}{\sin x \cos x}$ since $\sin^2 x + \cos^2 x = 1$.

    $\dfrac{2 \sin x \cos x + 1 - 1}{\sin x \cos x} = \dfrac{2 \sin x \cos x}{\sin x \cos x} = 2$.   Thus we have the equivalent sentence

    $2 = 2$ and the given sentence is an identity.

B 16. Begin with $\cos (\frac{\pi}{2} - x) \sin x + \sin (\frac{\pi}{2} - x) \cos x$.   This equals $\sin x \sin x +$
       $\cos x \cos x$ since $\cos (\frac{\pi}{2} - x) = \sin x$ and $\sin (\frac{\pi}{2} - x) = \cos x$.

$\sin x \sin x + \cos x \cos x = \sin^2 x + \cos^2 x = 1$.　　Thus we have the equivalent sentence $1 = 1$ and the given sentence is an identity.

17. $\sin^4 x - \cos^4 x = (\sin^2 x + \cos^2 x)(\sin^2 x - \cos^2 x) = 1(\sin^2 x - \cos^2 x)$.　　This gives the equivalent sentence $\sin^2 x - \cos^2 x = 2 \sin^2 x - 1$.　　By adding $\cos^2 x - \sin^2 x$ to each member, we have $0 = \sin^2 x + \cos^2 x - 1$, which becomes the known identity $1 = \sin^2 x + \cos^2 x$ when 1 is added to each member.

18. Apply the addition properties of cos to the left-hand member.　　$\cos (x - \pi) - \cos (x + \pi) = \cos x \cos \pi + \sin x \sin \pi - [\cos x \cos \pi - \sin x \sin \pi] = \cos x (-1) + \sin x \cdot 0 - \cos x (-1) + \sin x \cdot 0 = -\cos x + \cos x = 0$.　　Thus we have the equivalent sentence $0 = 0$, so the given sentence is an identity.

19. Apply the addition properties of cos to the left-hand member.
$\cos (x_2 - x_1) \cos (x_2 + x_1) =$
$[\cos x_2 \cos x_1 + \sin x_2 \sin x_1][\cos x_2 \cos x_1 - \sin x_2 \sin x_1] =$
$\cos^2 x_2 \cos^2 x_1 - \sin^2 x_2 \sin^2 x_1$.　　Since $\sin^2 x_1 + \cos^2 x_1 = 1$, we can replace $\cos^2 x_1$ with $1 - \sin^2 x_1$.　　Thus $\cos^2 x_2 \cos^2 x_1 - \sin^2 x_2 \sin^2 x_1 =$
$\cos^2 x_2 (1 - \sin^2 x_1) - \sin^2 x_2 \sin^2 x_1 = \cos^2 x_2 - \cos^2 x_2 \sin^2 x_1 - \sin^2 x_2 \sin^2 x_1 =$
$\cos^2 x_2 - \sin^2 x_1(\cos^2 x_2 + \sin^2 x_2) = \cos^2 x_2 - \sin^2 x_1 \cdot 1$.　　Thus we have the equivalent sentence $\cos^2 x_2 - \sin^2 x_1 = \cos^2 x_2 - \sin^2 x_1$ and the given sentence is an identity.

20. $\cos \frac{\pi}{6} \cos x + \sin \frac{\pi}{6} \sin x = \cos (\frac{\pi}{6} - x)$.　　Since $\cos 0 = 1$, $\frac{\pi}{6} - x = 0$ and $x = \frac{\pi}{6}$.

21. $\cos \frac{\pi}{6} \cos x + \sin \frac{\pi}{6} \sin x = \cos (\frac{\pi}{6} - x)$.　　$\cos \frac{\pi}{6} = \frac{\sqrt{3}}{2}$, so $\frac{\pi}{6} - x = \frac{\pi}{6}$ and $x = 0$.

22. $\cos \frac{\pi}{3} \cos x - \sin \frac{\pi}{3} \sin x = \cos (\frac{\pi}{3} + x)$.　　$\frac{\pi}{3} + x = \frac{\pi}{2}$, so $x = \frac{\pi}{6}$.

23. $\cos \frac{2\pi}{3} \cos (-\frac{\pi}{6}) - \sin \frac{2\pi}{3} \sin (-\frac{\pi}{6}) = \cos (\frac{2\pi}{3} + (-\frac{\pi}{6})) = \cos \frac{\pi}{2} = 0$, so $x = 0$.

24. $\sin (-\frac{\pi}{3}) \sin \frac{\pi}{2} + \cos (-\frac{\pi}{3}) \cos \frac{\pi}{2} = \cos (-\frac{\pi}{3} - \frac{\pi}{2}) = \cos (-\frac{5\pi}{6}) = \cos \frac{5\pi}{6} = -\frac{\sqrt{3}}{2}$.
Thus there is no non-negative value of x for which this is true.

Exercises 1-6 · Pages 30-31

<u>A</u>   <u>1</u>. $\sin \frac{5\pi}{6} = \sin (-\frac{\pi}{6} + \pi)$ <u>2</u>. $\sin \frac{13\pi}{12} = \sin (\frac{\pi}{12} + \pi)$

$= -\sin (-\frac{\pi}{6})$ (Iden. 6) $= -\sin \frac{\pi}{12}$ (Iden. 6)

$= -(-\sin \frac{\pi}{6})$ (Iden. 8)

$= \sin \frac{\pi}{6}$

<u>3</u>. $\sin \frac{11\pi}{3} = \sin (-\frac{\pi}{3} + 4\pi)$ <u>4</u>. $\cos \frac{5\pi}{3} = \cos (-\frac{\pi}{3} + 2\pi)$

$= \sin (-\frac{\pi}{3})$ (Iden. 5) $= \cos (-\frac{\pi}{3})$ (Iden. 1)

$= -\sin \frac{\pi}{3}$ (Iden. 8) $= \cos \frac{\pi}{3}$ (Iden. 4)

<u>5</u>. $\cos (-\frac{4\pi}{3}) = \cos (-\frac{\pi}{3} - \pi)$ <u>6</u>. $\cos \frac{11\pi}{2} = \cos (-\frac{\pi}{2} + 6\pi)$

$= -\cos (-\frac{\pi}{3})$ (Iden. 2) $= \cos (-\frac{\pi}{2})$ (Iden. 1)

$= -\cos \frac{\pi}{3}$ (Iden. 4) $= \cos \frac{\pi}{2}$ (Iden. 4)

<u>7</u>. $\sin (-\frac{5\pi}{4}) = \sin (-\frac{\pi}{4} - \pi)$ <u>8</u>. $\sin (-\frac{2\pi}{3}) = \sin (\frac{\pi}{3} - \pi)$

$= -\sin (-\frac{\pi}{4})$ (Iden. 6) $= -\sin \frac{\pi}{3}$ (Iden. 6)

$= \sin \frac{\pi}{4}$ (Iden. 8)

<u>9</u>. $\cos \frac{40\pi}{13} = \cos (-\frac{12\pi}{13} + 4\pi)$ <u>10</u>. $\cos (-\frac{8\pi}{3}) = \cos (\frac{4\pi}{3} - 4\pi)$

$= \cos (-\frac{12\pi}{13})$ (Iden. 1) $= \cos \frac{4\pi}{3}$ (Iden. 1)

$= \cos (\frac{\pi}{13} - \pi)$ $= \cos (\frac{\pi}{3} + \pi)$

$= -\cos \frac{\pi}{13}$ (Iden. 2) $= -\cos \frac{\pi}{3}$ (Iden. 2)

<u>11</u>. $\sin (-\frac{5\pi}{4}) = \sin (-\frac{\pi}{4} - \pi)$ <u>12</u>. $\cos \frac{7\pi}{3} = \cos (\frac{\pi}{3} + 2\pi)$

$= -\sin (-\frac{\pi}{4})$ (Iden. 6) $= \cos \frac{\pi}{3}$ (Iden. 1)

$= \sin \frac{\pi}{4}$ (Iden. 8)

<u>13</u>. If $x = \frac{\pi}{6}$, $\sin x = \frac{1}{2}$ and $\cos x = \frac{\sqrt{3}}{2}$. $\sin (\frac{\pi}{6} + \frac{\pi}{4}) = \sin \frac{\pi}{6} \cos \frac{\pi}{4} + \cos \frac{\pi}{6} \sin \frac{\pi}{4} =$

$\frac{1}{2} \cdot \frac{\sqrt{2}}{2} + \frac{\sqrt{3}}{2} \cdot \frac{\sqrt{2}}{2} = \frac{\sqrt{2} + \sqrt{6}}{4}$.   Hence $\sin \frac{5\pi}{12} = \frac{\sqrt{2} + \sqrt{6}}{4}$. $\sin (\frac{\pi}{6} + \frac{\pi}{2}) =$

$\sin \frac{\pi}{6} \cos \frac{\pi}{2} + \cos \frac{\pi}{6} \sin \frac{\pi}{2} = \frac{1}{2} \cdot 0 + \frac{\sqrt{3}}{2} \cdot 1 = \frac{\sqrt{3}}{2}$.   Hence $\sin \frac{2\pi}{3} = \frac{\sqrt{3}}{2}$.

$\sin\left(\frac{\pi}{6} + \frac{3\pi}{4}\right) = \sin\frac{\pi}{6}\cos\frac{3\pi}{4} + \cos\frac{\pi}{6}\sin\frac{3\pi}{4} = \frac{1}{2}\left(-\frac{\sqrt{2}}{2}\right) + \frac{\sqrt{3}}{2}\cdot\frac{\sqrt{2}}{2} = \frac{-\sqrt{2} + \sqrt{6}}{4}.$

Hence $\sin\frac{11\pi}{12} = \frac{-\sqrt{2} + \sqrt{6}}{4}.$

<u>14.</u> If $x = \frac{\pi}{3}$, $\sin x = \frac{\sqrt{3}}{2}$ and $\cos x = \frac{1}{2}$. $\sin\left(\frac{\pi}{3} + \frac{\pi}{4}\right) = \sin\frac{\pi}{3}\cos\frac{\pi}{4} + \cos\frac{\pi}{3}\sin\frac{\pi}{4} =$

$\frac{\sqrt{3}}{2}\cdot\frac{\sqrt{2}}{2} + \frac{1}{2}\cdot\frac{\sqrt{2}}{2} = \frac{\sqrt{6} + \sqrt{2}}{4}.$ Hence $\sin\frac{7\pi}{12} = \frac{\sqrt{6} + \sqrt{2}}{4}.$ $\sin\left(\frac{\pi}{3} + \frac{\pi}{2}\right) =$

$\sin\frac{\pi}{3}\cos\frac{\pi}{2} + \cos\frac{\pi}{3}\sin\frac{\pi}{2} = \frac{\sqrt{3}}{2}\cdot 0 + \frac{1}{2}\cdot 1 = \frac{1}{2}.$ Hence $\sin\frac{5\pi}{6} = \frac{1}{2}.$

$\sin\left(\frac{\pi}{3} + \frac{3\pi}{4}\right) = \sin\frac{\pi}{3}\cos\frac{3\pi}{4} + \cos\frac{\pi}{3}\sin\frac{3\pi}{4} = \frac{\sqrt{3}}{2}\left(-\frac{\sqrt{2}}{2}\right) + \frac{1}{2}\cdot\frac{\sqrt{2}}{2} = \frac{-\sqrt{6} + \sqrt{2}}{4}.$

Hence $\sin\frac{13\pi}{12} = \frac{-\sqrt{6} + \sqrt{2}}{4}.$

<u>15.</u> If $x = \frac{3\pi}{2}$, $\sin x = -1$ and $\cos x = 0$. $\sin\left(\frac{3\pi}{2} + \frac{\pi}{4}\right) = \sin\frac{3\pi}{2}\cos\frac{\pi}{4} +$

$\cos\frac{3\pi}{2}\sin\frac{\pi}{4} = -1\cdot\frac{\sqrt{2}}{2} + 0\cdot\frac{\sqrt{2}}{2} = -\frac{\sqrt{2}}{2}.$ Hence $\sin\frac{7\pi}{4} = -\frac{\sqrt{2}}{2}.$ $\sin\left(\frac{3\pi}{2} + \frac{\pi}{2}\right) =$

$\sin\frac{3\pi}{2}\cos\frac{\pi}{2} + \cos\frac{3\pi}{2}\sin\frac{\pi}{2} = -1\cdot 0 + 0\cdot 1 = 0.$ Hence $\sin 2\pi = 0.$

$\sin\left(\frac{3\pi}{2} + \frac{3\pi}{4}\right) = \sin\frac{3\pi}{2}\cos\frac{3\pi}{4} + \cos\frac{3\pi}{2}\sin\frac{3\pi}{4} = -1\left(-\frac{\sqrt{2}}{2}\right) + 0\cdot\frac{\sqrt{2}}{2} = \frac{\sqrt{2}}{2}.$

Hence $\sin\frac{9\pi}{4} = \frac{\sqrt{2}}{2}.$

<u>16.</u> If $x = \frac{2\pi}{3}$, $\sin x = \frac{\sqrt{3}}{2}$ and $\cos x = -\frac{1}{2}$. $\sin\left(\frac{2\pi}{3} + \frac{\pi}{4}\right) = \sin\frac{2\pi}{3}\cos\frac{\pi}{4} +$

$\cos\frac{2\pi}{3}\sin\frac{\pi}{4} = \frac{\sqrt{3}}{2}\cdot\frac{\sqrt{2}}{2} + \left(-\frac{1}{2}\right)\cdot\frac{\sqrt{2}}{2} = \frac{\sqrt{6} - \sqrt{2}}{4}.$ Hence $\sin\frac{11\pi}{12} = \frac{\sqrt{6} - \sqrt{2}}{4}.$

$\sin\left(\frac{2\pi}{3} + \frac{\pi}{2}\right) = \sin\frac{2\pi}{3}\cos\frac{\pi}{2} + \cos\frac{2\pi}{3}\sin\frac{\pi}{2} = \frac{\sqrt{3}}{2}\cdot 0 + \left(-\frac{1}{2}\right)\cdot 1 = -\frac{1}{2}.$

Hence $\sin\frac{7\pi}{6} = -\frac{1}{2}.$ $\sin\left(\frac{2\pi}{3} + \frac{3\pi}{4}\right) = \sin\frac{2\pi}{3}\cos\frac{3\pi}{4} + \cos\frac{2\pi}{3}\sin\frac{3\pi}{4} =$

$\frac{\sqrt{3}}{2}\left(-\frac{\sqrt{2}}{2}\right) + \left(-\frac{1}{2}\right)\left(\frac{\sqrt{2}}{2}\right) = \frac{-\sqrt{6} - \sqrt{2}}{4}.$ Hence $\sin\frac{17\pi}{12} = \frac{-\sqrt{6} - \sqrt{2}}{4}.$

<u>17.</u> If $\sin x_2 = \frac{12}{13}$, $\cos^2 x_2 + \left(\frac{12}{13}\right)^2 = 1$, so $\cos x_2 = \pm\frac{5}{13}.$ Since $\frac{\pi}{2} < x_2 < \pi$,

$\cos x_2 = -\frac{5}{13}.$ If $\sin x_1 = \frac{3}{5}$, $\cos^2 x_1 + \left(\frac{3}{5}\right)^2 = 1$, so $\cos x_1 = \pm\frac{4}{5}.$ Since

$0 < x_1 < \frac{\pi}{2}$, $\cos x_1 = \frac{4}{5}.$ $\sin(x_2 + x_1) = \sin x_2 \cos x_1 + \cos x_2 \sin x_1 =$

$\frac{12}{13}\cdot\frac{4}{5} + \left(-\frac{5}{13}\right)\cdot\frac{3}{5} = \frac{33}{65}.$ Since $\sin(x_2 + x_1) > 0$, and given the locations of

x_2 and x_1, $x_2 + x_1$ is in the second quadrant.

18. See Ex. 17 for values of $\cos x_2$ and $\cos x_1$. $\sin (x_2 - x_1) = \sin x_2 \cos x_1 -$
$\cos x_2 \sin x_1 = \frac{12}{13} \cdot \frac{4}{5} - (-\frac{5}{13}) \cdot \frac{3}{5} = \frac{63}{65}$. $x_2 - x_1$ is in Quadrant I or II.

19. $\sin (x_2 + \pi) + \sin (x_1 + \pi) = -\sin x_2 + (-\sin x_1) = -\frac{12}{13} + (-\frac{3}{5}) = -\frac{99}{65}$.

20. $\sin (-x_2) + \sin (-x_1) = -\sin x_2 + (-\sin x_1) = -\frac{12}{13} + (-\frac{3}{5}) = -\frac{99}{65}$.

21. See Ex. 17 for values of $\cos x_2$ and $\cos x_1$. $\cos (-x_2) + \cos (-x_1) =$
$\cos x_2 + \cos x_1 = -\frac{5}{13} + \frac{4}{5} = \frac{27}{65}$.

22. $\sin (\pi - x_2) = \sin x_2 = \frac{12}{13}$.

23. If $\sin x_2 = \frac{5}{13}$, $\cos^2 x_2 + (\frac{5}{13})^2 = 1$ so $\cos x_2 = \pm \frac{12}{13}$. Since $0 < x_2 < \frac{\pi}{2}$,
$\cos x_2 = \frac{12}{13}$. If $\sin x_1 = (-\frac{4}{5})$, $\cos^2 x_1 + (-\frac{4}{5})^2 = 1$ so $\cos x_1 = \pm \frac{3}{5}$. Since
$\pi < x_1 < \frac{3\pi}{2}$, $\cos x_1 = -\frac{3}{5}$. $\sin (x_2 + x_1) = \sin x_2 \cos x_1 + \cos x_2 \sin x_1 =$
$\frac{5}{13}(-\frac{3}{5}) + \frac{12}{13}(-\frac{4}{5}) = -\frac{63}{65}$. $x_2 + x_1$ is in Quadrant III or IV.

24. See Ex. 23 for values of $\cos x_2$ and $\cos x_1$. $\sin (x_2 - x_1) = \sin x_2 \cos x_1 -$
$\cos x_2 \sin x_1 = \frac{5}{13}(-\frac{3}{5}) - \frac{12}{13}(-\frac{4}{5}) = \frac{33}{65}$. $x_2 - x_1$ is in Quadrant II.

25. $\sin (x_2 + \pi) + \sin (x_1 + \pi) = -\sin x_2 + (-\sin x_1) = -\frac{5}{13} + (+\frac{4}{5}) = \frac{27}{65}$.

26. $\sin (-x_2) - \sin (-x_1) = -\sin x_2 - (-\sin x_1) = -\frac{5}{13} - \frac{4}{5} = -\frac{77}{65}$.

27. See Ex. 23 for values of $\cos x_2$ and $\cos x_1$. $\cos (-x_2) - \cos (-x_1) = \cos x_2 - \cos x_1$
$= \frac{12}{13} - (-\frac{3}{5}) = \frac{99}{65}$

28. $\sin (\pi - x_2) = \sin x_2 = \frac{5}{13}$.

29. If $\cos x_2 = -\frac{3}{5}$, $\sin^2 x_2 + (-\frac{3}{5})^2 = 1$ so $\sin x_2 = \pm\frac{4}{5}$. Since $\pi < x_2 < \frac{3\pi}{2}$, $\sin x_2 =$
$-\frac{4}{5}$. If $\cos x_1 = \frac{5}{13}$, $\sin^2 x_1 + (\frac{5}{13})^2 = 1$ so $\sin x_1 = \pm\frac{12}{13}$. Since $\frac{3\pi}{2} < x_1 < 2\pi$,
$\sin x_1 = -\frac{12}{13}$. $\sin (x_2 + x_1) = \sin x_2 \cos x_1 + \cos x_2 \sin x_1 =$
$(-\frac{4}{5}) \cdot \frac{5}{13} + (-\frac{3}{5})(-\frac{12}{13}) = -\frac{20}{65} + \frac{36}{65} = \frac{16}{65}$. $x_2 + x_1$ is in Quadrant II.

30. See Ex. 29 for values of $\sin x_2$ and $\sin x_1$. $\sin (x_2 - x_1) = \sin x_2 \cos x_1 -$

$\cos x_2 \sin x_1 = (-\frac{4}{5}) \cdot \frac{5}{13} - (-\frac{3}{5})(-\frac{12}{13}) = -\frac{20}{65} - \frac{36}{65} = -\frac{56}{65}.$ $x_2 - x_1$ is in

Quadrant III or IV.

31. See Ex. 29 for values of $\sin x_2$ and $\sin x_1$. $\sin (x_2 - \pi) + \sin (x_1 - \pi) =$

$-\sin x_2 + (-\sin x_1) = \frac{4}{5} + \frac{12}{13} = \frac{112}{65}.$

32. See Ex. 29 for value of $\sin x_2$. $\sin (\pi - x_2) + \cos (\pi - x_2) = \sin x_2 + (-\cos x_2) =$

$-\frac{4}{5} + \frac{3}{5} = -\frac{1}{5}.$

33. See Ex. 29 for values of $\sin x_2$ and $\sin x_1$. $\sin (-x_2) + \sin (-x_1) =$

$-\sin x_2 + (-\sin x_1) = \frac{4}{5} + \frac{12}{13} = \frac{112}{65}.$

34. $\cos (-x_2) - \cos (-x_1) = \cos x_2 - \cos x_1 = -\frac{3}{5} - \frac{5}{13} = -\frac{64}{65}.$

B 35.

x	cos x	sin x
$\frac{\pi}{12}$	$\frac{\sqrt{2} + \sqrt{6}}{4} \doteq 0.97$	$\frac{\sqrt{6} - \sqrt{2}}{4} \doteq 0.26$
$\frac{5\pi}{12}$	$\frac{\sqrt{6} - \sqrt{2}}{4} \doteq 0.26$	$\frac{\sqrt{2} + \sqrt{6}}{4} \doteq 0.97$

36. $\sin (x + a) = \sin x \cos a + \cos x \sin a.$ If $\sin (x + a) = \sin x$, $\cos a = 1$ and

$\sin a = 0.$ Thus $a = 0.$

37. $\sin (x + a) = \sin x \cos a + \cos x \sin a.$ If $\sin (x + a) = -\sin x$, $\cos a = -1$

and $\sin a = 0.$ Thus $a = \pi.$

38. $\sin (x + a) = \sin x \cos a + \cos x \sin a.$ If $\sin (x + a) = \cos x$, $\cos a = 0$ and

$\sin a = 1.$ Thus $a = \frac{\pi}{2}.$

39. $\sin (x + a) = \sin x \cos a + \cos x \sin a.$ If $\sin (x + a) = \frac{\sqrt{2}}{2} (\sin x + \cos x)$,

then $\cos a = \sin a = \frac{\sqrt{2}}{2}$ and $a = \frac{\pi}{4}.$

40. $\sin (x + a) = \sin x \cos a + \cos x \sin a.$ If $\sin (x + a) = \frac{\sqrt{2}}{2} (\sin x - \cos x)$,

$\sin a = -\frac{\sqrt{2}}{2}$ and $\cos a = \frac{\sqrt{2}}{2}.$ Thus $a = \frac{7\pi}{4}.$

C 41. If $x_1 + x_2 + x_3 = \pi$, $x_1 = \pi - (x_2 + x_3).$ $\sin x_1 = \sin (\pi - (x_2 + x_3)) = \sin (x_2 + x_3)$

by Iden. 7. Thus $\sin x_1 = \sin x_2 \cos x_3 + \cos x_2 \sin x_3$ by Iden. 10.

42. a. sin (x + π) = -sin x, sin (x + 2π) = sin x, sin (x + 3π) = -sin x, and so forth.

Thus if the coeff. of π is odd, sin (x + kπ) = -sin x, as does $(-1)^k$ sin x.   If the

coeff. of π is even, sin (x + kπ) = sin x as does $(-1)^k$ sin x.

b. sin (x + $\frac{\pi}{2}$) = cos x, for k = 0.   sin (x + $\frac{3\pi}{2}$) = -cos x, for k = 1.

sin (x + $\frac{5\pi}{2}$) = cos x, for k = 2.   If k is even, sin [x + (2k + 1)$\frac{\pi}{2}$] = cos x,

as does $(-1)^k$ cos x.   If k is odd, sin [x + (2k + 1)$\frac{\pi}{2}$] = -cos x, as does $(-1)^k$ cos x.

Exercises 1-7 · Pages 36-37

1.
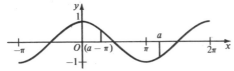

2.
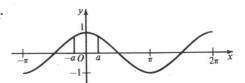

3.
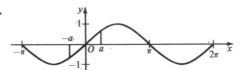

4.
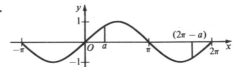

5. a.
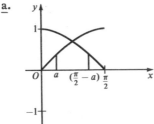

b.
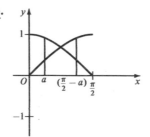

6. a.
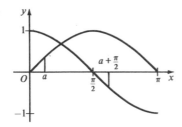

b.
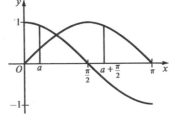

7.
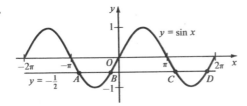

The coord. of the pts. of intersection are:

A = (-$\frac{5\pi}{6}$, -$\frac{1}{2}$),   B = (-$\frac{\pi}{6}$, -$\frac{1}{2}$),

C = ($\frac{7\pi}{6}$, -$\frac{1}{2}$),   D = ($\frac{11\pi}{6}$, -$\frac{1}{2}$)

8.

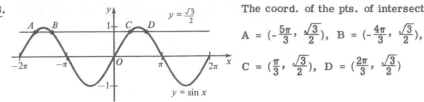

The coord. of the pts. of intersection are:

$A = (-\frac{5\pi}{3}, \frac{\sqrt{3}}{2})$, $B = (-\frac{4\pi}{3}, \frac{\sqrt{3}}{2})$,

$C = (\frac{\pi}{3}, \frac{\sqrt{3}}{2})$, $D = (\frac{2\pi}{3}, \frac{\sqrt{3}}{2})$

9.

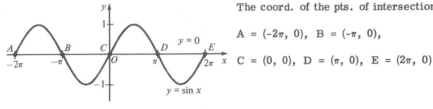

The coord. of the pts. of intersection are:

$A = (-2\pi, 0)$, $B = (-\pi, 0)$,

$C = (0, 0)$, $D = (\pi, 0)$, $E = (2\pi, 0)$

10.

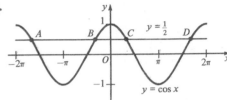

The coord. of the pts. of intersection are:

$A = (-\frac{5\pi}{3}, \frac{1}{2})$, $B = (-\frac{\pi}{3}, \frac{1}{2})$,

$C = (\frac{\pi}{3}, \frac{1}{2})$, $D = (\frac{5\pi}{3}, \frac{1}{2})$

11.

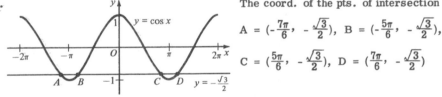

The coord. of the pts. of intersection are:

$A = (-\frac{7\pi}{6}, -\frac{\sqrt{3}}{2})$, $B = (-\frac{5\pi}{6}, -\frac{\sqrt{3}}{2})$,

$C = (\frac{5\pi}{6}, -\frac{\sqrt{3}}{2})$, $D = (\frac{7\pi}{6}, -\frac{\sqrt{3}}{2})$

12.

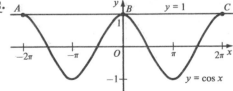

The coord. of the pts. of intersection are:

$A = (-2\pi, 1)$, $B = (0, 1)$, $C = (2\pi, 1)$

B   13.

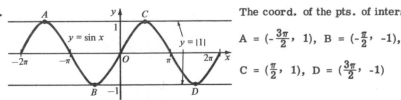

The coord. of the pts. of intersection are:

$A = (-\frac{3\pi}{2}, 1)$, $B = (-\frac{\pi}{2}, -1)$,

$C = (\frac{\pi}{2}, 1)$, $D = (\frac{3\pi}{2}, -1)$

14.

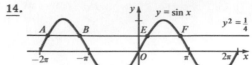

The coord. of the pts. of intersection are:

$A = (-\frac{11\pi}{6}, \frac{1}{2})$, $B = (-\frac{7\pi}{6}, \frac{1}{2})$,

$C = (-\frac{5\pi}{6}, -\frac{1}{2})$, $D = (-\frac{\pi}{6}, -\frac{1}{2})$,

$E = (\frac{\pi}{6}, \frac{1}{2})$, $F = (\frac{5\pi}{6}, \frac{1}{2})$,

$G = (\frac{7\pi}{6}, -\frac{1}{2})$, $F = (\frac{11\pi}{6}, -\frac{1}{2})$

15.

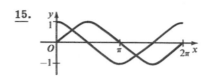

a.

b.

c.

d.

e.

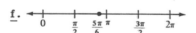

f.

16. a.

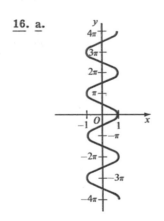

b.

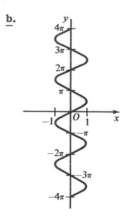

C   17. a. $\{-\frac{11\pi}{6}, -\frac{7\pi}{6}, -\frac{5\pi}{6}, -\frac{\pi}{6}, \frac{\pi}{6}, \frac{5\pi}{6}, \frac{7\pi}{6}, \frac{11\pi}{6}\}$ since $|\sin x| = \frac{1}{2} \Rightarrow \sin x = \frac{1}{2}$ or

$\sin x = -\frac{1}{2}$.

b. $\{-\frac{7\pi}{4}, -\frac{5\pi}{4}, -\frac{3\pi}{4}, -\frac{\pi}{4}, \frac{\pi}{4}, \frac{3\pi}{4}, \frac{5\pi}{4}, \frac{7\pi}{4}\}$ since $\cos^2 x = \frac{1}{2} \Rightarrow \cos x = \sqrt{\frac{1}{2}} = \frac{\sqrt{2}}{2}$

or $\cos x = -\sqrt{\frac{1}{2}} = -\frac{\sqrt{2}}{2}$.

c. $\{-\frac{7\pi}{4},\ -\frac{5\pi}{4},\ -\frac{3\pi}{4},\ -\frac{\pi}{4},\ \frac{\pi}{4},\ \frac{3\pi}{4},\ \frac{5\pi}{4},\ \frac{7\pi}{4}\}$ since $|\sin x| = |\cos x| \Rightarrow \sin x = \cos x$

or $\sin x = -\cos x$.

18. a.

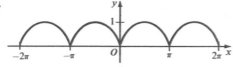

b.

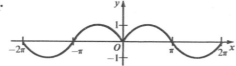

c.

Exercises 1-8 · Pages 39-40

A   1.

x	0	$\frac{\pi}{6}$	$\frac{\pi}{4}$	$\frac{3\pi}{4}$	$\frac{5\pi}{6}$	π
$\frac{x}{2}$	0	$\frac{\pi}{12}$	$\frac{\pi}{8}$	$\frac{3\pi}{8}$	$\frac{5\pi}{12}$	$\frac{\pi}{2}$
$\cos \frac{x}{2}$	1	$\sqrt{\frac{1}{2}(1 + \frac{\sqrt{3}}{2})}$ $\doteq 0.96$	$\sqrt{\frac{1}{2}(1 + \frac{\sqrt{2}}{2})}$ $\doteq 0.92$	$\sqrt{\frac{1}{2}(1 - \frac{\sqrt{2}}{2})}$ $\doteq 0.39$	$\sqrt{\frac{1}{2}(1 - \frac{\sqrt{3}}{2})}$ $\doteq 0.26$	0
$\sin \frac{x}{2}$	0	$\sqrt{\frac{1}{2}(1 - \frac{\sqrt{3}}{2})}$ $\doteq 0.26$	$\sqrt{\frac{1}{2}(1 - \frac{\sqrt{2}}{2})}$ $\doteq 0.39$	$\sqrt{\frac{1}{2}(1 + \frac{\sqrt{2}}{2})}$ $\doteq 0.92$	$\sqrt{\frac{1}{2}(1 + \frac{\sqrt{3}}{2})}$ $\doteq 0.96$	1

$\sin \frac{\pi}{12} = \sin \frac{1}{2}(\frac{\pi}{6}) = \sqrt{\frac{1}{2}(1 - \frac{\sqrt{3}}{2})} \doteq \sqrt{0.067} \doteq 0.26$

$\cos \frac{\pi}{8} = \cos \frac{1}{2}(\frac{\pi}{4}) = \sqrt{\frac{1}{2}(1 + \frac{\sqrt{2}}{2})} \doteq \sqrt{0.854} \doteq 0.92$

$\sin \frac{\pi}{8} = \sin \frac{1}{2}(\frac{\pi}{4}) = \sqrt{\frac{1}{2}(1 - \frac{\sqrt{2}}{2})} \doteq \sqrt{0.147} \doteq 0.385$

$\cos \frac{3\pi}{8} = \cos \frac{1}{2}(\frac{3\pi}{4}) = \sqrt{\frac{1}{2}(1 + (-\frac{\sqrt{2}}{2}))} \doteq \sqrt{0.147} \doteq 0.385$

$\sin \frac{3\pi}{8} = \sin \frac{1}{2}(\frac{3\pi}{4}) = \sqrt{\frac{1}{2}(1 - (-\frac{\sqrt{2}}{2}))} \doteq \sqrt{0.854} \doteq 0.92$

$\cos \frac{5\pi}{12} = \cos \frac{1}{2}(\frac{5\pi}{6}) = \sqrt{\frac{1}{2}(1 + (-\frac{\sqrt{3}}{2}))} \doteq \sqrt{0.067} \doteq 0.26$

$$\sin \frac{5\pi}{12} = \sin \frac{1}{2}(\frac{5\pi}{6}) = \sqrt{\frac{1}{2}(1 - (-\frac{\sqrt{3}}{2}))} \doteq \sqrt{0.933} \doteq 0.96$$

$$\cos \frac{\pi}{2} = \cos \frac{1}{2}(\pi) = \sqrt{\frac{1}{2}(1 + (-1))} = 0; \quad \sin \frac{\pi}{2} = \sin \frac{1}{2}(\pi) = \sqrt{\frac{1}{2}(1 + 1)} = 1$$

2.

x	$\frac{2\pi}{3}$	$\frac{3\pi}{4}$	$\frac{5\pi}{6}$	$\frac{7\pi}{6}$	$\frac{4\pi}{3}$
2x	$\frac{4\pi}{3}$	$\frac{3\pi}{2}$	$\frac{5\pi}{3}$	$\frac{7\pi}{3}$	$\frac{8\pi}{3}$
cos 2x	$1 - 2(-\frac{\sqrt{3}}{2})^2$ $= -0.5$	$1 - 2(\frac{1}{\sqrt{2}})^2$ $= 0$	$1 - 2(\frac{1}{2})^2$ $= 0.5$	$1 - 2(-\frac{1}{2})^2$ $= 0.5$	$1 - 2(-\frac{\sqrt{3}}{2})^2$ $= -0.5$
sin 2x	$2(-\frac{1}{2})(\frac{\sqrt{3}}{2})$ $\doteq -0.87$	$2(\frac{1}{\sqrt{2}})(-\frac{1}{\sqrt{2}})$ $= -1$	$2(-\frac{1}{2})(-\frac{\sqrt{3}}{2})$ $\doteq 0.87$	$2(-\frac{1}{2})(-\frac{\sqrt{3}}{2})$ $\doteq 0.87$	$2(-\frac{1}{2})(-\frac{\sqrt{3}}{2})$ $\doteq 0.87$

Since $\cos 2x = 1 - 2 \sin^2 x$:

$$\cos \frac{4\pi}{3} = 1 - 2 \sin^2 \frac{2\pi}{3} = 1 - 2(\frac{\sqrt{3}}{2})^2 = 1 - \frac{3}{2} = -\frac{1}{2} = -0.5,$$

$$\cos \frac{3\pi}{2} = 1 - 2 \sin^2 \frac{3\pi}{4} = 1 - 2(\frac{1}{\sqrt{2}})^2 = 1 - 1 = 0,$$

$$\cos \frac{5\pi}{3} = 1 - 2 \sin^2 \frac{5\pi}{6} = 1 - 2(\frac{1}{2})^2 = 1 - \frac{1}{2} = \frac{1}{2} = 0.5,$$

$$\cos \frac{7\pi}{3} = 1 - 2 \sin^2 (\frac{7\pi}{6}) = 1 - 2(-\frac{1}{2})^2 = 1 - \frac{1}{2} = \frac{1}{2} = 0.5,$$

$$\cos \frac{8\pi}{3} = 1 - 2 \sin^2 \frac{4\pi}{3} = 1 - 2(-\frac{\sqrt{3}}{2})^2 = 1 - \frac{3}{2} = -\frac{1}{2} = -0.5.$$

Since $\sin 2x = 2 \sin x \cos x$:

$$\sin \frac{4\pi}{3} = 2 \sin \frac{2\pi}{3} \cos \frac{2\pi}{3} = 2(-\frac{1}{2})(\frac{\sqrt{3}}{2}) = -\frac{\sqrt{3}}{2} \doteq -0.87,$$

$$\sin \frac{3\pi}{2} = 2 \sin \frac{3\pi}{4} \cos \frac{3\pi}{4} = 2(\frac{1}{\sqrt{2}})(-\frac{1}{\sqrt{2}}) = -2(\frac{1}{2}) = -1,$$

$$\sin \frac{5\pi}{3} = 2 \sin \frac{5\pi}{6} \cos \frac{5\pi}{6} = 2(\frac{1}{2})(-\frac{\sqrt{3}}{2}) = -\frac{\sqrt{3}}{2} \doteq -0.87,$$

$$\sin \frac{7\pi}{3} = 2 \sin \frac{7\pi}{6} \cos \frac{7\pi}{6} = 2(-\frac{1}{2})(-\frac{\sqrt{3}}{2}) = \frac{\sqrt{3}}{2} \doteq 0.87,$$

$$\sin \frac{8\pi}{3} = 2 \sin \frac{4\pi}{3} \cos \frac{4\pi}{3} = 2(-\frac{1}{2})(-\frac{\sqrt{3}}{2}) = \frac{\sqrt{3}}{2} \doteq 0.87.$$

3. a. If $\sin x = \frac{4}{5}$ and $\frac{\pi}{2} < x < \pi$, then $\cos x = -\frac{3}{5}$ since $\sin^2 x + \cos^2 x = 1$ implies

$$\cos x = \pm \sqrt{1 - \frac{16}{25}} = \pm \frac{3}{5}. \quad x \text{ is between } \frac{\pi}{2} \text{ and } \pi \text{ tells us } \sin x > 0 \text{ and } \cos x < 0,$$

so $\cos x = -\frac{3}{5}$.

b. $\sin 2x = 2 \sin x \cos x = 2(\frac{4}{5})(-\frac{3}{5}) = -\frac{24}{25}$

c. If $\frac{\pi}{2} < x < \pi$, $\frac{\pi}{4} < \frac{x}{2} < \frac{\pi}{2}$, so $\sin \frac{x}{2} > 0$. $\sin \frac{x}{2} = \sqrt{\frac{1}{2}(1 - (-\frac{3}{5}))} = \sqrt{\frac{4}{5}} = \frac{2}{\sqrt{5}}$.

d. $\cos 2x = \cos^2 x - \sin^2 x = (\frac{9}{25}) - (\frac{16}{25}) = -\frac{7}{25}$.

e. $\cos \frac{x}{2} > 0$ since $\frac{\pi}{2} < x < \pi$ implies $\frac{\pi}{4} < \frac{x}{2} < \frac{\pi}{2}$. $\cos \frac{x}{2} = \sqrt{\frac{1}{2}(1 + (-\frac{3}{5}))} =$

$\sqrt{\frac{1}{5}} = \frac{1}{\sqrt{5}}$.

4. a. Since $\pi < x < \frac{3\pi}{2}$, $\sin x < 0$.   If $\cos x = -\frac{2}{3}$, $\sin^2 x + \cos^2 x = 1$ implies

$\sin x = -\sqrt{1 - \frac{4}{9}} = -\frac{\sqrt{5}}{3}$.

b. $\sin 2x = 2 \sin x \cos x = 2(-\frac{\sqrt{5}}{3})(-\frac{2}{3}) = \frac{4\sqrt{5}}{9}$.

c. Since $\pi < x < \frac{3\pi}{2}$, $\frac{\pi}{2} < \frac{x}{2} < \frac{3\pi}{4}$ and $\sin \frac{x}{2} > 0$. $\sin \frac{x}{2} = \sqrt{\frac{1}{2}(1 - (-\frac{2}{3}))} =$

$\sqrt{\frac{5}{6}} = \frac{\sqrt{30}}{6}$.

d. $\cos 2x = \cos^2 x - \sin^2 x = \frac{4}{9} - \frac{5}{9} = -\frac{1}{9}$.

e. Since $\pi < x < \frac{3\pi}{2}$, $\frac{\pi}{2} < \frac{x}{2} < \frac{3\pi}{4}$ and $\cos \frac{x}{2} < 0$. $\cos \frac{x}{2} = \sqrt{\frac{1}{2}(1 + (-\frac{2}{3}))} =$

$-\sqrt{\frac{1}{6}} = -\frac{\sqrt{6}}{6}$.

5.

x	Quadrant			
	I	II	III	IV
$\frac{x}{2}$	I	I	II	II
2x	I or II	III or IV	I or II	III or IV

If $\frac{\pi}{2} < x < \pi$, then $\frac{\pi}{4} < \frac{x}{2} < \frac{\pi}{2}$ and $\pi < 2x < 2\pi$.

If $\pi < x < \frac{3\pi}{2}$, then $\frac{\pi}{2} < \frac{x}{2} < \frac{3\pi}{4}$ and $2\pi < 2x < 3\pi$.

If $\frac{3\pi}{2} < x < 2\pi$, then $\frac{3\pi}{4} < \frac{x}{2} < \pi$ and $3\pi < 2x < 4\pi$.

B   6. a. $\cos^2 x + \sin^2 x = 1$, so $\cos x = \sqrt{1 - \frac{1}{9}} = \sqrt{\frac{8}{9}} = \frac{2\sqrt{2}}{3}$.

$\cos x > 0$ since $0 < x < \frac{\pi}{2}$.

b. $\cos 2x = 1 - 2 \sin^2 x = 1 - 2(\frac{1}{3})^2 = 1 - \frac{2}{9} = \frac{7}{9}$; $\sin 2x = 2 \sin x \cos x =$

$2 \cdot \frac{1}{3} \cdot \frac{2\sqrt{2}}{3} = \frac{4\sqrt{2}}{9}$.

c. $\cos^2 (2x) + \sin^2 (2x) = \frac{49}{81} + \frac{32}{81} = \frac{81}{81} = 1$.

<u>7</u>. $\cos 2x + 2 \sin^2 x = 1 - 2 \sin^2 x + 2 \sin^2 x = 1$

<u>8</u>. $\sin 2x \cos x - \cos 2x \sin x = \sin (2x - x) = \sin x$

<u>9</u>. $(\sin x + \cos x)^2 = \sin^2 x + 2 \sin x \cos x + \cos^2 x = 1 + \sin 2x$

<u>10</u>. $(\sin x - \cos x)^2 = \sin^2 x - 2 \sin x \cos x + \cos^2 x = 1 - \sin 2x$

<u>11</u>. $\cos^4 x - \sin^4 x = (\cos^2 x + \sin^2 x)(\cos^2 x - \sin^2 x) = \cos 2x$

<u>12</u>. $2 \sin \frac{x}{2} \cos \frac{x}{2} = \sin 2(\frac{x}{2}) = \sin x.$

<u>C</u>   <u>13</u>. $\sin 3x = \sin (2x + x) = \sin 2x \cos x + \cos 2x \sin x = 2 \sin x \cos^2 x +$

$(1 - 2 \sin^2 x) \sin x = 2 \sin x - 2 \sin^3 x + \sin x - 2 \sin^3 x = 3 \sin x - 4 \sin^3 x$

<u>14</u>. $\cos 3x = \cos (2x + x) = \cos 2x \cos x - \sin 2x \sin x = (2 \cos^2 x - 1) \cos x -$

$2 \sin^2 x \cos x = 2 \cos^3 x - \cos x - 2(1 - \cos^2 x)\cos x = 2 \cos^3 x - \cos x -$

$2 \cos x + 2 \cos^3 x = 4 \cos^3 x - 3 \cos x$

<u>15</u>. $\sin 4x = 2 \sin 2x \cos 2x = 4 \sin x \cos x (1 - 2 \sin^2 x) = 4 \sin x \cos x -$

$8 \sin^3 x \cos x$

<u>16</u>. $\cos 4x = \cos^2 2x - \sin^2 2x = (1 - 2 \sin^2 x)^2 - 4 \sin^2 x \cos^2 x =$

$1 - 4 \sin^2 x + 4 \sin^4 x - 4 \sin^2 x \cos^2 x = 1 - 4 \sin^2 x(1 - \sin^2 x) -$

$4 \sin^2 x \cos^2 x = 1 - 4 \sin^2 x \cos^2 x - 4 \sin^2 x \cos^2 x = 1 - 8 \sin^2 x \cos^2 x$

<u>Exercises 1-9 · Pages 42-43</u>

<u>A</u>   <u>1</u>. $\sin (\frac{0.68\pi}{2}) \doteq 0.876$; $\cos (\frac{0.68\pi}{2}) \doteq 0.482$

<u>2</u>. $\sin (\frac{0.78\pi}{2}) \doteq 0.941$; $\cos (\frac{0.78\pi}{2}) \doteq 0.339$

<u>3</u>. $\sin (\frac{0.28\pi}{2}) \doteq 0.426$; $\cos (\frac{0.28\pi}{2}) \doteq 0.905$

<u>4</u>. $\sin (\frac{0.42\pi}{2}) \doteq 0.613$; $\cos (\frac{0.42\pi}{2}) \doteq 0.790$

<u>5</u>. $\sin (-\frac{2.24\pi}{2}) = -\sin (\frac{2.24\pi}{2}) = -\sin (\frac{2\pi}{2} + \frac{0.24\pi}{2}) = \sin (\frac{0.24\pi}{2}) \doteq 0.368$;

$\cos (-\frac{2.24\pi}{2}) = \cos (\frac{2.24\pi}{2}) = \cos (\frac{2\pi}{2} + \frac{0.24\pi}{2}) = -\cos (\frac{0.24\pi}{2}) \doteq -0.930$

<u>6</u>. $\sin (-\frac{2.18\pi}{2}) = -\sin (\frac{2.18\pi}{2}) = -\sin (\frac{2\pi}{2} + \frac{0.18\pi}{2}) = \sin (\frac{0.18\pi}{2}) \doteq 0.279$;

$\cos (-\frac{2.18\pi}{2}) = \cos (\frac{2.18\pi}{2}) = \cos (\frac{2\pi}{2} + \frac{0.18\pi}{2}) = -\cos (\frac{0.18\pi}{2}) \doteq -0.960$

7. $\sin (1.23\pi) = \sin (\frac{2.46\pi}{2}) = \sin (\frac{2\pi}{2} + \frac{0.46\pi}{2}) = -\sin (\frac{0.46\pi}{2}) \doteq -0.661$;

$\cos (1.23\pi) = \cos (\frac{2.46\pi}{2}) = \cos (\frac{2\pi}{2} + \frac{0.46\pi}{2}) = -\cos (\frac{0.46\pi}{2}) \doteq -0.750$

8. $\sin (0.93\pi) = \sin (\frac{1.86\pi}{2}) = \sin (\frac{\pi}{2} + \frac{0.86\pi}{2}) = \cos (\frac{0.86\pi}{2}) \doteq 0.218$;

$\cos (0.93\pi) = \cos (\frac{1.86\pi}{2}) = \cos (\frac{\pi}{2} + \frac{0.86\pi}{2}) = -\sin (\frac{0.86\pi}{2}) \doteq -0.976$

9. $\sin (-1.11\pi) = \sin (-\frac{2.22\pi}{2}) = -\sin (\frac{2.22\pi}{2}) = \sin (\frac{0.22\pi}{2}) \doteq 0.339$;

$\cos (-1.11\pi) = \cos (-\frac{2.22\pi}{2}) = \cos (\frac{2.22\pi}{2}) = -\cos (\frac{0.22\pi}{2}) \doteq -0.941$

10. $\sin (-3.52\pi) = -\sin (\frac{7.04\pi}{2}) = -\sin (\frac{8\pi}{2} - \frac{0.96\pi}{2}) = \sin (\frac{0.96\pi}{2}) \doteq 0.998$;

$\cos (-3.52\pi) = \cos (\frac{7.04\pi}{2}) = \cos (\frac{8\pi}{2} - \frac{0.96\pi}{2}) = \cos (\frac{0.96\pi}{2}) \doteq 0.063$

11. $\sin (\frac{5.76\pi}{2}) = \sin (\frac{4\pi}{2} + \frac{\pi}{2} + \frac{0.76\pi}{2}) = \cos (\frac{0.76\pi}{2}) \doteq 0.368$;

$\cos (\frac{5.76\pi}{2}) = \cos (\frac{4\pi}{2} + \frac{\pi}{2} + \frac{0.76\pi}{2}) = -\sin (\frac{0.76\pi}{2}) \doteq -0.930$

12. $\sin (\frac{7.38\pi}{2}) = \sin (\frac{8\pi}{2} - \frac{0.62\pi}{2}) = -\sin (\frac{0.62\pi}{2}) \doteq -0.827$;

$\cos (\frac{7.38\pi}{2}) = \cos (\frac{8\pi}{2} - \frac{0.62\pi}{2}) = \cos (\frac{0.62\pi}{2}) \doteq 0.562$

13. $x \doteq \frac{0.21\pi}{2}$ 14. $x \doteq \frac{0.42\pi}{2}$

15. $x \doteq \frac{0.66\pi}{2}$ 16. $x \doteq \frac{0.82\pi}{2}$

17. Since $\cos x = \cos (-x)$, $\cos x = 0.946$ and $\sin x < 0$ imply $x \doteq \frac{-0.21\pi}{2}$.

$\cos [-0.21(\frac{\pi}{2}) + 4(\frac{\pi}{2})] = \cos \frac{3.79\pi}{2}$, so $x \doteq \frac{3.79\pi}{2}$

18. Since $\sin (\pi - x) = \sin x$, $\sin x = 0.613$ and $\cos x < 0$ imply $x \doteq \pi - \frac{0.42\pi}{2} \doteq$

0.79π $(\frac{1.58\pi}{2})$

19. Since $\sin (\pi - x) = \sin x$, $\sin x = 0.861$ and $\cos x < 0$ imply $x \doteq \pi - \frac{0.66\pi}{2} \doteq$

0.67π $(\frac{1.34\pi}{2})$

20. Since $\cos x = \cos (-x)$, $\cos x = 0.279$ and $\sin x < 0$ imply $x \doteq -\frac{0.82\pi}{2}$.

$\cos [-0.82 (\frac{\pi}{2}) + 4(\frac{\pi}{2})] = \cos \frac{3.18\pi}{2}$, so $x \doteq 1.59\pi$

Chapter Test · Page 45

$\underline{1}$. $f(2p) = f(p + p) = f(p)$; $f(0) = f(0 + p) = f(p)$

$\underline{2}$. $\cos^2 x + (-\frac{12}{13})^2 = 1$, so $\cos^2 x = \frac{25}{169}$ and $\cos x = -\frac{5}{13}$ since $\pi < x < \frac{3\pi}{3}$.

$\underline{3}$.

x	cos x	sin x
$\frac{2\pi}{3}$	$-\frac{1}{2}$	$\frac{\sqrt{3}}{2}$
$\frac{3\pi}{4}$	$-\frac{\sqrt{2}}{2}$	$\frac{\sqrt{2}}{2}$
$\frac{5\pi}{6}$	$-\frac{\sqrt{3}}{2}$	$\frac{1}{2}$

$\underline{4}$. $\cos \frac{\pi}{6} \cos \frac{\pi}{3} - \sin \frac{\pi}{6} \sin \frac{\pi}{3} = \cos (\frac{\pi}{6} + \frac{\pi}{3}) = \cos \frac{\pi}{2} = 0$

$\underline{5}$. $\cos (\frac{\pi}{4} + x) = \cos \frac{\pi}{4} \cos x - \sin \frac{\pi}{4} \sin x = \frac{\sqrt{2}}{2} (\cos x - \sin x)$

$\underline{6}$. $\sin (x - \frac{\pi}{2}) = \sin x \cos \frac{\pi}{2} - \cos x \sin \frac{\pi}{2} = 0 - \cos x \cdot 1 = -\cos x$

$\underline{7}$.

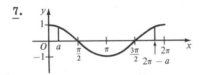

$\underline{8}$. $\cos^2 x + (-\frac{2}{3})^2 = 1$, so $\cos x = -\frac{\sqrt{5}}{3}$ since $\pi < x < \frac{3\pi}{2}$.

$\cos 2x = \cos^2 x - \sin^2 x = \frac{5}{9} - \frac{4}{9} = \frac{1}{9}$

$\underline{9}$. Since $\sin (\pi + z) = -\sin z$, $\sin z = -0.941$ and $\cos z < 0$ imply $z \doteq$

$\pi + \frac{0.78\pi}{2} = 1.39\pi$

Exercises 2-1 · Pages 53-55

<u>A</u> <u>1</u>. $\tan \frac{5\pi}{4} = \tan (\pi + \frac{\pi}{4}) = \tan \frac{\pi}{4} = 1$ <u>2</u>. $\tan \frac{10\pi}{3} = \tan (3\pi + \frac{\pi}{3}) = \tan \frac{\pi}{3} = \sqrt{3}$

<u>3</u>. $\tan (-\frac{\pi}{3}) = -\tan \frac{\pi}{3} = -\sqrt{3}$

<u>4</u>. $\tan (-\frac{10\pi}{3}) = -\tan \frac{10\pi}{3} = -\tan (3\pi + \frac{\pi}{3}) = -\tan \frac{\pi}{3} = -\sqrt{3}$

<u>5</u>. $\tan \frac{2\pi}{3} = \tan (-\frac{\pi}{3} + \pi) = -\tan \frac{\pi}{3} = -\sqrt{3}$ <u>6</u>. $\tan \frac{22\pi}{3} = \tan (7\pi + \frac{\pi}{3}) = \tan \frac{\pi}{3} = \sqrt{3}$

<u>7</u>. $\tan (-\frac{7\pi}{4}) = \tan (\frac{\pi}{4} + (-2\pi)) = \tan \frac{\pi}{4} = 1$ <u>8</u>. $\tan \frac{5\pi}{6} = \tan (-\frac{\pi}{6} + \pi) = -\tan \frac{\pi}{6} = -\frac{1}{\sqrt{3}}$

<u>9</u>. $\cos^2 x + (\frac{5}{13})^2 = 1$, so $\cos^2 x = \frac{144}{169}$ and $\cos x = \frac{12}{13}$ or $-\frac{12}{13}$. For $\sin x = \frac{5}{13}$

and $\cos x = \frac{12}{13}$ (Quadrant I), $\tan x = \frac{\frac{5}{13}}{\frac{12}{13}} = \frac{5}{12}$. For $\sin x = \frac{5}{13}$ and $\cos x =$

$-\frac{12}{13}$ (Quadrant II), $\tan x = \frac{\frac{5}{13}}{-\frac{12}{13}} = -\frac{5}{12}$.

<u>10</u>. $\sin^2 x + (\frac{4}{5})^2 = 1$, so $\sin^2 x = \frac{9}{25}$ and $\sin = \frac{3}{5}$ or $-\frac{3}{5}$. For $\sin x = \frac{3}{5}$ (Quadrant I),

$\tan x = \frac{\frac{3}{5}}{\frac{4}{5}} = \frac{3}{4}$. For $\sin x = -\frac{3}{5}$ (Quadrant IV), $\tan x = \frac{-\frac{3}{5}}{\frac{4}{5}} = -\frac{3}{4}$.

<u>11</u>. $\sin^2 x + (-\frac{12}{13})^2 = 1$, so $\sin^2 x = \frac{25}{169}$ and $\sin x = \frac{5}{13}$ or $-\frac{5}{13}$. For $\sin x = \frac{5}{13}$

(Quadrant II), $\tan x = \frac{\frac{5}{13}}{-\frac{12}{13}} = -\frac{5}{12}$. For $\sin x = -\frac{5}{13}$ (Quadrant III), $\tan x =$

$\frac{-\frac{5}{13}}{-\frac{12}{13}} = +\frac{5}{12}$.

<u>12</u>. $\cos^2 x + (-\frac{1}{2})^2 = 1$, so $\cos^2 x = \frac{3}{4}$ and $\cos x = \frac{\sqrt{3}}{2}$ or $-\frac{\sqrt{3}}{2}$. For $\cos x = \frac{\sqrt{3}}{2}$

(Quadrant IV), $\tan x = \frac{-\frac{1}{2}}{\frac{\sqrt{3}}{2}} = -\frac{1}{\sqrt{3}}$. For $\cos x = -\frac{\sqrt{3}}{2}$ (Quadrant III),

$\tan x = \frac{-\frac{1}{2}}{-\frac{\sqrt{3}}{2}} = \frac{1}{\sqrt{3}}$.

<u>13</u>.
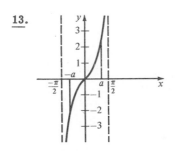

<u>14</u>.
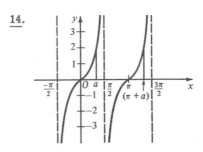

<u>15.</u>

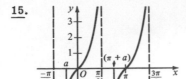

<u>16.</u>

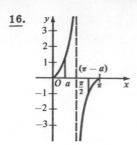

<u>17.</u> tan $(\pi - x)$ = tan $(\pi + (-x))$ = tan $(-x)$, using tan $(x + \pi)$ = tan x.   tan $(-x)$ = -tan x, $x \neq \frac{\pi}{2} + k\pi$.

<u>18.</u> $A(\frac{\pi}{3}, \sqrt{3})$

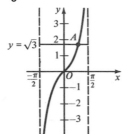

<u>19.</u> $A(-\frac{\pi}{4}, -1)$

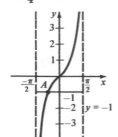

<u>20.</u> $A(-\frac{\pi}{6}, -\frac{\sqrt{3}}{3})$

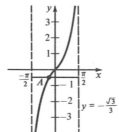

<u>21.</u> <u>a.</u> If tan x = 1, x = $\frac{\pi}{4}$ and sin x = cos x = $\frac{\sqrt{2}}{2}$

   <u>b.</u> If tan x = -1, x = -$\frac{\pi}{4}$ and sin x = -$\frac{\sqrt{2}}{2}$, cos x = $\frac{\sqrt{2}}{2}$

   <u>c.</u> If tan x = $\sqrt{3}$, x = $\frac{\pi}{3}$ and sin x = $\frac{\sqrt{3}}{2}$, cos x = $\frac{1}{2}$

   <u>d.</u> If tan x = -$\frac{\sqrt{3}}{3}$, x = -$\frac{\pi}{6}$ and sin x = -$\frac{1}{2}$, cos x = $\frac{\sqrt{3}}{2}$.

B   <u>22.</u> <u>a.</u> If tan x = 2, $\frac{\sin x}{\cos x}$ = 2 and sin x = 2 cos x.   Thus $\sin^2 x = 4 \cos^2 x$,

   $\sin^2 x = 4 - 4 \sin^2 x$, $5 \sin^2 x = 4$, and sin x = $\frac{2}{\sqrt{5}}$ since $0 < x < \frac{\pi}{2}$.

   cos x = $\sqrt{1 - \frac{4}{5}} = \frac{1}{\sqrt{5}}$

b. Since $\pi < x < \frac{3\pi}{2}$, $\sin x = -\frac{2}{\sqrt{5}}$ and $\cos x = -\frac{1}{\sqrt{5}}$

23. a. $x = \frac{\pi}{3}, \frac{4\pi}{3}$ b. $x = -\frac{\pi}{3}, \frac{2\pi}{3}, \frac{5\pi}{3}$

c. $\tan x = \sqrt{3}$ or $-\sqrt{3}$, so $x = -\frac{\pi}{3}, \frac{\pi}{3}, \frac{2\pi}{3}, \frac{4\pi}{3}, \frac{5\pi}{3}$

d. $\tan x = \pm\sqrt{3}$, so $x = -\frac{\pi}{3}, \frac{\pi}{3}, \frac{2\pi}{3}, \frac{4\pi}{3}, \frac{5\pi}{3}$

e. $2x = \frac{\pi}{4}, \frac{5\pi}{4}, \frac{9\pi}{4}, \frac{13\pi}{4}$ and $x = \frac{\pi}{8}, \frac{5\pi}{8}, \frac{9\pi}{8}, \frac{13\pi}{8}$

f. $\tan^2 x = 1$ so $\tan x = \pm 1$ and $x = -\frac{\pi}{4}, \frac{\pi}{4}, \frac{3\pi}{4}, \frac{5\pi}{4}, \frac{7\pi}{4}$

24.    25.

26.

C   27.

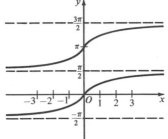

28. a.   b.

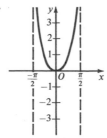   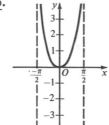

29. If $0 < x < \frac{\pi}{2}$, $\sin x > 0$ and $0 < \cos x < 1$. $\cos x < 1$ implies $1 < \frac{1}{\cos x}$.

Thus $\sin x < \frac{\sin x}{\cos x}$ or $\tan x > \sin x$.

Exercises 2-2 · Pages 59-60

A 1.

x	sec x	csc x	cot x
0	$\dfrac{1}{\cos 0} = 1$	$\dfrac{1}{\sin 0}$ undef.	$\dfrac{\cos 0}{\sin 0}$ undef.
$\dfrac{\pi}{6}$	$\dfrac{1}{\cos \frac{\pi}{6}} = \dfrac{2}{\sqrt{3}}$	$\dfrac{1}{\sin \frac{\pi}{6}} = 2$	$\dfrac{\cos \frac{\pi}{6}}{\sin \frac{\pi}{6}} = \sqrt{3}$
$\dfrac{\pi}{4}$	$\dfrac{1}{\cos \frac{\pi}{4}} = \sqrt{2}$	$\dfrac{1}{\sin \frac{\pi}{4}} = \sqrt{2}$	$\dfrac{\cos \frac{\pi}{4}}{\sin \frac{\pi}{4}} = 1$
$\dfrac{\pi}{3}$	$\dfrac{1}{\cos \frac{\pi}{3}} = 2$	$\dfrac{1}{\sin \frac{\pi}{3}} = \dfrac{2}{\sqrt{3}}$	$\dfrac{\cos \frac{\pi}{3}}{\sin \frac{\pi}{3}} = \dfrac{1}{\sqrt{3}}$
$\dfrac{\pi}{2}$	$\dfrac{1}{\cos \frac{\pi}{2}}$ undef.	$\dfrac{1}{\sin \frac{\pi}{2}} = 1$	$\dfrac{\cos \frac{\pi}{2}}{\sin \frac{\pi}{2}} = 0$

2. $\sec 6\pi = \dfrac{1}{\cos 6\pi} = \dfrac{1}{\cos 0} = 1$

3. $\sec \dfrac{4\pi}{3} = \dfrac{1}{\cos \frac{4\pi}{3}} = \dfrac{1}{-\frac{1}{2}} = -2$

4. $\csc \left(-\dfrac{\pi}{6}\right) = \dfrac{1}{\sin \left(-\frac{\pi}{6}\right)} = -\dfrac{1}{\sin \frac{\pi}{6}} = -\dfrac{1}{\frac{1}{2}} = -2$

5. $\csc \dfrac{19\pi}{2} = \dfrac{1}{\sin \frac{19\pi}{2}} = \dfrac{1}{\sin \left(8\pi + \frac{3\pi}{2}\right)} = \dfrac{1}{\sin \frac{3\pi}{2}} = \dfrac{1}{-1} = -1$

6. $\cot \dfrac{8\pi}{3} = \dfrac{\cos \frac{8\pi}{3}}{\sin \frac{8\pi}{3}} = \dfrac{\cos \frac{2\pi}{3}}{\sin \frac{2\pi}{3}} = \dfrac{-\frac{1}{2}}{\frac{\sqrt{3}}{2}} = -\dfrac{1}{\sqrt{3}}$

7. $\cot \left(-\dfrac{2\pi}{3}\right) = \dfrac{\cos \left(-\frac{2\pi}{3}\right)}{\sin \left(-\frac{2\pi}{3}\right)} = \dfrac{\cos \frac{2\pi}{3}}{-\sin \frac{2\pi}{3}} = \dfrac{-\frac{1}{2}}{-\frac{\sqrt{3}}{2}} = \dfrac{1}{\sqrt{3}}$

8. $\sec \dfrac{5\pi}{6} = \dfrac{1}{\cos \frac{5\pi}{6}} = \dfrac{1}{-\frac{\sqrt{3}}{2}} = -\dfrac{2}{\sqrt{3}}$

9. $\sec \left(-\dfrac{4\pi}{3}\right) = \dfrac{1}{\cos \left(-\frac{4\pi}{3}\right)} = \dfrac{1}{\cos \frac{4\pi}{3}} = \dfrac{1}{-\frac{1}{2}} = -2$

10. $\csc \left(-\dfrac{5\pi}{4}\right) = \dfrac{1}{\sin \left(-\frac{5\pi}{4}\right)} = \dfrac{1}{-\sin \frac{5\pi}{4}} = \dfrac{1}{-\left(-\frac{1}{\sqrt{2}}\right)} = \sqrt{2}$

11. $\csc \left(\dfrac{\pi}{2}\right) = \dfrac{1}{\sin \frac{\pi}{2}} = \dfrac{1}{1} = 1$

12.

Function	Domain	Range
sin x	−1 0 1	−1 0 1
csc x = $\dfrac{1}{\sin x}$	−2π −π 0 π 2π	−1 0 1
tan x	$-\dfrac{3\pi}{2}$ $-\dfrac{\pi}{2}$ $\dfrac{\pi}{2}$ $\dfrac{3\pi}{2}$ $\dfrac{5\pi}{2}$	−1 0 1
cot x = $\dfrac{1}{\tan x}$	−2π −π 0 π 2π	−1 0 1

13. $\sec (x + \pi) = \dfrac{1}{\cos (x + \pi)} = \dfrac{1}{-\cos x} = -\sec x$

14. $\sec (\pi - x) = \dfrac{1}{\cos (\pi - x)} = \dfrac{1}{-\cos x} = -\sec x$

15. $\csc (x + \pi) = \dfrac{1}{\sin (x + \pi)} = \dfrac{1}{-\sin x} = -\csc x$

16. $\csc (\pi - x) = \dfrac{1}{\sin (\pi - x)} = \dfrac{1}{\sin x} = \csc x$

17. $\cot (x + \pi) = \dfrac{\cos (x + \pi)}{\sin (x + \pi)} = \dfrac{-\cos x}{-\sin x} = \cot x$

18. $\cot (\pi - x) = \dfrac{\cos (\pi - x)}{\sin (\pi - x)} = \dfrac{-\cos x}{\sin x} = -\cot x$

19. $\sec (-x) = \dfrac{1}{\cos (-x)} = \dfrac{1}{\cos x} = \sec x$

20. $\csc (-x) = \dfrac{1}{\sin (-x)} = \dfrac{1}{-\sin x} = -\csc x$

21. $\cot (-x) = \dfrac{\cos (-x)}{\sin (-x)} = \dfrac{\cos x}{-\sin x} = -\cot x$

22. $\sec x \cot x = \dfrac{1}{\cos x} \cdot \dfrac{\cos x}{\sin x} = \dfrac{1}{\sin x} = \csc x$

B 23. $\sec^2 x + \csc^2 x = \dfrac{1}{\cos^2 x} + \dfrac{1}{\sin^2 x} = \dfrac{\sin^2 x + \cos^2 x}{\cos^2 x \cdot \sin^2 x} = \dfrac{1}{\cos^2 x \sin^2 x} = \sec^2 x \csc^2 x$

24. $\sec x = 2$ implies $\cos x = \dfrac{1}{2}$, so $x = -\dfrac{\pi}{3}, \dfrac{\pi}{3}, \dfrac{5\pi}{3}$

25. $\sec x = -\sqrt{2}$ implies $\cos x = -\dfrac{1}{\sqrt{2}}$, so $x = -\dfrac{3\pi}{4}, \dfrac{3\pi}{4}, \dfrac{5\pi}{4}$

26. $\csc x = \dfrac{\sqrt{2}}{3}$ implies $\sin x = \dfrac{3}{\sqrt{2}}$. $\dfrac{3}{\sqrt{2}} > 1$, so no x satisfy the condition.

27. $\csc x = 1$ implies $\sin x = 1$, so $x = \dfrac{\pi}{2}$

28. $\cot x = -1$ implies $\dfrac{\sin x}{\cos x} = \tan x = -1$, so $x = -\dfrac{\pi}{4}, \dfrac{3\pi}{4}, \dfrac{7\pi}{4}$

29. $\cot x = \sqrt{3}$ implies $\dfrac{\sin x}{\cos x} = \tan x = \dfrac{1}{\sqrt{3}}$ so $x = -\dfrac{5\pi}{6}, \dfrac{\pi}{6}, \dfrac{7\pi}{6}$

<u>C</u>   <u>30.</u> $\tan^2 x + 1 = \dfrac{\sin^2 x}{\cos^2 x} + 1 = \dfrac{\sin^2 x + \cos^2 x}{\cos^2 x} = \dfrac{1}{\cos^2 x} = \sec^2 x;\ 1 + \cot^2 x =$

$1 + \dfrac{\cos^2 x}{\sin^2 x} = \dfrac{\sin^2 x + \cos^2 x}{\sin^2 x} = \dfrac{1}{\sin^2 x} = \csc^2 x$

Exercises 2-3 · Page 63

<u>A</u>   <u>1.</u>

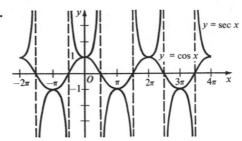

<u>2.</u> $x = \dfrac{\pi}{2},\ x = \dfrac{3\pi}{2}$

<u>3.</u> $x = 0,\ x = \pi,\ x = 2\pi$

<u>4.</u> $x = \dfrac{\pi}{2},\ x = \dfrac{3\pi}{2}$

<u>5.</u> $x = 0,\ x = \pi,\ x = 2\pi$

<u>6.</u>

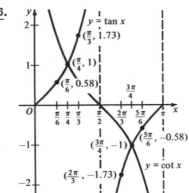

<u>7.</u>

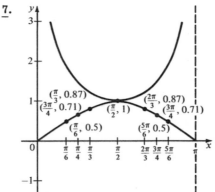

<u>B</u>   <u>8.</u>

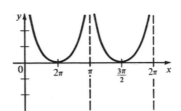

<u>9.</u>

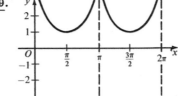

<u>10.</u>

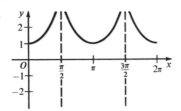

<u>11.</u> <u>a.</u> All $x \neq k\pi,\ k \in J$

   <u>b.</u> All $x \neq k\dfrac{\pi}{2},\ k \in J$

Exercises 2-4 · Pages 67-68

__A__   __1.__ $\tan(x - \pi) = \dfrac{\tan x - \tan \pi}{1 + \tan x \tan \pi} = \dfrac{\tan x - 0}{1 + 0} = \tan x$, $x \neq k \cdot \dfrac{\pi}{2}$, $k \in J$

__2.__ $\tan \dfrac{x}{2} = \dfrac{\sin x}{1 + \cos x} = \dfrac{\sin x(1 - \cos x)}{1 - \cos^2 x} = \dfrac{\sin x(1 - \cos x)}{\sin^2 x} = \dfrac{(1 - \cos x)}{\sin x}$,

$x \neq (2k + 1)\pi$, $k \in J$

__3.__ $\tan(x + \dfrac{\pi}{4}) = \dfrac{\tan x + \tan \dfrac{\pi}{4}}{1 - \tan x \tan \dfrac{\pi}{4}} = \dfrac{1 + \tan x}{1 - \tan x}$, $x \neq (4k + 1)\dfrac{\pi}{4}$

__4.__ $\cot(\dfrac{\pi}{2} - x) = \dfrac{\cos(\dfrac{\pi}{2} - x)}{\sin(\dfrac{\pi}{2} - x)} = \dfrac{\sin x}{\cos x} = \tan x$, $x \neq (2k + 1)\dfrac{\pi}{2}$, $k \in J$

__5.__ $\sec(x - \pi) = \dfrac{1}{\cos(x - \pi)} = \dfrac{1}{-\cos x} = -\sec x$, $x \neq (2k + 1)\dfrac{\pi}{2}$, $k \in J$

__6.__ $\sec(\dfrac{\pi}{2} - x) = \dfrac{1}{\cos(\dfrac{\pi}{2} - x)} = \dfrac{1}{\sin x} = \csc x$, $x \neq k\pi$, $k \in J$

__7.__ $\csc(\dfrac{\pi}{2} - x) = \dfrac{1}{\sin(\dfrac{\pi}{2} - x)} = \dfrac{1}{\cos x} = \sec x$, $x \neq (2k + 1)\dfrac{\pi}{2}$, $k \in J$

__8.__ $\sec x \csc x = \dfrac{1}{\sin x \cos x} = \dfrac{2}{2 \sin x \cos x} = \dfrac{2}{\sin 2x} = 2 \csc 2x$, $x \neq k \cdot \dfrac{\pi}{2}$, $k \in J$

__9.__ $\sin^2 x \cot^2 x + \cos^2 x \tan^2 x = \sin^2 x \cdot \dfrac{\cos^2 x}{\sin^2 x} + \cos^2 x \cdot \dfrac{\sin^2 x}{\cos^2 x} =$

$\cos^2 x + \sin^2 x = 1$, $x \neq k \cdot \dfrac{\pi}{2}$, $k \in J$

__10.__ $\dfrac{\sec x}{\tan x + \cot x} = \dfrac{\dfrac{1}{\cos x}}{\dfrac{\sin x}{\cos x} + \dfrac{\cos x}{\sin x}} = \dfrac{\dfrac{1}{\cos x}}{\dfrac{1}{\sin x \cos x}} = \sin x$, $x \neq k \cdot \dfrac{\pi}{2}$, $k \in J$

__B__   __11.__ $\dfrac{1 - \cos 2x}{\sin 2x} = \dfrac{1 - (1 - 2 \sin^2 x)}{2 \sin x \cos x} = \dfrac{2 \sin^2 x}{2 \sin x \cos x} = \dfrac{\sin x}{\cos x} = \tan x$, $x \neq k \cdot \dfrac{\pi}{2}$, $k \in J$

__12.__ $\dfrac{\cos(x_1 + x_2)}{\cos x_1 \cos x_2} = \dfrac{\cos x_1 \cos x_2 - \sin x_1 \sin x_2}{\cos x_1 \cos x_2} = 1 - \dfrac{\sin x_1}{\cos x_1} \cdot \dfrac{\sin x_2}{\cos x_2} =$

$1 - \tan x_1 \tan x_2$, $x_1 \neq (2k + 1)\dfrac{\pi}{2}$, $x_2 \neq (2k + 1)\dfrac{\pi}{2}$, $k \in J$

__13.__ $\cot(x_2 + x_1) = \dfrac{\cos(x_2 + x_1)}{\sin(x_2 + x_1)} = \dfrac{\cos x_2 \cos x_1 - \sin x_2 \sin x_1}{\sin x_2 \cos x_1 + \cos x_2 \sin x_1} =$

$\dfrac{\dfrac{\cos x_2 \cos x_1}{\sin x_2 \sin x_1} - 1}{\dfrac{\sin x_2 \cos x_1 + \cos x_2 \sin x_1}{\sin x_2 \sin x_1}} = \dfrac{\cot x_2 \cot x_1 - 1}{\dfrac{\cos x_1}{\sin x_1} + \dfrac{\cos x_2}{\sin x_2}} = \dfrac{\cot x_2 \cot x_1 - 1}{\cot x_2 + \cot x_1}$,

$x_1 \neq k\pi$, $x_2 \neq k\pi$, $x_1 + x_2 \neq k\pi$, $k \in J$

<u>14.</u> $\tan 2x = \dfrac{2 \tan x}{1 - \tan^2 x} = \dfrac{2}{\dfrac{1}{\tan x} - \tan x} = \dfrac{2}{\cot x - \tan x}, \quad x \neq k \cdot \dfrac{\pi}{4}, \ k \in J$

<u>15.</u> $\cot 2x = \dfrac{\cos 2x}{\sin 2x} = \dfrac{\cos^2 x - \sin^2 x}{2 \sin x \cos x} = \dfrac{\dfrac{\cos^2 x}{\sin^2 x} - 1}{\dfrac{2 \sin x \cos x}{\sin^2 x}} = \dfrac{\cot^2 x - 1}{2 \cot x}, \quad x \neq k \cdot \dfrac{\pi}{2}, \ k \in J$

<u>16.</u> $\sec (x_2 - x_1) = \dfrac{1}{\cos (x_2 - x_1)} = \dfrac{1}{\cos x_2 \cos x_1 + \sin x_2 \sin x_1} =$

$\dfrac{\dfrac{1}{\cos x_2 \cos x_1}}{1 + \dfrac{\sin x_2 \sin x_1}{\cos x_2 \cos x_1}} = \dfrac{\sec x_2 \sec x_1}{1 + \tan x_2 \tan x_1}, \quad x_1 \neq (2k + 1) \cdot \dfrac{\pi}{2}, \ x_2 \neq (2k + 1) \cdot \dfrac{\pi}{2},$

$x_2 - x_1 \neq (2k + 1) \cdot \dfrac{\pi}{2}, \ k \in J$

<u>C</u>   <u>17.</u> $\left| \sec x - \tan x \right| = \left| \dfrac{1}{\cos x} - \dfrac{\sin x}{\cos x} \right| = \left| \dfrac{1 - \sin x}{\cos x} \right| =$

$\sqrt{\dfrac{(1 - \sin x)^2}{\cos^2 x}} = \sqrt{\dfrac{(1 - \sin x)^2}{1 - \sin^2 x}} = \sqrt{\dfrac{1 - \sin x}{1 + \sin x}}, \quad x \neq (2k + 1) \cdot \dfrac{\pi}{2}, \ k \in J$

<u>18.</u> <u>a.</u> Since circle O is a unit circle, $\sin x = d(\overline{CD})$ and $\cos x = d(\overline{OD})$.

     $\triangle ODC \sim \triangle OAB$, so $\dfrac{d(\overline{OD})}{d(\overline{OA})} = \dfrac{d(\overline{CD})}{d(\overline{AB})}$.   This is equivalent to $\dfrac{d(\overline{CD})}{d(\overline{OD})} = \dfrac{d(\overline{AB})}{d(\overline{AO})}$.

     Substituting, $\dfrac{\sin x}{\cos x} = \dfrac{d(\overline{AB})}{1}$ or $d(\overline{AB}) = \tan x$.

   <u>b.</u> For $0 < x < \dfrac{\pi}{2}$, $d(\overline{OA}) = 1$ and so $d(\overline{OD}) < 1 = d(\overline{OA})$.   Since $\triangle ODC \sim \triangle OAB$,

     $d(\overline{OA}) > d(\overline{OD})$ implies $d(\overline{AB}) > d(\overline{DC})$.   Thus, since $d(\overline{AB}) = \tan x$ and

     $d(\overline{DC}) = \sin x$, $\tan x > \sin x$.

<u>19.</u> Drop a $\perp$ from C to some point Y on $\overline{OA}$. $\triangle COY \sim \triangle BOA \sim \triangle EOD$.

     $\dfrac{d(\overline{OC})}{d(\overline{OY})} = \dfrac{d(\overline{OB})}{d(\overline{OA})}$. $d(\overline{OC}) = 1 = d(\overline{OA})$ and $d(\overline{OY}) = \cos x$, so $d(\overline{OB}) = \dfrac{1}{\cos x} = \sec x$.

     $\dfrac{d(\overline{OC})}{d(\overline{CY})} = \dfrac{d(\overline{OE})}{d(\overline{ED})}$ and $d(\overline{ED}) = 1$, so $d(\overline{OE}) = \dfrac{1}{\sin x}$ and $d(\overline{OE}) = \csc x$.

     $\dfrac{d(\overline{CY})}{d(\overline{OY})} = \dfrac{d(\overline{ED})}{d(\overline{OD})}$, so $\dfrac{\sin x}{\cos x} = \dfrac{1}{d(\overline{OD})}$ and $d(\overline{OD}) = \cot x$

<u>20.</u> $\tan x_1 + \tan x_2 + \tan x_3 = \tan (x_1 + x_2)(1 - \tan x_1 \tan x_2) + \tan x_3 =$

     $\tan (\pi - x_3)(1 - \tan x_1 \tan x_2) + \tan x_3 =$

     $-\tan x_3 (1 - \tan x_1 \tan x_2) + \tan x_3 = -\tan x_3 + \tan x_1 \tan x_2 \tan x_3 + \tan x_3 =$

     $\tan x_1 \tan x_2 \tan x_3$

Exercises 2-5 · Pages 71-72

<u>A</u>   <u>1</u>. $\sin x + \cos x \cot x = \sin x + \dfrac{\cos x \cdot \cos x}{\sin x} = \dfrac{\sin^2 x + \cos^2 x}{\sin x} = \dfrac{1}{\sin x} = \csc x,$

$x \neq k\pi, \ k \in J$

<u>2</u>. $\cos x \csc x = \cos x \cdot \dfrac{1}{\sin x} = \cot x, \ x \neq k\pi, \ k \in J$

<u>3</u>. $\tan x (\sin x + \cot x \cos x) = \tan x \sin x + \cos x = \dfrac{\sin^2 x}{\cos x} + \dfrac{\cos^2 x}{\cos x} = \dfrac{1}{\cos x} =$

$\sec x, \ x \neq (2k + 1)\dfrac{\pi}{2}, \ k \in J$

<u>4</u>. $2 \cos^2 x - \sin^2 x + 1 = 2 \cos^2 x - \sin^2 x + \sin^2 x + \cos^2 x = 3 \cos^2 x, \ x \in \mathcal{R}$

<u>5</u>. $\sin x \tan x + \cos x = \dfrac{\sin^2 x}{\cos x} + \dfrac{\cos^2 x}{\cos x} = \dfrac{1}{\cos x} = \sec x, \ x \neq (2k + 1)\dfrac{\pi}{2}, \ k \in J$

<u>6</u>. $\sin x (\sec x - \csc x) = \dfrac{\sin x}{\cos x} - \dfrac{\sin x}{\sin x} = \tan x - 1, \ x \neq k \cdot \dfrac{\pi}{2}, \ k \in J$

<u>7</u>. $\cos x (\csc x - \sec x) = \dfrac{\cos x}{\sin x} - \dfrac{\cos x}{\cos x} = \cot x - 1, \ x \neq k \cdot \dfrac{\pi}{2}, \ k \in J$

<u>8</u>. $\dfrac{1}{1 + \sin x} + \dfrac{1}{1 - \sin x} = \dfrac{1 - \sin x + 1 + \sin x}{1 - \sin^2 x} = \dfrac{2}{\cos^2 x} = 2 \sec^2 x,$

$x \neq (2k + 1)\dfrac{\pi}{2}, \ k \in J$

<u>9</u>. $\dfrac{1 + \sin x}{\cos x} - \dfrac{\cos x}{1 - \sin x} = \dfrac{1 - \sin^2 x - \cos^2 x}{\cos x(1 - \sin x)} = \dfrac{1 - (\sin^2 x + \cos^2 x)}{\cos x(1 - \sin x)} = 0,$

$x \neq (2k + 1)\dfrac{\pi}{2}, \ k \in J$

<u>10</u>. $\dfrac{\sin^2 x}{1 + \cos x} + \cos x = \dfrac{\sin^2 x + \cos x + \cos^2 x}{1 + \cos x} = \dfrac{1 + \cos x}{1 + \cos x} = 1,$

$x \neq (2k + 1)\pi, \ k \in J$

<u>11</u>. $\dfrac{\sin x \cot x + \cos x}{\sin x} = \cot x + \dfrac{\cos x}{\sin x} = \cot x + \cot x = 2 \cot x, \ x \neq k\pi, \ k \in J$

<u>12</u>. $\dfrac{1 + 2 \sin x \cos x}{\sin x + \cos x} = \dfrac{\sin^2 x + \cos^2 x + 2 \sin x \cos x}{\sin x + \cos x} = \dfrac{(\sin x + \cos x)^2}{\sin x + \cos x} =$

$\sin x + \cos x, \ x \neq (4k - 1)\dfrac{\pi}{4}, \ k \in J$

<u>13</u>. a. $\dfrac{1}{\sec^2 x} + \dfrac{1}{\csc^2 x} = \cos^2 x + \sin^2 x = 1$

   b. $\dfrac{\sec x}{\tan x + \cot x} = \dfrac{\sec x}{\tan x + \cot x} \cdot \dfrac{\sin x \cos x}{\sin x \cos x} = \dfrac{\sin x}{\sin^2 x + \cos^2 x} = \sin x$

   c. $\dfrac{\csc^2 x - 1}{\cot^2 x} = \dfrac{\csc^2 x - 1}{\cot^2 x} \cdot \dfrac{\sin^2 x}{\sin^2 x} = \dfrac{1 - \sin^2 x}{\cos^2 x} = \dfrac{\cos^2 x}{\cos^2 x} = 1$

d. $(1 + \tan^2 x)(1 - \sin^2 x) = (1 + \tan^2 x)\cos^2 x = \cos^2 x + \sin^2 x = 1$

e. $\sec x - \sin x \tan x = \dfrac{1}{\cos x} - \dfrac{\sin^2 x}{\cos x} = \dfrac{1 - \sin^2 x}{\cos x} = \dfrac{\cos^2 x}{\cos x} = \cos x$

f. $\sin x \cos x \tan x \cot x \sec x \csc x = (\sin x \cdot \csc x)(\cos x \cdot \sec x)(\tan x \cdot \cot x)$

   $= 1 \cdot 1 \cdot 1 = 1$

g. $\dfrac{\sin x + \tan x}{\tan x (\csc x + \cot x)} = \dfrac{\sin x + \tan x}{\dfrac{1}{\cos x} + 1} = \dfrac{\dfrac{\sin x \cos x + \sin x}{\cos x}}{\dfrac{1 + \cos x}{\cos x}} =$

   $\dfrac{\sin x(\cos x + 1)}{1 + \cos x} = \sin x$

h. $\dfrac{1 + \tan x}{1 + \cot x} = \dfrac{(1 + \tan x)}{(1 + \cot x)} \cdot \dfrac{\tan x}{\tan x} = \dfrac{\tan x(1 + \tan x)}{\tan x + 1} = \tan x$

i. $\dfrac{\sin 2x}{2 \sin x} = \dfrac{2 \sin x \cos x}{2 \sin x} = \cos x$

j. $\sin x \sec x = \sin x \cdot \dfrac{1}{\cos x} = \tan x$

k. $\sec^2 x (1 - \sin^2 x) = \sec^2 x \cdot \cos^2 x = \dfrac{1}{\cos^2 x} \cdot \cos^2 x = 1$

l. $\csc^2 x (1 - \cos^2 x) = \dfrac{1}{\sin^2 x} \cdot \sin^2 x = 1$

m. $\sec x - \sin x \tan x = \dfrac{1}{\cos x} - \dfrac{\sin^2 x}{\cos x} = \dfrac{1 - \sin^2 x}{\cos x} = \dfrac{\cos^2 x}{\cos x} = \cos x$

B 14. $\dfrac{\sin x \cos x}{1 - 2 \sin^2 x} = \dfrac{\sin x \cos x}{\cos^2 x - \sin^2 x} = \dfrac{1}{\dfrac{\cos^2 x}{\sin x \cos x} - \dfrac{\sin^2 x}{\sin x \cos x}} = \dfrac{1}{\cot x - \tan x}$,

   $x \neq k \cdot \dfrac{\pi}{4}, \ k \in J$

15. $\dfrac{1 + \tan^2 x}{\tan^2 x} = \cot^2 x + 1 = \csc^2 x, \ x \neq k \cdot \dfrac{\pi}{2}, \ k \in J$

16. $\cos^4 x - \sin^4 x = (\cos^2 x + \sin^2 x)(\cos^2 x - \sin^2 x) = \cos^2 x - \sin^2 x$

17. $\dfrac{1 - \tan x}{1 + \tan x} = \dfrac{\dfrac{1 - \tan x}{\tan x}}{\dfrac{1 + \tan x}{\tan x}} = \dfrac{\cot x - 1}{\cot x + 1}, \ x \neq k \cdot \dfrac{\pi}{2}, \ (4k + 3)\dfrac{\pi}{4}, \ k \in J$

18. $\dfrac{1 - \sin^2 x}{1 - \cos^2 x} = \dfrac{\cos^2 x}{\sin^2 x} = \cot^2 x, \ x \neq k\pi, \ k \in J$

19. $\dfrac{\sin (x - y)}{\sin (x + y)} = \dfrac{\sin x \cos y - \cos x \sin y}{\sin x \cos y + \cos x \sin y} = \dfrac{\dfrac{\sin x \cos y - \cos x \sin y}{\cos x \cos y}}{\dfrac{\sin x \cos y + \cos x \sin y}{\cos x \cos y}} =$

   $\dfrac{\tan x - \tan y}{\tan x + \tan y}, \ x + y \neq k\pi, \ x, y \neq k\dfrac{\pi}{2}, \ k \in J$

20. $\dfrac{\cos\,(x - y)}{\cos\,(x + y)} = \dfrac{\cos x \cos y + \sin x \sin y}{\cos x \cos y - \sin x \sin y} = \dfrac{\dfrac{\cos x \cos y + \sin x \sin y}{\sin x \cos y}}{\dfrac{\cos x \cos y - \sin x \sin y}{\sin x \cos y}} =$

$\dfrac{\cot x + \tan y}{\cot x - \tan y}$, $x + y \neq (2k + 1)\dfrac{\pi}{2}$, $x, y \neq k\dfrac{\pi}{2}$, $k \in J$

21. $\sin\,(x + y)\,\sin\,(x - y) = (\sin x \cos y + \cos x \sin y)(\sin x \cos y - \cos x \sin y) =$

$\sin^2 x \cos^2 y - \cos^2 x \sin^2 y = \sin^2 x \cos^2 y + \sin^2 x \sin^2 y - \sin^2 x \sin^2 y -$

$\cos^2 x \sin^2 y = \sin^2 x\,(\cos^2 y + \sin^2 y) - \sin^2 y\,(\sin^2 x + \cos^2 x) = \sin^2 x - \sin^2 y$

22. $\sin 2x = 2 \sin x \cos x = \dfrac{2 \sin x \cos x \cdot \dfrac{1}{\cos^2 x}}{\dfrac{1}{\cos^2 x}} = \dfrac{2 \tan x}{\sec^2 x} = \dfrac{2 \tan x}{1 + \tan^2 x}$,

$x \neq (2k + 1)\dfrac{\pi}{2}$, $k \in J$

23. $\sec 2x \tan 2x = \dfrac{1}{\cos 2x} + \dfrac{\sin 2x}{\cos 2x} = \dfrac{1 + \sin 2x}{\cos 2x} = \dfrac{\sin^2 x + 2 \sin x \cos x + \cos^2 x}{\cos^2 x - \sin^2 x} =$

$\dfrac{(\sin x + \cos x)^2}{(\sin x + \cos x)(\cos x - \sin x)} = \dfrac{\cos x + \sin x}{\cos x - \sin x}$, $x \neq (2k + 1)\dfrac{\pi}{4}$

C 24. $\cos\,(x_2 + x_1) + \cos\,(x_2 - x_1) = \cos x_2 \cos x_1 - \sin x_2 \sin x_1 + \cos x_2 \cos x_1 +$

$\sin x_2 \sin x_1 = 2 \cos x_2 \cos x_1$

25. $\cos\,(x_2 + x_1) - \cos\,(x_2 - x_1) = \cos x_2 \cos x_1 - \sin x_2 \sin x_1 -$

$(\cos x_2 \cos x_1 + \sin x_2 \sin x_1) = -2 \sin x_2 \sin x_1$

26. $\sin\,(x_2 + x_1) + \sin\,(x_2 - x_1) = \sin x_2 \cos x_1 + \cos x_2 \sin x_1 + \sin x_2 \cos x_1 -$

$\cos x_2 \sin x_1 = 2 \sin x_2 \cos x_1$

27. If $x = x_2 + x_1$ and $y = x_2 - x_1$, $2x_2 = x + y$ and $x_2 = \dfrac{x + y}{2}$, and $2x_1 = x - y$ and

$x_1 = \dfrac{x - y}{2}$.   Using Ex. 24, we have $\cos x + \cos y = 2 \cos \dfrac{x + y}{2} \cos \dfrac{x - y}{2}$.

28. Let $x = x_2 + x_1$ and $y = x_2 - x_1$ and use Ex. 25.   Then $x_2 = \dfrac{x + y}{2}$ and $x_1 = \dfrac{x - y}{2}$,

so $\cos x - \cos y = -2 \sin \dfrac{x + y}{2} \sin \dfrac{x - y}{2}$.

29. Let $x = x_2 + x_1$ and $y = x_2 - x_1$ and use Ex. 26.   Then $x_2 = \dfrac{x + y}{2}$, $x_1 = \dfrac{x - y}{2}$

and $\sin x + \sin y = 2 \sin \dfrac{x + y}{2} \cos \dfrac{x - y}{2}$

30. $\dfrac{\sin 2x}{1 + \cos 2x} = \dfrac{\sin 2x(1 - \cos 2x)}{(1 + \cos 2x)(1 - \cos 2x)} = \dfrac{\sin 2x(1 - \cos 2x)}{1 - \cos^2 2x} = \dfrac{\sin 2x(1 - \cos 2x)}{\sin^2 2x} =$

$\dfrac{1 - \cos 2x}{\sin 2x}$, $x \neq k \cdot \dfrac{\pi}{2}$, $k \in J$

31. $\sin 3x + \sin x = \sin 2x \cos x + \cos 2x \sin x + \sin x = \sin 2x \cos x +$

    $(2 \cos^2 x - 1) \sin x + \sin x = \sin 2x \cos x + 2 \cos^2 x \sin x = \sin 2x \cos x +$

    $\sin 2x \cos x = 2 \sin 2x \cos x$

32. $\tan 3x = \dfrac{\tan 2x + \tan x}{1 - \tan 2x \tan x} = \dfrac{\dfrac{2 \tan x}{1 - \tan^2 x} + \tan x}{1 - \dfrac{2 \tan x}{1 - \tan^2 x} \cdot \tan x} = \dfrac{2 \tan x + \tan x - \tan^3 x}{1 - \tan^2 x - 2 \tan^2 x} =$

    $\dfrac{3 \tan x - \tan^3 x}{1 - 3 \tan^2 x}$, $x \neq (2k + 1)\dfrac{\pi}{6}$, $k \in J$

33. $\sin (x + y) \sin (x - y) = (\sin x \cos y + \cos x \sin y)(\sin x \cos y - \cos x \sin y) =$

    $\sin^2 x \cos^2 y - \cos^2 x \sin^2 y = \sin^2 x \cos^2 y + \sin^2 x \sin^2 y - \sin^2 x \sin^2 y -$

    $\cos^2 x \sin^2 y = \sin^2 x (\cos^2 y + \sin^2 y) - \sin^2 y (\sin^2 x + \cos^2 x) = \sin^2 x - \sin^2 y$

34. $\cos^4 \left(\dfrac{x}{2}\right) - \sin^4 \left(\dfrac{x}{2}\right) = \left(\pm\sqrt{\dfrac{1}{2}(1 + \cos x)}\right)^4 - \left(\pm\sqrt{\dfrac{1}{2}(1 - \cos x)}\right)^4 =$

    $\dfrac{1}{4}(1 + \cos x)^2 - \dfrac{1}{4}(1 - \cos x)^2 = \dfrac{1}{4}[1 + 2 \cos x + \cos^2 x - 1 + 2 \cos x - \cos^2 x] =$

    $\dfrac{1}{4} \cdot 4 \cos x = \cos x$

Chapter Test · Page 73

1. If $\cos x = -\dfrac{3}{5}$, $\sin x = \sqrt{1 - \dfrac{9}{25}} = \dfrac{4}{5}$ since $\dfrac{\pi}{2} < x < \pi$.   Hence $\tan x = \dfrac{\frac{4}{5}}{-\frac{3}{5}} = -\dfrac{4}{3}$

2. a. $\sec \dfrac{5\pi}{4} = \dfrac{1}{\cos \frac{5\pi}{4}} = -\sqrt{2}$ b. $\cot \left(-\dfrac{\pi}{6}\right) = \dfrac{\cos \left(-\frac{\pi}{6}\right)}{\sin \left(-\frac{\pi}{6}\right)} = -\dfrac{\cos \frac{\pi}{6}}{\sin \frac{\pi}{6}} = -\sqrt{3}$

3.

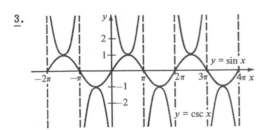

4. $\tan (2\pi - x) = \dfrac{\tan 2\pi - \tan x}{1 + \tan 2\pi \tan x} = \dfrac{0 - \tan x}{1 + 0} = -\tan x$

5. $\dfrac{2 \sin x}{\sin x \cot x + \cos x} = \dfrac{2 \sin x}{\cos x + \cos x} = \dfrac{2 \sin x}{2 \cos x} = \tan x$

Exercises 3-1 · Pages 82-83

<u>A</u>   <u>1.</u>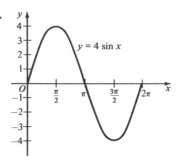
$y = 4 \sin x$

<u>2.</u>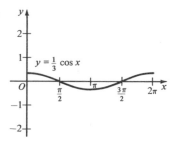
$y = \frac{1}{3} \cos x$

<u>3.</u>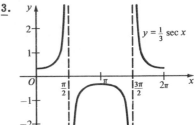
$y = \frac{1}{3} \sec x$

<u>4.</u>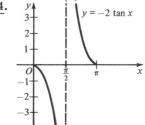
$y = -2 \tan x$

<u>5.</u>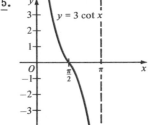
$y = 3 \cot x$

<u>6.</u>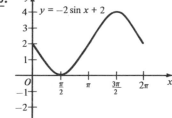
$y = -2 \sin x + 2$

<u>7.</u>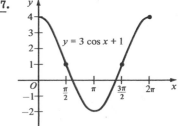
$y = 3 \cos x + 1$

<u>8.</u>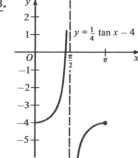
$y = \frac{1}{4} \tan x - 4$

9.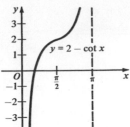

$y = 2 - \cot x$

10.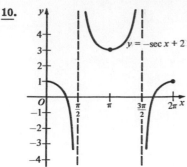

$y = -\sec x + 2$

B **11.** $\{y: y \in R \text{ and } 1 \leq y \leq 5\}$

12. $\{y: y \in R \text{ and } -8 \leq y \leq 0\}$

13. $\{y: y \in R \text{ and } 1 \leq y \leq 2\}$

14. $\{y: y \in R\}$

15. $\{y: y \in R\}$

16. $\{y: y \in R \text{ and } y \geq -1 \text{ or } y \leq -3\}$

17. $\sin \frac{\pi}{2} + d = 1 + d = 4$, so $d = 3$

18. $-3 \sin \frac{\pi}{2} + d = -3 + d = -6$, so $d = -3$

19. $-\cos \theta + d = -1 + d = -6$, so $d = -5$

20. $\tan (-\frac{\pi}{4}) + d = -1 + d = -3$, so $d = -2$

C **21.**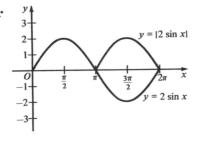

$y = |2 \sin x|$

$y = 2 \sin x$

22.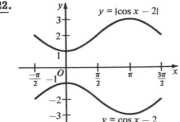

$y = |\cos x - 2|$

$y = \cos x - 2$

Exercises 3-2 · Page 88

1.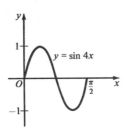

$y = \sin 4x$

2.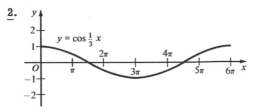

$y = \cos \frac{1}{3} x$

3.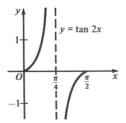

$y = \tan 2x$

4.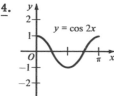

$y = \cos 2x$

5.

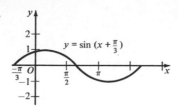

$y = \sin\left(x + \frac{\pi}{3}\right)$

6.

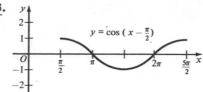

$y = \cos\left(x - \frac{\pi}{2}\right)$

7.

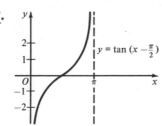

$y = \tan\left(x - \frac{\pi}{2}\right)$

8.

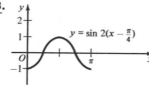

$y = \sin 2\left(x - \frac{\pi}{4}\right)$

9.

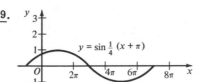

$y = \sin\frac{1}{4}(x + \pi)$

10.

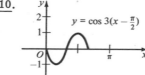

$y = \cos 3\left(x - \frac{\pi}{2}\right)$

11.

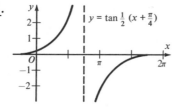

$y = \tan\frac{1}{2}\left(x + \frac{\pi}{4}\right)$

12.

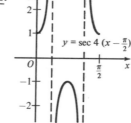

$y = \sec 4\left(x - \frac{\pi}{2}\right)$

B   **13.**

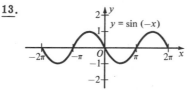

$y = \sin(-x)$

14.

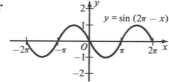

$y = \sin(2\pi - x)$

15.

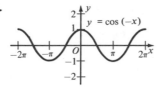

$y = \cos(-x)$

16.

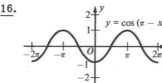

$y = \cos(\pi - x)$

17.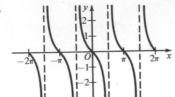

18.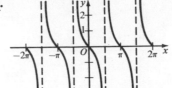

19.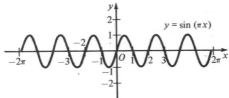

$y = \sin (\pi x)$

20.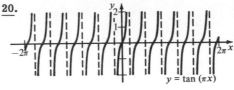

$y = \tan (\pi x)$

21. $y = \sin (x - c)$, $-1 = \sin (0 - c)$, $-1 = \sin (-c)$, $-1 = -\sin c$, $1 = \sin c$,

   $c = \frac{\pi}{2} \pm 2\pi k$, $k \in J$.

22. $y = 2 \tan (x - c)$, $2\sqrt{3} = 2 \tan (\frac{\pi}{6} - c)$, $\sqrt{3} = \dfrac{\frac{1}{\sqrt{3}} - \tan c}{1 + \frac{1}{\sqrt{3}} \tan c}$,

   $\sqrt{3} + \tan c = \frac{1}{\sqrt{3}} - \tan c$, $2 \tan c = -\frac{2\sqrt{3}}{3}$,

   $\tan c = \frac{\sqrt{3}}{3}$, $c = -\frac{\pi}{6}$, $c = \frac{5}{6}\pi + 2\pi k$, $k \in J$.

C   23. $y = -\cos x = \cos (x + \pi)$; phase shift $\frac{\pi}{1} = \pi$

24. $\cos (x - \frac{\pi}{2}) = \sin x$, so a shift of $\frac{\pi}{2}$ to the right will make cos coincide with sin;

   yes, for any c such that $\cos (x - c) = \sin x$ and $c = \frac{\pi}{2} + 2\pi k$, $k \in J$.

Exercises 3-3 · Page 92

A   1.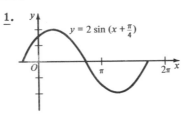

$y = 2 \sin (x + \frac{\pi}{4})$

amplitude: 2; period: 2π;

phase shift: $-\frac{\pi}{4}$

2.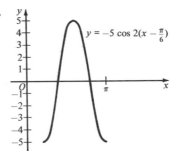

$y = -5 \cos 2(x - \frac{\pi}{6})$

amplitude: 5; period: π;

phase shift: $\frac{\pi}{6}$

3.

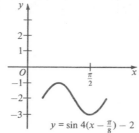

$$y = \sin 4\left(x - \tfrac{\pi}{8}\right) - 2$$

amplitude: 1; period: $\frac{\pi}{2}$;

phase shift: $\frac{\pi}{8}$

4.

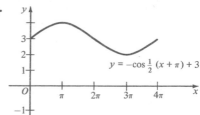

$$y = -\cos \tfrac{1}{2}(x + \pi) + 3$$

amplitude: 1; period: 4π

phase shift: $-\pi$

5.

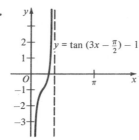

$$y = \tan\left(3x - \tfrac{\pi}{2}\right) - 1$$

amplitude: not defined;

period: $\frac{\pi}{3}$; phase shift: $\frac{\pi}{6}$

6.

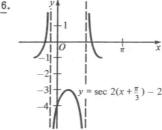

$$y = \sec 2\left(x + \tfrac{\pi}{3}\right) - 2$$

amplitude: not defined;

period: π; phase shift: $\frac{\pi}{3}$

B **7.**

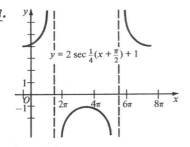

$$y = 2 \sec \tfrac{1}{4}\left(x + \tfrac{\pi}{2}\right) + 1$$

amplitude: not defined;

period: 8π; phase shift: $-\frac{\pi}{2}$

8.

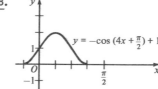

$$y = -\cos\left(4x + \tfrac{\pi}{2}\right) + 1$$

amplitude: 1; period: $\frac{\pi}{2}$;

phase shift: $-\frac{\pi}{8}$

9.

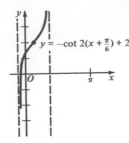

amplitude: not defined;

period: $\frac{\pi}{2}$; phase shift: $-\frac{\pi}{6}$

10.

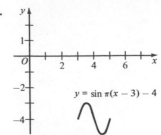

amplitude: 1; period: 2;

phase shift: 3

11.

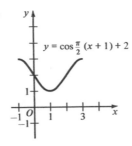

amplitude: 1; period: 4;

phase shift: -1

12.

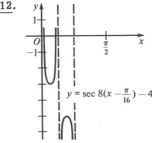

amplitude: not defined;

period: $\frac{\pi}{4}$; phase shift: $\frac{\pi}{16}$

C   **13.** Let a = -1 and c = $\frac{\pi}{2}$. $-\tan\left(x - \frac{\pi}{2}\right) = \tan\left(-x + \frac{\pi}{2}\right) = \dfrac{\sin\left(\frac{\pi}{2} - x\right)}{\cos\left(\frac{\pi}{2} - x\right)} = \dfrac{\cos x}{\sin x}$

= cot x.

14. $x = \frac{\pi}{4}$, $x = \frac{3\pi}{4}$, $x = \frac{5\pi}{4}$, $x = \frac{7\pi}{4}$ since $2x = (2k + 1)\frac{\pi}{2}$ or $x = (2k + 1)\frac{\pi}{4}$ for

k = 0, 1, 2, 3

15. $4x - \frac{\pi}{2} = (2k + 1)\frac{\pi}{2}$, $4x = (2k + 2)\frac{\pi}{2}$, $x = (k + 1)\frac{\pi}{4}$ for $0 \le k \le 8$ and $k \in J$.

Exercises 3-4 · Page 94

A   **1.**

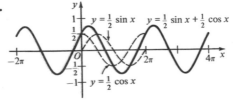

2.

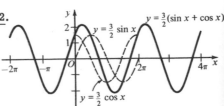

<u>3.</u>
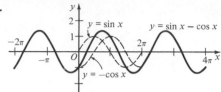

<u>4.</u>
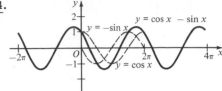

<u>5.</u>
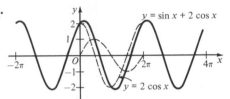

<u>6.</u>
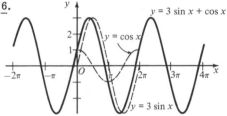

<u>7.</u>
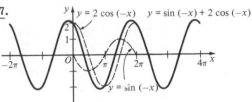

<u>8.</u>
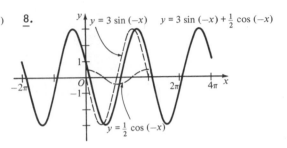

<u>9.</u>
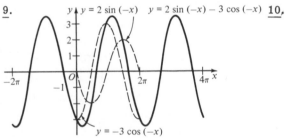

<u>10.</u>
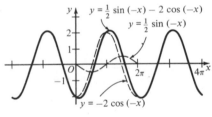

B <u>11.</u>
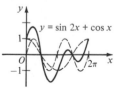

<u>12.</u>
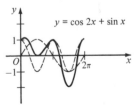

<u>13.</u>
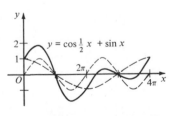

<u>14.</u>
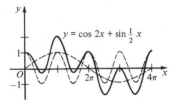

15.

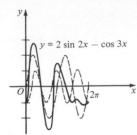

$y = 2 \sin 2x - \cos 3x$

16.

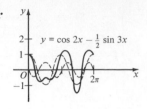

$y = \cos 2x - \frac{1}{2} \sin 3x$

17.

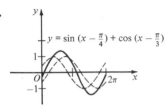

$y = \sin (x - \frac{\pi}{4}) + \cos (x - \frac{\pi}{3})$

18.

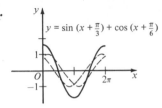

$y = \sin (x + \frac{\pi}{3}) + \cos (x + \frac{\pi}{6})$

C **19.** $\sqrt{2} \cos (x - \frac{\pi}{4}) = \sqrt{2} [\cos x \cos \frac{\pi}{4} + \sin x \sin \frac{\pi}{4}] = \sqrt{2} \cos x \cdot \frac{\sqrt{2}}{2} + \sqrt{2} \sin x \cdot \frac{\sqrt{2}}{2}$

$\cos x + \sin x$

20. a cos $(2x - c)$ = a(cos 2x cos c + sin 2x sin c) = a cos c · cos 2x +

a sin c · sin 2x.   Thus we want a cos c = 1 and a sin c = 1. $a^2 \cos^2 c$ = 1 and

a $\sin^2$ c = 1, so $a^2(\cos^2 c + \sin^2 c)$ = 2, a^2 = 2, and a = $\pm\sqrt{2}$.   If a = $\sqrt{2}$,

sin c = $\frac{1}{\sqrt{2}}$ and cos c = $\frac{1}{\sqrt{2}}$.   If a = $-\sqrt{2}$, sin c = $-\frac{1}{\sqrt{2}}$ and cos c = $-\frac{1}{\sqrt{2}}$.

Thus a = $\sqrt{2}$ and c = $\frac{\pi}{4}$ + 2πk or a = $-\sqrt{2}$ and c = $\frac{5\pi}{4}$ + 2πk satisfy the given

conditions.

21. a cos $(x - c)$ = a cos x cos c + a sin x sin c.   Thus we want a cos c = 4 and

a sin c = 3. $a^2 \cos^2 c$ = 16 and $a^2 \sin^2 c$ = 9, so $a^2(\cos^2 c + \sin^2 c)$ = 25,

a^2 = 25, and a = ± 5.   If a = 5, cos c = $\frac{4}{5}$ and sin c = $\frac{3}{5}$.   If a = -5,

cos c = $-\frac{4}{5}$ and sin c = $-\frac{3}{5}$.   Thus a = 5 and c $\doteq$ 0.41 · $\frac{\pi}{2}$ + 2πk or a = -5

and c = $0.41(\frac{\pi}{2})$ + (2k + 1)π satisfy the given conditions.

Exercises 3-5 · Pages 98-100

A **1.** rωt = 3 · $\frac{\pi}{6}$ · 2 = π

2. rωt = 4 · $\frac{3\pi}{2}$ · 3 = 18π

3. rωt = 9 · $\frac{2\pi}{3}$ · $\frac{1}{2}$ = 3π

4. rωt = 6 · $\frac{7\pi}{12}$ · $\frac{1}{3}$ = $\frac{7\pi}{6}$

5. rωt = $\frac{4}{3}$ · $\frac{5\pi}{4}$ · 2 = $\frac{10\pi}{3}$

6. rωt = $\frac{6}{5}$ · $\frac{11\pi}{12}$ · 5 = $\frac{11\pi}{2}$

7. $\omega = \frac{2\pi}{T} = \frac{2\pi}{\frac{2}{3}} = 3\pi$ sec.; speed $= \frac{2\pi r}{T} = \frac{2\pi \cdot 5}{\frac{2}{3}} = 15\pi$ in./sec.

8. $\omega = \frac{2\pi}{\frac{1}{120}} = 240\pi/\text{min.}$; $3(120)2\pi \cdot 6 = 4320$ in.

9. $2\pi\left(\frac{15}{60}\right) \cdot 2 = \pi$ in.; $\omega = -\frac{2\pi}{T} = -2\pi/\text{hr.} = -\frac{\pi}{30}/\text{min.}$

10. speed of belt $= \omega a = \frac{2\pi \cdot 4}{\frac{1}{120}} = 960\pi$ in./min.; $\omega = \frac{960\pi}{3} = 320\pi$ /min.

B   11. a. At t = 3, (a cos ωt, a sin ωt) = (6 cos $\frac{5\pi}{6} \cdot 3$, 6 sin $\frac{5\pi}{6} \cdot 3$) =

   (6 cos $\frac{5\pi}{2}$, 6 sin $\frac{5\pi}{2}$) = (6 cos $\frac{\pi}{2}$, 6 sin $\frac{\pi}{2}$) = (0, 6)

   b. At t = -3, (a cos ωt, a sin ωt) = (6 cos $\frac{5\pi}{6}$(-3), 6 cos $\frac{5\pi}{6}$(-3)) =

   (6 cos (-$\frac{5\pi}{2}$), 6 sin (-$\frac{5\pi}{2}$)) = (6 cos $\frac{\pi}{2}$, -6 sin $\frac{\pi}{2}$) = (0, -6)

12. $3(\frac{5\pi}{6})t = 3(\frac{\pi}{3})t + 2\pi \cdot 3$, $\frac{5\pi}{2}t = \pi t + 6\pi$, $\frac{3\pi}{2}t = 6\pi$, t = 4; $P_1 =$

   (3 cos $\frac{5\pi}{6} \cdot 4$, 3 sin $\frac{5\pi}{6} \cdot 4$) = (3 cos $\frac{10\pi}{3}$, 3 sin $\frac{10\pi}{3}$) = (3 cos $\frac{4\pi}{3}$, 3 sin $\frac{4\pi}{3}$) =

   $(-\frac{3}{2}, -\frac{3\sqrt{3}}{2})$

13. $3(\frac{4\pi}{3})t = 3(\frac{\pi}{2})t + 2\pi \cdot 3$, $4\pi t = \frac{3\pi}{2} + 6\pi$, $\frac{5\pi}{2}t = 6\pi$, t = $\frac{12}{5}$;

   $P_1 = $ (3 cos $\frac{4\pi}{3} \cdot \frac{12}{5}$, 3 sin $\frac{4\pi}{3} \cdot \frac{12}{5}$) = (3 cos $\frac{16\pi}{5}$, 3 sin $\frac{16\pi}{5}$) =

   (3 cos $\frac{6\pi}{5}$, 3 sin $\frac{6\pi}{5}$)

14. $3(\frac{\pi}{4})t = 3(-\frac{\pi}{8})t + 2\pi \cdot 3$, $\frac{3\pi}{4}t = -\frac{3\pi}{8}t + 6\pi$, $\frac{9}{8}t = 6\pi$, t = $\frac{16}{3}$; $P_1 =$

   (3 cos $(-\frac{\pi}{8}) \cdot \frac{16}{3}$, 3 sin $(-\frac{\pi}{8}) \cdot \frac{16}{3}$) = (3 cos $(-\frac{2\pi}{3})$, 3 sin $(-\frac{2\pi}{3})$) =

   (3 cos $\frac{2\pi}{3}$, -3 sin $\frac{2\pi}{3}$) = $(-\frac{3}{2}, -\frac{3\sqrt{3}}{2})$

C   15. $3(\frac{\pi}{3})t = 3(\frac{\pi}{4})t + 3(\frac{\pi}{6})$, $\pi t = \frac{3\pi}{4}t + \frac{\pi}{2}$, $\frac{\pi}{4}t = \frac{\pi}{2}$, t = 2;

   P = (3 cos 2 $\cdot \frac{\pi}{3}$, 3 sin 2 $\cdot \frac{\pi}{3}$) = $(-\frac{3}{2}, \frac{3\sqrt{3}}{2})$; $\frac{2\pi}{3} \cdot 3 = 2\pi$ units or $\frac{1}{3}$ revolution

16. $3(-\frac{\pi}{3})t - \frac{\pi}{6}(3) = 3(-\frac{\pi}{4})t - 2\pi(3)$, $-\pi t - \frac{\pi}{2} = -\frac{3\pi}{4}t - 6\pi$, $\frac{11\pi}{2} = \frac{\pi}{4}t$, t = 22;

   P = (3 cos $(-\frac{\pi}{3}) \cdot 22$, 3 sin $(-\frac{\pi}{3}) \cdot 22$) = (3 cos $\frac{22\pi}{3}$, -3 sin $\frac{22\pi}{3}$) =

   (3 cos $\frac{4\pi}{3}$, -3 sin $\frac{4\pi}{3}$) = $(-\frac{3}{2}, \frac{3\sqrt{3}}{2})$; $3(-\frac{\pi}{3}) \cdot 22 = -22\pi$ units or $3\frac{2}{3}$ revolutions

   in a clockwise direction

17. $u = 5 \cos (\omega t + \frac{\pi}{6})$, $v = 5 \sin (\omega t + \frac{\pi}{6})$

18. $u = 2 \cos (\omega t + \frac{\pi}{4})$, $v = 2 \sin (\omega t + \frac{\pi}{4})$

Exercises 3-6 · Page 103

A   1. $f = \frac{\omega}{2\pi} = 60$, so $\omega = 120\pi$; $E_{max} = E_{max} \sin (120\pi \cdot \frac{1}{4} + b)$, $1 = \sin (30\pi + b) =$

   $\sin b$, so $b = \frac{\pi}{2}$; $E = 120 \sin (120\pi \cdot \frac{1}{120} + \frac{\pi}{2}) = 120 \sin (\pi + \frac{\pi}{2}) = -120$ volts

   2. $E = 120 \sin (120\pi \cdot \frac{1}{360} + \frac{\pi}{2}) = 120 \sin (\frac{\pi}{3} + \frac{\pi}{2}) = 120 \sin \frac{5\pi}{6} = 120 \cdot \frac{1}{2} = 60$ volts

   3. $f = \frac{\omega}{2\pi} = 50$, so $\omega = 100\pi$; $E = 115 \sin (100\pi \cdot \frac{1}{200}) = 115 \sin \frac{\pi}{2} = 115$ volts

   4. $E = 115 \sin (100\pi \cdot \frac{1}{50}) = 115 \sin 2\pi = 0$ volts

B   5. $y = a \cos (\omega t + b)$, $2\pi = \frac{2\pi}{\omega} = t$, so $\omega = 1$, $a = 3$; when $t = 0$, $y = 3$ so

   $3 = 3 \cos (0 + b)$; $b = 0$; $y = 3 \cos t$

   6. $1 = 2 \cos (0 + b)$; $\cos b = \frac{1}{2}$, $b = \frac{\pi}{3}$; $y = 2 \cos (t + \frac{\pi}{3})$

C   7. When $y = 0$, $0 = \sin (\omega t + b)$, so $\omega t + b = k\pi$ and $v = a\omega \cos (k\pi) = \pm a\omega$;

   when $y = a$, $a = a \sin (\omega t + b)$ and $\sin (\omega t + b) = 1$, $\omega t + b = (4k + 1)\frac{\pi}{2}$, and

   $v = a\omega \cos (4k + 1)\frac{\pi}{2} = 0$; Speed maximum when $\cos (\omega t + b) = 1$ or $\omega t + b = 2k\pi$,

   and $t = \frac{2k\pi - b}{\omega}$; Speed minimum when $\cos (\omega t + b) = -1$ or $\omega t + b = (2k + 1)\pi$,

   and $t = \frac{(2k + 1)\pi - b}{\omega}$; Speed zero when $\cos (\omega t + b) = 0$ or $\omega t + b = (2k + 1)\frac{\pi}{2}$,

   and $t = \frac{(2k + 1)\frac{\pi}{2} - b}{\omega}$

   8. $v = 3\sqrt{2} \cos 3t$, so $v = 3\sqrt{2} \cos \frac{\pi}{4} = 3$; v maximum when $\cos 3t = 1$ and $v = 3\sqrt{2}$

Chapter Test · Pages 104-105

1.
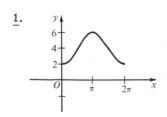

2.
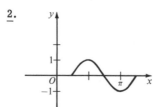

3.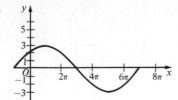

4.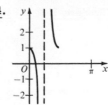

5.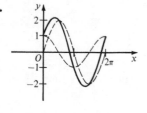

6. $r\omega t = 4 \cdot \dfrac{5\pi}{6} \cdot 3 = 10\pi$

7. $y = a \cos (\omega t + b);$

   $y = a \sin (\omega t + b)$

Cumulative Review: Chapters 1-3 · Pages 106-107

Chapter 1

1. $\cos^2 x + \sin^2 x = 1$, $\cos^2 x + \frac{1}{36} = 1$, $\cos^2 x = \frac{35}{36}$, $\cos x = \pm\frac{\sqrt{35}}{6}$; since $\pi < x < \frac{3\pi}{2}$, then $\cos x = -\frac{\sqrt{35}}{6}$

2. $\sin^2 x \cos x + \cos^3 x = \cos x (\sin^2 x + \cos^2 x) = \cos x$

3. $\cos \frac{2\pi}{3} \cos \frac{\pi}{6} + \sin \frac{2\pi}{3} \sin \frac{\pi}{6} = \cos (\frac{2\pi}{3} - \frac{\pi}{6}) = \cos \frac{\pi}{2} = 0$

4. $\cos \pi \sin (-\frac{\pi}{3}) - \sin \pi \cos (-\frac{\pi}{3}) = \sin (-\frac{\pi}{3} - \pi) = \sin (-\frac{4\pi}{3}) = -\sin \frac{4\pi}{3} = \frac{\sqrt{3}}{2}$

5. $\cos x_1 = \frac{3}{4}$, $\sin x_1 = \pm\sqrt{1 - \frac{9}{16}} = \pm\frac{\sqrt{7}}{4}$, since $0 < x_1 < \frac{\pi}{2}$, then $\sin x_1 = \frac{\sqrt{7}}{4}$;

$\sin x_2 = \frac{1}{2}$, $\cos x_2 = \pm\sqrt{1 - \frac{1}{4}} = \pm\frac{\sqrt{3}}{2}$, since $\frac{\pi}{2} < x_2 < \pi$, then $\cos x_2 = -\frac{\sqrt{3}}{2}$;

$\sin (x_1 + x_2) = \sin x_1 \cos x_2 + \cos x_1 \sin x_2 = (\frac{\sqrt{7}}{4})(-\frac{\sqrt{3}}{2}) + (\frac{3}{4})(\frac{1}{2}) = \frac{-\sqrt{21} + 3}{8} = \frac{3 - \sqrt{21}}{8}$

6. $\cos^2 x - \sin^2 x + 1 = \cos^2 x - \sin^2 x + \sin^2 x + \cos^2 x = 2 \cos^2 x$

7. $\cos 2x + 2 \sin^2 x = (1 - 2 \sin^2 x) + 2 \sin^2 x = 1$

8.
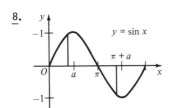

9. The solution set $= \{(\frac{\pi}{3}, \frac{\sqrt{3}}{2}), (\frac{2\pi}{3}, \frac{\sqrt{3}}{2})\}$
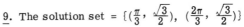
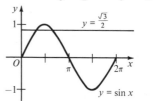

10. The solution set $= \{(\frac{2\pi}{3}, -\frac{1}{2}), (\frac{4\pi}{3}, -\frac{1}{2})\}$
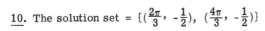
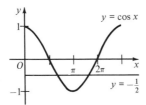

Chapter 2

11. $\sin x = -\frac{4}{5}$, $\cos x = \pm\sqrt{1 - \frac{16}{25}} = \pm\frac{3}{5}$, since $\cos x > 0$, so $\cos x = \frac{3}{5}$, $\tan x = \frac{\sin x}{\cos x} = \frac{-\frac{4}{5}}{\frac{3}{5}} = -\frac{4}{3}$

12. a. $\tan \frac{23\pi}{3} = \tan (6\pi + \frac{5\pi}{3}) = \tan \frac{5\pi}{3} = -\sqrt{3};$

   b. $\cot (-\frac{5\pi}{6}) = \cot (\frac{1}{6}\pi - \pi) = \cot \frac{\pi}{6} = \sqrt{3};$

   c. $\sec \frac{17\pi}{4} = \sec (4\pi + \frac{\pi}{4}) = \sec \frac{\pi}{4} = \sqrt{2}$

13. $\csc x = \frac{5}{4};$ $\sin x = \frac{1}{\csc x} = \frac{4}{5};$ $\cos x = \pm \sqrt{1 - \frac{16}{25}} = \pm \frac{3}{5},$ since $\frac{\pi}{2} < x < \pi,$ then

   $\cos x = -\frac{3}{5};$ $\sec x = \frac{1}{\cos x} = -\frac{5}{3};$ $\tan x = \frac{\sin x}{\cos x} = -\frac{4}{3};$ $\cot x = \frac{1}{\tan x} = -\frac{3}{4}$

14. $\sin x \sec x = \sin x (\frac{1}{\cos x}) = \frac{\sin x}{\cos x} = \tan x$

15. $\sin x + \cos x \cot x = \sin x + \frac{\cos^2 x}{\sin x} = \frac{\sin^2 x + \cos^2 x}{\sin x} = \frac{1}{\sin x} = \csc x$

16. $\csc^2 x (1 - \cos^2 x) = \csc^2 x \sin^2 x = \frac{1}{\sin^2 x}(\sin^2 x) = 1$

17. $\sin x \tan x + \cos x = \frac{\sin^2 x}{\cos x} + \cos x = \frac{\sin^2 x + \cos^2 x}{\cos x} = \frac{1}{\cos x} = \sec x$

18. $\sec x - \cos x = \frac{1}{\cos x} - \cos x = \frac{1 - \cos^2 x}{\cos x} = \frac{\sin^2 x}{\cos x} = \sin x \tan x$

19. $\frac{\sin x}{\sin x \cot x + \cos x} = \frac{\sin x}{\cos x + \cos x} = \frac{\sin x}{2 \cos x} = \frac{1}{2} \frac{\sin x}{\cos x} = \frac{1}{2} \tan x$

Chapter 3

21.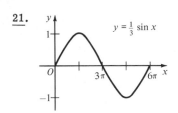
   $y = \frac{1}{3} \sin x$

22.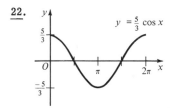
   $y = \frac{5}{3} \cos x$

23.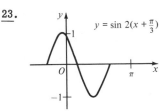
   $y = \sin 2(x + \frac{\pi}{3})$

24.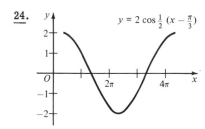
   $y = 2 \cos \frac{1}{2} (x - \frac{\pi}{3})$

25.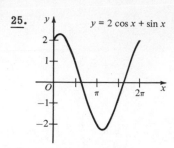

26.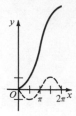

27. $\frac{2\pi}{3} \cdot 3 \cdot 4 = 8\pi$

28. $\frac{\pi}{3} \cdot 10 \cdot \frac{9}{2} = 15\pi$

29. $\frac{2\pi}{T} = \omega, \quad \frac{2\pi}{\frac{3}{4}} = \omega, \quad \frac{8\pi}{3} = \omega$

30. speed $= \frac{2\pi a}{T} = \frac{2\pi(3)}{\frac{3}{4}} = 8\pi$

Exercises 4-1 · Pages 115-117

A 1. $y = 0$ 2. $y = \frac{5\pi}{6}$ 3. $y = \frac{\pi}{4}$ 4. $y = \frac{\pi}{2}$ 5. $y = \frac{\pi}{3}$

6. $y = -\frac{\pi}{4}$ 7. $y = \frac{\pi}{6}$ 8. $y = -\frac{\pi}{2}$ 9. $\cos^{-1} \frac{1}{2} = \frac{\pi}{3}$

10. $\cos^{-1}(-1) = \pi$ 11. $\sin^{-1} \frac{\sqrt{3}}{2} = \frac{\pi}{3}$ 12. $\sin^{-1}(-\frac{1}{2}) = -\frac{\pi}{6}$

13. $\cos^{-1} 0.853 \doteq 0.35 \cdot \frac{\pi}{2} \doteq 0.175\pi$

14. $\cos^{-1}(-0.924) \doteq \pi - 0.25\frac{\pi}{2} \doteq 1.75\frac{\pi}{2} \doteq 0.875\pi$

15. $\sin^{-1} 0.800 \doteq 0.59\frac{\pi}{2} \doteq 0.295\pi$ 16. $\sin^{-1}(-0.203) \doteq -0.13\frac{\pi}{2} \doteq -0.065\pi$

17. $\cos(\cos^{-1} 0.800) = 0.800$ 18. $\cos(\cos^{-1} \frac{3}{5}) = \frac{3}{5}$

19. $\sin(\sin^{-1} \frac{\sqrt{2}}{2}) = \frac{\sqrt{2}}{2}$ 20. $\sin(\sin^{-1}(-\frac{3}{5})) = -\frac{3}{5}$

21. $\cos^{-1}(\cos \frac{\pi}{6}) = \cos^{-1} \frac{\sqrt{3}}{2} = \frac{\pi}{6}$ 22. $\cos[\cos^{-1}(-0.426)] = -0.426$

B 23.

x	$\cos^{-1} x$	$\cos(\cos^{-1} x)$	$\sin^{-1} x$	$\sin(\sin^{-1} x)$
-1	π	-1	$-\frac{\pi}{2}$	-1
$-\frac{\sqrt{3}}{2}$	$\frac{5\pi}{6}$	$-\frac{\sqrt{3}}{2}$	$-\frac{\pi}{3}$	$-\frac{\sqrt{3}}{2}$
$-\frac{1}{2}$	$\frac{2\pi}{3}$	$-\frac{1}{2}$	$-\frac{\pi}{6}$	$-\frac{1}{2}$
0	$\frac{\pi}{2}$	0	0	0
$\frac{1}{2}$	$\frac{\pi}{3}$	$\frac{1}{2}$	$\frac{\pi}{6}$	$\frac{1}{2}$
$\frac{\sqrt{3}}{2}$	$\frac{\pi}{6}$	$\frac{\sqrt{3}}{2}$	$\frac{\pi}{3}$	$\frac{\sqrt{3}}{2}$
1	0	1	$\frac{\pi}{2}$	1

24. a. $\cos(\cos^{-1} x) = x$ b. $\sin(\sin^{-1} x) = x$

25. $\cos^{-1}(-1) + \sin^{-1}(-1) = \pi + (-\frac{\pi}{2}) = \frac{\pi}{2}$; $\cos^{-1}(-\frac{\sqrt{3}}{2}) + \sin^{-1}(-\frac{\sqrt{3}}{2}) = \frac{5\pi}{6} + (-\frac{\pi}{3}) = \frac{\pi}{2}$; $\cos^{-1}(-\frac{1}{2}) + \sin^{-1}(-\frac{1}{2}) = \frac{2\pi}{3} + (-\frac{\pi}{6}) = \frac{\pi}{2}$; $\cos^{-1} 0 + \sin^{-1} 0 = \frac{\pi}{2} + 0 = \frac{\pi}{2}$; $\cos^{-1} \frac{1}{2} + \sin^{-1} \frac{1}{2} = \frac{\pi}{3} + \frac{\pi}{6} = \frac{\pi}{2}$; $\cos^{-1} \frac{\sqrt{3}}{2} + \sin^{-1} \frac{\sqrt{3}}{2} = \frac{\pi}{6} + \frac{\pi}{3} = \frac{\pi}{2}$; $\cos^{-1} 1 + \sin^{-1} 1 = 0 + \frac{\pi}{2} = \frac{\pi}{2}$; $\cos^{-1} x + \sin^{-1} x = \frac{\pi}{2}$, $-1 \leq x \leq 1$

<u>C</u> <u>26.</u> <u>a.</u>

x	cos x	Cos⁻¹ (cos x)
0	1	0
$\frac{\pi}{2}$	0	$\frac{\pi}{2}$
π	-1	π
$\frac{3\pi}{2}$	0	$\frac{\pi}{2}$
2π	1	0
$\frac{5\pi}{2}$	0	$\frac{\pi}{2}$
3π	-1	π
$\frac{7\pi}{2}$	0	$\frac{\pi}{2}$
4π	1	0

<u>b.</u>

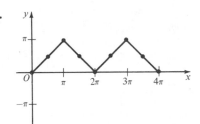

<u>c.</u> yes, 2π

<u>d.</u> $0 \le x \le \pi$

<u>e.</u> $0 \le x \le \pi$; $0 \le x \le \pi$

<u>27.</u> <u>a.</u>

x	sin x	Sin⁻¹ (sin x)
$-\frac{\pi}{2}$	-1	$-\frac{\pi}{2}$
0	0	0
$\frac{\pi}{2}$	1	$\frac{\pi}{2}$
π	0	0
$\frac{3\pi}{2}$	-1	$-\frac{\pi}{2}$
2π	0	0
$\frac{5\pi}{2}$	1	$\frac{\pi}{2}$
3π	0	0
$\frac{7\pi}{2}$	-1	$-\frac{\pi}{2}$

b.

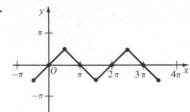

<u>c</u>. yes, 2π

<u>d</u>. $-\frac{\pi}{2} \le x \le \frac{\pi}{2}$

<u>e</u>. $-\frac{\pi}{2} \le x \le \frac{\pi}{2}$; $-\frac{\pi}{2} \le x \le \frac{\pi}{2}$

Exercises 4-2 · Pages 120-122

<u>A</u> <u>1</u>. $y = \frac{\pi}{3}$ <u>2</u>. $y = \frac{3\pi}{4}$ <u>3</u>. $y = \frac{2\pi}{3}$ <u>4</u>. $y = \frac{\pi}{4}$ <u>5</u>. $y = \frac{\pi}{2}$

<u>6</u>. $y = -\frac{\pi}{6}$ <u>7</u>. $\text{Tan}^{-1} 1 = \frac{\pi}{4}$ <u>8</u>. $\text{Sec}^{-1} 2 = \frac{\pi}{3}$ <u>9</u>. $\text{Sec}^{-1} \sqrt{2} = \frac{\pi}{4}$

<u>10</u>. $\text{Cot}^{-1} (-\sqrt{3}) = \frac{5\pi}{6}$ <u>11</u>. $\text{Csc}^{-1} (-\frac{2}{\sqrt{3}}) = -\frac{\pi}{3}$ <u>12</u>. $\text{Tan}^{-1} 0 = 0$

<u>13</u>. $\text{Csc}^{-1} 2 = \frac{\pi}{6}$ <u>14</u>. $\text{Csc}^{-1} (-\sqrt{2}) = -\frac{\pi}{4}$ <u>15</u>. $\tan (\text{Tan}^{-1} 1) = 1$

<u>16</u>. $\sec (\text{Sec}^{-1} \frac{2}{\sqrt{3}}) = \frac{2}{\sqrt{3}}$ <u>17</u>. $\sec (\text{Sec}^{-1} 1) = 1$ <u>18</u>. $\cot [\text{Cot}^{-1} (-1)] = -1$

<u>19</u>. $\tan (\text{Cot}^{-1} \frac{1}{\sqrt{3}}) = \tan \frac{\pi}{3} = \sqrt{3}$ <u>20</u>. $\cot [\text{Tan}^{-1} (-\sqrt{3})] = \cot (-\frac{\pi}{3}) = -\frac{1}{\sqrt{3}}$

<u>21</u>. $\sin (\text{Csc}^{-1} \frac{2}{\sqrt{3}}) = \sin \frac{\pi}{3} = \frac{\sqrt{3}}{2}$

<u>B</u> <u>22</u>. <u>a</u>.

x	$-\sqrt{3}$	-1	$-\frac{1}{\sqrt{3}}$	0	$\frac{1}{\sqrt{3}}$	1	$\sqrt{3}$
(i) $\text{Tan}^{-1} x$	$-\frac{\pi}{3}$	$-\frac{\pi}{4}$	$-\frac{\pi}{6}$	0	$\frac{\pi}{6}$	$\frac{\pi}{4}$	$\frac{\pi}{3}$
(ii) $\text{Cot}^{-1} x$	$\frac{5\pi}{6}$	$\frac{3\pi}{4}$	$\frac{2\pi}{3}$	$\frac{\pi}{2}$	$\frac{\pi}{3}$	$\frac{\pi}{4}$	$\frac{\pi}{6}$
(iii) $\text{Tan}^{-1} x + \text{Cot}^{-1} x$	$\frac{\pi}{2}$	$\frac{\pi}{2}$	$\frac{\pi}{2}$	$\frac{\pi}{2}$	$\frac{\pi}{2}$	$\frac{\pi}{2}$	$\frac{\pi}{2}$
(iv) $\text{Tan}^{-1} \frac{1}{x}$	$-\frac{\pi}{6}$	$-\frac{\pi}{4}$	$-\frac{\pi}{3}$	$-$	$\frac{\pi}{3}$	$\frac{\pi}{4}$	$\frac{\pi}{6}$
(v) $\text{Tan}^{-1} x + \text{Tan}^{-1} \frac{1}{x}$	$-\frac{\pi}{2}$	$-\frac{\pi}{2}$	$-\frac{\pi}{2}$	$-$	$\frac{\pi}{2}$	$\frac{\pi}{2}$	$\frac{\pi}{2}$

<u>b</u>. $\text{Tan}^{-1} x + \text{Cot}^{-1} x = \frac{\pi}{2}$

<u>23</u>. $\text{Tan}^{-1} (-\frac{1}{\sqrt{3}}) + \pi = -\frac{\pi}{6} + \pi = \frac{5\pi}{6} = \text{Cot}^{-1} (-\sqrt{3})$; $\text{Tan}^{-1} (-1) + \pi = -\frac{\pi}{4} + \pi = \frac{3\pi}{4} =$

$\text{Cot}^{-1} (-1)$; $\text{Tan}^{-1} (-\sqrt{3}) + \pi = -\frac{\pi}{3} + \pi = \frac{2\pi}{3} = \text{Cot}^{-1} (-\frac{1}{\sqrt{3}})$; $\text{Tan}^{-1} \sqrt{3} = \frac{\pi}{3} =$

$\text{Cot}^{-1} \frac{1}{\sqrt{3}}$; $\text{Tan}^{-1} 1 = \frac{\pi}{4} = \text{Cot}^{-1} 1$; $\text{Tan}^{-1} \frac{1}{\sqrt{3}} = \frac{\pi}{6} = \text{Cot}^{-1} \sqrt{3}$

<u>C</u>   <u>24.</u> $\text{Tan}^{-1} x + \text{Tan}^{-1} \frac{1}{x} = -\frac{\pi}{2}$ for $x < 0$; $\text{Tan}^{-1} x + \text{Tan}^{-1} \frac{1}{x} = \frac{\pi}{2}$ for $x > 0$;

$\text{Tan}^{-1} \frac{1}{6} + \text{Tan}^{-1} 6 = \frac{\pi}{2}$, $\text{Tan}^{-1} \frac{1}{6} + 1.406 \doteq 1.571$, $\text{Tan}^{-1} \frac{1}{6} \doteq 0.165$;

$\text{Tan}^{-1} (-7) + \text{Tan}^{-1} (-\frac{1}{7}) = -\frac{\pi}{2}$, $\text{Tan}^{-1} (-7) - 0.1435 \doteq -1.571$, $\text{Tan}^{-1} (-7) \doteq -1.427$

<u>25.</u> $\text{Cot} (\text{Cot}^{-1} x) = x,\ x \in \mathcal{R}$; $\text{Cot}^{-1} (\text{Cot } x) = x,\ 0 < x < \pi$; $\text{Sec} (\text{Sec}^{-1} x) = x,$

$|x| \geq 1$; $\text{Sec}^{-1} (\text{Sec } x) = x,\ 0 \leq x \leq \pi,\ x \neq \frac{\pi}{2}$; $\text{Csc} (\text{Csc}^{-1} x) = x,\ |x| \geq 1$;

$\text{Csc}^{-1} (\text{Csc } x) = x,\ -\frac{\pi}{2} \leq x \leq \frac{\pi}{2},\ x \neq 0$

<u>26.</u> $\tan (\text{Tan}^{-1} \frac{1}{2} + \text{Tan}^{-1} \frac{1}{3}) = \dfrac{\tan (\text{Tan}^{-1} \frac{1}{2}) + \tan (\text{Tan}^{-1} \frac{1}{3})}{1 - \tan (\text{Tan}^{-1} \frac{1}{2})(\text{Tan}^{-1} \frac{1}{3})} = \dfrac{\frac{1}{2} + \frac{1}{3}}{1 - \frac{1}{6}} = \dfrac{\frac{5}{6}}{\frac{5}{6}} = 1$;

$\text{Tan}^{-1} \frac{1}{2} + \text{Tan}^{-1} \frac{1}{3} = \frac{\pi}{4}$

<u>27.</u>   yes, π

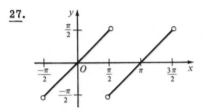

<u>Exercises 4-3</u> · <u>Pages 126-127</u>

<u>A</u>   <u>1.</u> $\sin (\text{Cos}^{-1} \frac{1}{3}) = \sqrt{1 - (\frac{1}{3})^2} = \sqrt{\frac{8}{9}} = \frac{2\sqrt{2}}{3}$

<u>2.</u> $\cos (\text{Sin}^{-1} \frac{2}{3}) = \sqrt{1 - (\frac{2}{3})^2} = \sqrt{\frac{5}{9}} = \frac{\sqrt{5}}{3}$

<u>3.</u> $\sin [\text{Arccos} (-\frac{1}{2})] = \sqrt{1 - (-\frac{1}{2})^2} = \sqrt{\frac{3}{4}} = \frac{\sqrt{3}}{2}$

<u>4.</u> $\cos [\text{Arcsin} (-\frac{1}{4})] = \sqrt{1 - (-\frac{1}{4})^2} = \sqrt{\frac{15}{16}} = \frac{\sqrt{15}}{4}$

<u>5.</u> $\text{Arcsin } 1 + \text{Arcsin } 0 = \frac{\pi}{2} + 0 = \frac{\pi}{2}$

<u>6.</u> $\text{Arcsin } \frac{\sqrt{2}}{2} + \text{Arcsin } \frac{\sqrt{3}}{2} = \frac{\pi}{4} + \frac{\pi}{3} = \frac{7\pi}{12}$

<u>7.</u> $\text{Sin}^{-1} \frac{1}{2} + \text{Sin}^{-1} \frac{\sqrt{3}}{2} = \frac{\pi}{6} + \frac{\pi}{3} = \frac{\pi}{2}$

<u>8.</u> $\text{Cos}^{-1} (-\frac{\sqrt{3}}{2}) + \text{Cos}^{-1} (-\frac{1}{2}) = \frac{5\pi}{6} + \frac{2\pi}{3} = \frac{3\pi}{2}$

<u>9.</u> $\tan (\text{Sin}^{-1} \frac{1}{2}) = \dfrac{\sin (\text{Sin}^{-1} \frac{1}{2})}{\cos (\text{Sin}^{-1} \frac{1}{2})} = \dfrac{\frac{1}{2}}{\sqrt{1 - (\frac{1}{2})^2}} = \dfrac{\frac{1}{2}}{\frac{\sqrt{3}}{2}} = \dfrac{1}{\sqrt{3}}$

<u>10.</u> $\cot [\text{Cos}^{-1} (-\frac{1}{2})] = \dfrac{\cos [\text{Cos}^{-1} (-\frac{1}{2})]}{\sin [\text{Cos}^{-1} (-\frac{1}{2})]} = \dfrac{-\frac{1}{2}}{\sqrt{1 - (-\frac{1}{2})^2}} = \dfrac{-\frac{1}{2}}{\frac{\sqrt{3}}{2}} = -\dfrac{1}{\sqrt{3}}$

<u>11.</u> $\csc (\text{Sin}^{-1} \frac{\sqrt{3}}{2}) = \dfrac{1}{\sin (\text{Sin}^{-1} \frac{\sqrt{3}}{2})} = \dfrac{2}{\sqrt{3}}$

<u>12.</u> $\tan (\text{Cos}^{-1} \frac{1}{3}) = \dfrac{\sin (\text{Cos}^{-1} \frac{1}{3})}{\cos (\text{Cos}^{-1} \frac{1}{3})} = \dfrac{\sqrt{1 - (\frac{1}{3})^2}}{\frac{1}{3}} = \dfrac{\frac{2\sqrt{2}}{3}}{\frac{1}{3}} = 2\sqrt{2}$

<u>13.</u> $\sec (\text{Cos}^{-1} \frac{3}{5}) = \dfrac{1}{\cos (\text{Cos}^{-1} \frac{3}{5})} = \dfrac{5}{3}$

<u>14.</u> $\text{Arccos } \frac{1}{2}$, $\text{Arccos } 0$, $\text{Arccos } (-\frac{\sqrt{3}}{2})$ <u>15.</u> $\text{Arcsin } (-\frac{\sqrt{3}}{2})$, $\text{Arcsin } 0$, $\text{Arcsin } \frac{1}{2}$

<u>16.</u> $\text{Arctan } (-\sqrt{3})$, $\text{Arctan } 0$, $\text{Arctan } 1$ <u>17.</u> $\text{Arccos } 1$, $\text{Arctan } 1$, $\text{Arcsin } 1$

<u>18.</u> $\text{Arcsin } (-1)$, $\text{Arctan } (-1)$, $\text{Arccos } (-1)$

<u>19.</u> $\sin (\text{Arcsin } \frac{1}{2} + \text{Arccos } \frac{\sqrt{3}}{2}) = \sin (\text{Arcsin } \frac{1}{2}) \cos (\text{Arccos } \frac{\sqrt{3}}{2}) +$

$\cos (\text{Arcsin } \frac{1}{2}) \sin (\text{Arccos } \frac{\sqrt{3}}{2}) = \frac{1}{2} \cdot \frac{\sqrt{3}}{2} + \frac{\sqrt{3}}{2} \cdot \frac{1}{2} = \frac{\sqrt{3}}{2}$

<u>20.</u> $\cos (\text{Arcsin } \frac{1}{2} + \text{Arccos } \frac{\sqrt{3}}{2}) = \cos (\text{Arcsin } \frac{1}{2}) \cos (\text{Arccos } \frac{\sqrt{3}}{2}) -$

$\sin (\text{Arcsin } \frac{1}{2}) \sin (\text{Arccos } \frac{\sqrt{3}}{2}) = \frac{\sqrt{3}}{2} \cdot \frac{\sqrt{3}}{2} - \frac{1}{2} \cdot \frac{1}{2} = \frac{1}{2}$

<u>21.</u> $\sin (\text{Arccos } \frac{1}{2} + \text{Arccos } \frac{\sqrt{2}}{2}) = \sin (\text{Arccos } \frac{1}{2}) \cos (\text{Arccos } \frac{\sqrt{2}}{2}) +$

$\cos (\text{Arccos } \frac{1}{2}) \sin (\text{Arccos } \frac{\sqrt{2}}{2}) = \frac{\sqrt{3}}{2} \cdot \frac{\sqrt{2}}{2} + \frac{1}{2} \cdot \frac{\sqrt{2}}{2} = \frac{\sqrt{6} + \sqrt{2}}{4}$

B   <u>22.</u> $\sin (\text{Arcsin } \frac{3}{5} + \text{Arcsin } \frac{3}{5}) = \sin (\text{Arcsin } \frac{3}{5}) \cos (\text{Arcsin } \frac{3}{5}) +$

$\cos (\text{Arcsin } \frac{3}{5}) \sin (\text{Arcsin } \frac{3}{5}) = \frac{3}{5} \cdot \frac{4}{5} + \frac{4}{5} \cdot \frac{3}{5} = \frac{24}{25}$

<u>23.</u> $\sin (2 \text{ Arcsin } \frac{4}{5}) = 2 \sin (\text{Arcsin } \frac{4}{5}) \cos (\text{Arcsin } \frac{4}{5}) = 2 \cdot \frac{4}{5} \cdot \frac{3}{5} = \frac{24}{25}$

<u>24.</u> $\cos (2 \text{ Arcsin } \frac{1}{5}) = 1 - 2 \sin^2 (\text{Arcsin } \frac{1}{5}) = 1 - \frac{2}{25} = \frac{23}{25}$

<u>25.</u> $\tan (\text{Tan}^{-1} \sqrt{3} + \text{Tan}^{-1} 1) = \dfrac{\tan (\text{Tan}^{-1} \sqrt{3}) + \tan (\text{Tan}^{-1} 1)}{1 - \tan (\text{Tan}^{-1} \sqrt{3}) \tan (\text{Tan}^{-1} 1)} = \dfrac{\sqrt{3} + 1}{1 - \sqrt{3}} =$
$-(2 + \sqrt{3})$

<u>26.</u> $\tan (\text{Tan}^{-1} \frac{1}{4} + \frac{\pi}{3}) = \dfrac{\tan (\text{Tan}^{-1} \frac{1}{4}) + \tan \frac{\pi}{3}}{1 - \tan (\text{Tan}^{-1} \frac{1}{4}) \tan \frac{\pi}{3}} = \dfrac{\frac{1}{4} + \sqrt{3}}{1 - \frac{\sqrt{3}}{4}} = \dfrac{1 + 4\sqrt{3}}{4 - \sqrt{3}} = \dfrac{16 + 17\sqrt{3}}{13}$

27. $\tan (\text{Tan}^{-1} \frac{1}{2} + \text{Tan}^{-1} \frac{1}{3}) = \dfrac{\tan (\text{Tan}^{-1} \frac{1}{2}) + \tan (\text{Tan}^{-1} \frac{1}{3})}{1 - \tan (\text{Tan}^{-1} \frac{1}{2}) \tan (\text{Tan}^{-1} \frac{1}{3})} = \dfrac{\frac{1}{2} + \frac{1}{3}}{1 - \frac{1}{2} \cdot \frac{1}{3}} = 1$

28. a. $x_1 < x_2$, T   b. $x_1 < x_2$, T

   c. $\sqrt{1 - x_1^2} < \sqrt{1 - x_2^2}$, $-x_1^2 < -x_2^2$, $x_1^2 > x_2^2$, F

   d. $\sqrt{1 - x_1^2} < \sqrt{1 - x_2^2}$, F   e. T

C   29. a. Choose any $x \in \{x: x \in R$ and $-\frac{\pi}{2} \le x \le \frac{\pi}{2}\}$

   b. Choose any $x \in \{x: x \in R$ and $|x| > \frac{\pi}{2}\}$

30. a. Choose any $x \in \{x: x \in R$ and $0 \le x \le \pi\}$

   b. Choose any $x \in \{x: x \in R$ and $x < 0$ or $x > \pi\}$

31. $\sin (\text{Arcsin } x + \text{Arccos } x) = \sin (\text{Arcsin } x) \cos (\text{Arccos } x) +$
    $\cos (\text{Arcsin } x) \sin (\text{Arccos } x) = x \cdot x + \sqrt{1 - x^2} \sqrt{1 - x^2}$ for $|x| \le 1$;
    $x^2 + (\sqrt{1 - x^2})^2 = 1$

32. $\cos (\text{Arcsin } x + \text{Arccos } x) = \cos (\text{Arcsin } x) \cos (\text{Arccos } x) -$
    $\sin (\text{Arcsin } x) \sin (\text{Arccos } x) = (\sqrt{1 - x^2}) \cdot x - x\sqrt{1 - x^2} = 0 = \cos \frac{\pi}{2}$;
    $\therefore \text{Arcsin } x + \text{Arccos } x = \frac{\pi}{2}$

33.

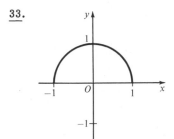

34. Let $\omega = \text{Cos}^{-1} x$ so $x = \cos \omega$. $\sin \omega = \pm\sqrt{1 - \cos^2 \omega} = \pm\sqrt{1 - x^2}$. Since
    $0 \le \text{Cos}^{-1} x \le \pi$ for all $|x| \le 1$, $\sin (\text{Cos}^{-1} x) \ge 0$ for all $|x| \le 1$ because
    $\sin x \ge 0$ for $0 \le x \le \pi$. Thus $\sin \omega = \sin (\text{Cos}^{-1} x) = \sqrt{1 - x^2}$ for $|x| \le 1$

35. a. $\sin (\frac{1}{2} \text{Cos}^{-1} x) = \sqrt{\frac{1}{2}(1 - \cos (\text{Cos}^{-1} x))} = \sqrt{\frac{1 - x}{2}}$

   b. $\cos (\frac{1}{2} \text{Sin}^{-1} x) = \sqrt{\frac{1}{2}(1 + \cos (\text{Sin}^{-1} x))} = \sqrt{\frac{1 + \sqrt{1 - x^2}}{2}}$

Exercises 4-4 · Page 131

A   1. $\sin x - 1 = 0$, $\sin x = 1$; $\{\frac{\pi}{2}\}$

     2. $\sqrt{2} \cos x + 1 = 0$, $\cos x = -\frac{1}{\sqrt{2}}$; $\{\frac{3\pi}{4}, \frac{5\pi}{4}\}$

     3. $2 + \sec x = 0$, $\sec x = -2$; $\{\frac{2\pi}{3}, \frac{4\pi}{3}\}$

     4. $4 \sin^2 x - 1 = 0$, $\sin^2 x = \frac{1}{4}$, $\sin x = \pm\frac{1}{2}$; $\{\frac{\pi}{6}, \frac{5\pi}{6}, \frac{7\pi}{6}, \frac{11\pi}{6}\}$

     5. $\tan^2 x - 1 = 0$, $\tan^2 x = 1$, $\tan x = \pm 1$; $\{\frac{\pi}{4}, \frac{3\pi}{4}, \frac{5\pi}{4}, \frac{7\pi}{4}\}$

     6. $\cot^2 x - \sqrt{3} \cot x = 0$, $\cot x (\cot x - \sqrt{3}) = 0$, $\cot x = 0$ or $\cot x = \sqrt{3}$;

       $\{\frac{\pi}{6}, \frac{\pi}{2}, \frac{7\pi}{6}, \frac{3\pi}{2}\}$

     7. $\sin^2 3x = 0$, $\sin 3x = 0$; $3x = 0 + k\pi$ and $k \in J$; $x = 0 + \frac{1}{3}k\pi$, $k \in J$;

       $\{0, \frac{\pi}{3}, \frac{2\pi}{3}, \pi, \frac{4\pi}{3}, \frac{5\pi}{3}\}$

     8. $\sin^2 x - 2 \sin x + 1 = 0$, $(\sin x - 1)^2 = 0$, $\sin x = 1$; $\{\frac{\pi}{2}\}$

     9. $2 \cos^2 x + 3 \cos x + 1 = 0$, $(2 \cos + 1)(\cos x + 1) = 0$, $\cos x = -\frac{1}{2}$ or $\cos x = -1$;

       $\{\frac{2\pi}{3}, \pi, \frac{4\pi}{3}\}$

   10. $\tan x (1 - \tan x) = 0$, $\tan x = 0$ or $\tan x = 1$; $\{0, \pi, 2\pi, \frac{\pi}{4}, \frac{5\pi}{4}\}$

   11. $\cot^2 x + 2 \cot x + 1 = 0$, $(\cot x + 1)^2 = 0$, $\cot x = -1$; $\{\frac{3\pi}{4}, \frac{7\pi}{4}\}$

   12. $2 \sec^2 x = 2 - 3 \sec x$, $2 \sec^2 x + 3 \sec x - 2 = 0$, $(2 \sec x - 1)(\sec x + 2) = 0$,

       $\sec x = \frac{1}{2}$ or $\sec x = -2$; $\{\frac{2\pi}{3}, \frac{4\pi}{3}\}$

   13. $3 \tan^2 x - 1 = 0$, $\tan^2 x = \frac{1}{3}$, $\tan x = \pm\frac{1}{\sqrt{3}}$; $\{\frac{\pi}{6}, \frac{5\pi}{6}, \frac{7\pi}{6}, \frac{11\pi}{6}\}$

   14. $2 \cos^2 x = 1 + \cos^2 x$, $\cos^2 x = 1$, $\cos x = \pm 1$; $\{0, \pi\}$

   15. $2 \cot^2 x + 2 \cot x = 0$, $2 \cot x (\cot x + 1) = 0$, $\cot x = 0$ or $\cot x = -1$;

       $\{\frac{\pi}{2}, \frac{3\pi}{4}, \frac{3\pi}{2}, \frac{7\pi}{4}\}$

   16. $\tan^2 x - \sqrt{3} \tan x = 0$, $\tan x (\tan x - \sqrt{3}) = 0$, $\tan x = 0$ or $\tan x = \sqrt{3}$;

       $\{0, \frac{\pi}{3}, \pi, \frac{4\pi}{3}\}$

B   17. $\cos^2 4x + \cos 4x = 0$, $\cos 4x (\cos 4x + 1) = 0$, $\cos 4x = 0$ or $\cos 4x = -1$;

       $4x = \frac{\pi}{2} + k\pi$ or $4x = \pi + 2k\pi$; $\{\frac{\pi}{8}, \frac{\pi}{4}, \frac{3\pi}{8}, \frac{5\pi}{8}, \frac{3\pi}{4}, \frac{7\pi}{8}, \frac{9\pi}{8}, \frac{5\pi}{4}, \frac{11\pi}{8}, \frac{13\pi}{8},$

       $\frac{7\pi}{4}, \frac{15\pi}{8}\}$

18. $\tan^2 6x = 1$, $\tan 6x = \pm 1$; $6x = \frac{\pi}{4} + k \cdot \frac{\pi}{2}$; $\{\frac{\pi}{24}, \frac{\pi}{8}, \frac{5\pi}{24}, \frac{7\pi}{24}, \frac{3\pi}{8}, \frac{11\pi}{24}, \frac{13\pi}{24}, \frac{5\pi}{8},$

$\frac{17\pi}{24}, \frac{19\pi}{24}, \frac{7\pi}{8}, \frac{23\pi}{24}, \frac{25\pi}{24}, \frac{9\pi}{8}, \frac{29\pi}{24}, \frac{31\pi}{24}, \frac{11\pi}{8}, \frac{35\pi}{24}, \frac{37\pi}{24}, \frac{13\pi}{8}, \frac{41\pi}{24}, \frac{43\pi}{24}, \frac{15\pi}{8}, \frac{47\pi}{24}\}$

19. $\cos^2 \frac{x}{2} = \frac{1}{4}$, $\cos^2 \frac{x}{2} = \pm \frac{1}{2}$; $\frac{x}{2} = \frac{\pi}{3}$ or $\frac{x}{2} = \frac{2\pi}{3}$; $\{\frac{2\pi}{3}, \frac{4\pi}{3}\}$

20. $\sin^2 (x + 4) = 1$, $\sin (x + 4) = \pm 1$; $x + 4 = \frac{\pi}{2} + k\pi$; $\{\frac{3\pi}{2} - 4, \frac{5\pi}{2} - 4\}$

21. $(1 + \sin \frac{x}{2})(\sin \frac{x}{2}) = 0$, $\sin \frac{x}{2} = -1$ or $\sin \frac{x}{2} = 0$; $\frac{x}{2} = \frac{3\pi}{2}$ or $\frac{x}{2} = 0$, π; $\{0\}$

22. $2 \cos^2 (\frac{2\pi}{3} + 2k\pi) - \cos (\frac{2\pi}{3} + 2k\pi) = 2 \cos^2 \frac{2\pi}{3} - \cos \frac{2\pi}{3} = 2(-\frac{1}{2})^2 - (-\frac{1}{2}) = 1$;

$2 \cos^2 (\frac{4\pi}{3} + 2k\pi) - \cos (\frac{4\pi}{3} + 2k\pi) = 2 \cos^2 \frac{4\pi}{3} - \cos \frac{4\pi}{3} = 2(-\frac{1}{2})^2 - (-\frac{1}{2}) = 1$;

$2 \cos^2 (2k\pi) - \cos (2k\pi) = 2 \cos^2 0 - \cos 0 = 2 - 1 = 1$; $\tan^2 (\frac{\pi}{3} + k\pi) - 3 =$

$\tan^2 \frac{\pi}{3} - 3 = (\sqrt{3})^2 - 3 = 0$; $\tan^2 (\frac{2\pi}{3} + k\pi) - 3 = \tan^2 \frac{2\pi}{3} - 3 = (-\sqrt{3})^2 - 3 = 0$

C 23. $4 \cos^3 x - \cos x = 0$, $\cos x (4 \cos^2 x - 1) = 0$, $\cos x = 0$ or $\cos^2 x = \frac{1}{4}$;

$\cos x = 0$ or $\cos x = \pm \frac{1}{2}$; $\{x: x = \frac{\pi}{2} + k\pi, k \in J\} \cup$

$\{x: x = \frac{\pi}{3} + k\pi$ or $x = \frac{2\pi}{3} + k\pi, k \in J\}$

24. $\sin^2 |x - \frac{\pi}{3}| = \frac{1}{4}$, $\sin |x - \frac{\pi}{3}| = \pm \frac{1}{2}$, $|x - \frac{\pi}{3}| = \arcsin (\pm \frac{1}{2})$;

$\{x: x = \pm \frac{\pi}{6} + k\pi$ or $x = \pm \frac{5\pi}{6} + 2k\pi$ or $x = \frac{\pi}{2} + k\pi\}$

$\{x: x = \pm \frac{\pi}{6} + k\pi$ or $x = \frac{\pi}{2} + k\pi, k \in J\}$

25. $|\sin \frac{x}{\pi}| = 1$, $\sin \frac{x}{\pi} = \pm 1$; $\frac{x}{\pi} = (2k + 1)\frac{\pi}{2}$, $k \in J$; $\{x: x = (2k + 1)\frac{\pi^2}{2}, k \in J\}$

Exercises 4-5 · Page 136

A 1. $\sin 2x = \sin x$, $2 \sin x \cos x - \sin x = 0$, $\sin x (2 \cos x - 1) = 0$, $\sin x = 0$ or
$\cos x = \frac{1}{2}$; $\{0, \frac{\pi}{3}, \pi, \frac{5\pi}{3}\}$

2. $\sin 2x = \cos x$, $2 \sin x \cos x - \cos x = 0$, $\cos x (2 \sin x - 1) = 0$, $\cos x = 0$ or
$\sin x = \frac{1}{2}$; $\{\frac{\pi}{6}, \frac{\pi}{2}, \frac{5\pi}{6}, \frac{3\pi}{2}\}$

3. $\cos 2x + \sin x - 1 = 0$, $1 - 2 \sin^2 x + \sin x - 1 = 0$, $\sin x (-2 \sin x + 1) = 0$,
$\sin x = 0$ or $\sin x = \frac{1}{2}$; $\{0, \frac{\pi}{6}, \frac{5\pi}{6}, \pi\}$

4. $\sin x - \sqrt{3} \cos x = 1$, $\sin^2 x = 1 + 2\sqrt{3} \cos x + 3 \cos^2 x$, $4 \cos^2 x + 2\sqrt{3} \cos x$
$= 0$, $2 \cos x (2 \cos x + \sqrt{3}) = 0$, $\cos x = 0$ or $\cos x = -\frac{\sqrt{3}}{2}$; $\{\frac{\pi}{2}, \frac{7\pi}{6}\}$

5. $\sin 2x + \cos 2x - 1 = 0$, $2 \sin x \cos x + \cos^2 x - \sin^2 x - (\cos^2 x + \sin^2 x) = 0$,

$2 \sin x \cos x - 2 \sin^2 x = 0$, $2 \sin x (\cos x - \sin x) = 0$, $\sin x = 0$ or $\tan x = 1$;

$\{0, \frac{\pi}{4}, \pi, \frac{5\pi}{4}\}$

6. $\sqrt{3} \cos x + \sin x = -1$, $\sin^2 x = 1 + 2\sqrt{3} \cos x + 3 \cos^2 x$, $4 \cos^2 x + 2\sqrt{3} \cos x$

$= 0$, $2 \cos x (2 \cos x + \sqrt{3}) = 0$, $\cos x = 0$ or $\cos x = -\frac{\sqrt{3}}{2}$; $\{\frac{5\pi}{6}, \frac{3\pi}{2}\}$

7. $\sin 2x + \cos x + 2 \sin x = -1$, $2 \sin x \cos x + \cos x + 2 \sin x + 1 = 0$,

$\cos x (2 \sin x + 1) + (2 \sin x + 1) = 0$, $(\cos x + 1)(2 \sin x + 1) = 0$, $\cos x = -1$

or $\sin x = -\frac{1}{2}$; $\{\pi, \frac{7\pi}{6}, \frac{11\pi}{6}\}$

8. $2 \sin x \sec x = \sec x$, $2 \sin x \sec x - \sec x = 0$, $\sec x (2 \sin x - 1) = 0$,

$\sec x = 0$ or $2 \sin x - 1 = 0$, $\sec x = 0$ or $\sin x = \frac{1}{2}$, $x = \frac{\pi}{6}$ or $\frac{5\pi}{6}$; $\{\frac{\pi}{6}, \frac{5\pi}{6}\}$

9. $3 \csc x - \sin x = 2$, $3 - \sin^2 x = 2 \sin x$, $(\sin x + 3)(\sin x - 1) = 0$, $\sin x = -3$

or $\sin x = 1$; $\{\frac{\pi}{2}\}$

10. $3 \tan \frac{x}{2} = \cot \frac{x}{2}$, $3 \tan^2 \frac{x}{2} = 1$, $\tan^2 \frac{x}{2} = \frac{1}{3}$, $\tan \frac{x}{2} = \pm \frac{1}{\sqrt{3}}$; $\{\frac{\pi}{3}, \frac{5\pi}{3}\}$

11. $\frac{1}{2} \tan 2x = \cos x$, $\frac{\sin 2x}{2 \cos 2x} = \cos x$, $(\cos 2x \neq 0)$, $\frac{2 \sin x \cos x}{2(1 - 2 \sin^2 x)} = \cos x$,

$\cos x (\frac{\sin x}{1 - 2 \sin^2 x} - 1) = 0$, $\cos x = 0$ or $\sin x = 1 - 2 \sin^2 x$,

$(2 \sin x - 1)(\sin x + 1) = 0$, $\sin x = \frac{1}{2}$ or $\sin x = -1$; $\{\frac{\pi}{6}, \frac{\pi}{2}, \frac{5\pi}{6}, \frac{3\pi}{2}\}$

12. $\sec^2 2x = 1 - \tan 2x$, $1 + \tan^2 2x = 1 - \tan 2x$, $\tan^2 2x + \tan 2x = 0$,

$\tan 2x (\tan 2x + 1) = 0$, $\tan 2x = 0$ or $\tan 2x = -1$; $2x = 0$, π, 2π, 3π,

$\frac{3\pi}{4}$, $\frac{7\pi}{4}$, $\frac{11\pi}{4}$, $\frac{15\pi}{4}$; $\{0, \frac{3}{8}\pi, \frac{\pi}{2}, \frac{7}{8}\pi, \pi, \frac{11}{8}\pi, \frac{3\pi}{2}, \frac{15\pi}{8}\}$

13. $\tan^2 x = \sin 2x$, $\frac{\sin^2 x}{\cos^2 x} = 2 \sin x \cos x$, $\sin^2 x = 2 \sin x \cos^3 x$, $\sin x = 0$ or

$\sin x = 2 \cos^3 x$, $\sin^2 x = 4 \cos^6 x$, $1 - \cos^2 x = 4 \cos^6 x$, $4 \cos^6 x + \cos^2 x - 1$

$= 0$, $(2 \cos^2 x - 1)(2 \cos^4 x + \cos^2 x + 1) = 0$, $\cos^2 x = \frac{1}{2}$ or $2 \cos^4 x + \cos^2 x +$

$1 = 0$, $\cos x = \pm \frac{\sqrt{2}}{2}$ or $2 \cos^4 x + \cos^2 x + 1 = 0$, no real solution;

$\{0, \frac{\pi}{4}, \pi, \frac{5\pi}{4}\}$

14. $\sin x + \cos x \sin x = 2$ has no solution since $\sin x \leq 1$ and $\cos x \sin x \leq 1$, their

sum is 2 only if $\sin x = 1$ and $\cos x \sin x = 1$. But if $\sin x = 1$, $x = \frac{\pi}{2}$ and

$\cos x = 0$. Thus $\sin x + \cos x \sin x = 1 + 1 \cdot 0 = 1 \neq 2$.

15. $\cos 2x = \cos^2 x - 1$, $2 \cos^2 x - 1 = \cos^2 x - 1$, $\cos^2 x = 0$, $\cos x = 0$; $\{\frac{\pi}{2}, \frac{3\pi}{2}\}$

16. $1 - \sin^2 x = \cos 2x$, $1 - \sin^2 x = 1 - 2 \sin^2 x$, $\sin^2 x = 0$, $\sin x = 0$; $\{0, \pi\}$

B   17. $\text{Arctan } x = \frac{\pi}{6}$, $\tan (\text{Arctan } x) = \tan \frac{\pi}{6}$, $x = \frac{1}{\sqrt{3}}$

18. $\text{Arctan } \sqrt{x} = \frac{\pi}{3}$, $\tan (\text{Arctan } \sqrt{x}) = \tan \frac{\pi}{3}$, $\sqrt{x} = \sqrt{3}$, $x = 3$

19. $\text{Arctan } (x - 1) = -\frac{\pi}{4}$, $\tan (\text{Arctan } (x - 1)) = \tan (-\frac{\pi}{4})$, $x - 1 = -1$, $x = 0$

20. $\sin (\pi - 2x) = \cos x$, $\sin 2x = \cos x$, $2 \sin x \cos x - \cos x = 0$,

  $\cos x (2 \sin x - 1) = 0$, $\cos x = 0$ or $\sin x = \frac{1}{2}$; $\{x: x = \frac{\pi}{2} + k\pi, \; k \in J\} \cup$

  $\{x: x = \frac{\pi}{6} + 2k\pi$ or $x = \frac{5\pi}{6} + 2k\pi, \; k \in J\}$

21. $\sin (2x + \frac{2\pi}{3}) + \sin (2x - \frac{2\pi}{3}) = \frac{\sqrt{2}}{2}$, $\sin 2x \cos \frac{2\pi}{3} + \cos 2x \sin \frac{2\pi}{3} +$

  $\sin 2x \cos \frac{2\pi}{3} - \cos 2x \sin \frac{2\pi}{3} = \frac{\sqrt{2}}{2}$, $2 \sin 2x \, (-\frac{1}{2}) = \frac{\sqrt{2}}{2}$, $\sin 2x = -\frac{\sqrt{2}}{2}$,

  $2x = (\frac{5\pi}{4} + 2\pi k)$ or $(\frac{7\pi}{4} + 2\pi k)$; $\{x: x = \frac{5\pi}{8} + k\pi$ or $x = \frac{7\pi}{8} + k\pi, \; k \in J\}$

22. $\sin 2x \tan^2 2x - \tan 2x = \sin 2x$, $\sin 2x (\tan^2 2x - \sec 2x - 1) = 0$, $\sin 2x = 0$ or

  $\tan^2 2x - \sec 2x - 1 = 0$, $(\sec^2 2x - 1) - \sec 2x - 1 = 0$, $\sec^2 2x - \sec 2x - 2$

  $= 0$, $(\sec 2x - 2)(\sec 2x + 1) = 0$, $\sec 2x = 2$ or $\sec 2x = -1$; $2x = \frac{\pi}{3} + 2k\pi$ or

  $\frac{5\pi}{3} + 2k\pi$ or $\pi + 2k\pi$ or $k\pi$; $\{x: x = \frac{\pi}{6} + k\pi$ or $x = \frac{5\pi}{6} + k\pi, \; k \in J\} \cup$

  $\{x: x = (2k + 1)\frac{\pi}{2}$ or $x = k \cdot \frac{\pi}{2}, \; k \in J\}$; $\{x: x = \frac{\pi}{6} + k\pi$ or $x = \frac{5\pi}{6} + k\pi, \; k \in J\} \cup$

  $\{x: x = k\frac{\pi}{2}, \; k \in J\}$

23. $\sin^4 x - \cos^4 x = 1$, $(\sin^2 x - \cos^2 x)(\sin^2 x + \cos^2 x) = 1$, $\sin^2 x - \cos^2 x = 1$,

  $1 - \cos^2 x - \cos^2 x = 1$, $\cos^2 x = 0$, $\{x: x = (2k + 1)\frac{\pi}{2}, \; k \in J\}$

C   24. $2 \tan x + \sec x = 1$, $\frac{2 \sin x}{\cos x} + \frac{1}{\cos x} = 1$, $2 \sin x + 1 = \cos x$,

  $4 \sin^2 x + 4 \sin x + 1 = \cos^2 x$, $4 \sin^2 x + 4 \sin x + 1 = 1 - \sin^2 x$,

  $5 \sin^2 x + 4 \sin x = 0$, $\sin x (5 \sin x + 4) = 0$, $\sin x = 0$ or $\sin x = -0.8$;

  $\{x: x = 2k\pi, \; k \in J\} \cup \{x: x \doteq (2k + 1)\pi + 0.59(\frac{\pi}{2})$ or $2\pi(k + 1) - 0.59(\frac{\pi}{2}), \; k \in J\}$

25. $3 \cot x + 5 \csc x = 4$, $\frac{3 \cos x}{\sin x} + \frac{5}{\sin x} = 4$, $3 \cos x + 5 = 4 \sin x$,

  $9 \cos^2 x + 30 \cos x + 25 = 16 \sin^2 x$, $9 \cos^2 x + 30 \cos x + 25 = 16 - 16 \cos^2 x$,

  $25 \cos^2 x + 30 \cos x + 9 = 0$, $(5 \cos x + 3)^2 = 0$, $\cos x = -0.6$; $x \doteq \pi \pm \frac{0.59\pi}{2} +$

  $2k\pi$; $\{x: x \doteq 1.41 \frac{\pi}{2} + 2k\pi$ or $x \doteq 2.59(\frac{\pi}{2}) + 2k\pi, \; k \in J\}$

$\underline{26}$. $\sin x \cos x = 0$, $\sin x = 0$ or $\cos x = 0$; $\{x : x = k\frac{\pi}{2}, k \in J\}$

$\underline{27}$. $\sin 2x - \sin 4x = 2 \sin x$, $\sin 2x - 2 \sin 2x \cos 2x = 2 \sin x$,

$2 \sin x \cos x (1 - 2 \cos 2x) - 2 \sin x = 0$, $\sin x = 0$ or

$\cos x (1 - 2(2 \cos^2 x - 1)) = 0$, $3 \cos x - 4 \cos^3 x - 1 = 0$,

$(\cos x + 1)(-4 \cos^2 x + 4 \cos x - 1) = 0$, $\cos x = -1$ or $(2 \cos x - 1)^2 = 0$;

$\sin x = 0$ or $\cos x = -1$ or $\cos x = \frac{1}{2}$; $\{x : x = k\pi, \frac{\pi}{3} + 2k\pi, \frac{5\pi}{3} + 2k\pi; k \in J\}$

$\underline{28}$. $\cos 3x + \cos x = \cos 2x \cos x - \sin 2x \sin x + \cos x = 2 \cos^3 x - \cos x -$

$2 \sin^2 x \cos x + \cos x = 2 \cos^3 x - 2 \sin^2 x \cos x = 2 \cos x (\cos^2 x - \sin^2 x) =$

$2 \cos 2x \cos x$.   Thus $\cos 3x + \cos x = 2 \cos 2x \cos x = \cos 2x$,

$\cos 2x (2 \cos x - 1) = 0$, $\cos 2x = 0$ or $\cos x = \frac{1}{2}$, $2x = (2k + 1)\frac{\pi}{2}$ or

$x = \frac{\pi}{3} + 2k\pi, \frac{5\pi}{3} + 2k\pi$; $\{x : x = (2k + 1)\frac{\pi}{4}$ or $x = \frac{\pi}{3} + 2k\pi$ or $x = \frac{5\pi}{3} + 2k\pi, k \in J\}$

Chapter Test · Page 138

$\underline{1}$. $y = -\frac{\pi}{6}$ $\underline{2}$. $\text{Cos}^{-1} \frac{\sqrt{3}}{2} = \frac{\pi}{6}$ $\underline{3}$. $\cos (\text{Cos}^{-1} 1) = 1$

$\underline{4}$. $\text{Tan}^{-1} (\sin \pi) = \text{Tan}^{-1} 0 = 0$ $\underline{5}$. $\text{Sec}^{-1} (-2) = \frac{2\pi}{3}$

$\underline{6}$. $\text{Tan}^{-1} \frac{\sqrt{3}}{2} + \text{Cot}^{-1} \frac{\sqrt{3}}{2} = \frac{\pi}{2}$ $\underline{7}$. $\cos (\text{Sin}^{-1} \frac{3}{5}) = \sqrt{1 - (\frac{3}{5})^2} = \frac{4}{5}$

$\underline{8}$. $4 \sin^2 x = 3$, $\sin^2 x = \frac{3}{4}$, $\sin x = \pm \frac{\sqrt{3}}{2}$, $\{x : x = \frac{\pi}{3} + k\pi$ or $x = \frac{2\pi}{3} + k\pi, k \in J\}$

$\underline{9}$. $\tan^2 x - \tan x = 0$, $\tan x (\tan x - 1) = 0$, $\tan x = 0$ or $\tan x = 1$;

$\{x : x = k\pi$ or $x = \frac{\pi}{4} + k\pi, k \in J\}$

$\underline{10}$. $\cos 2x = \sin x$, $1 - 2 \sin^2 x = \sin x$, $2 \sin^2 x + \sin x - 1 = 0$,

$(2 \sin x - 1)(\sin x + 1) = 0$, $\sin x = \frac{1}{2}$ or $\sin x = -1$; $\{\frac{\pi}{6}, \frac{5\pi}{6}, \frac{3\pi}{2}\}$

$\underline{11}$. $3 \sin^2 x = 1 - \frac{5}{2} \cos x$, $3 - 3 \cos^2 x = 1 - \frac{5}{2} \cos x$, $6 \cos^2 x - 5 \cos x - 4 = 0$,

$(3 \cos x - 4)(2 \cos x + 1) = 0$, $\cos x = -\frac{1}{2}$; $\{\frac{2\pi}{3}, \frac{4\pi}{3}\}$

$\underline{12}$. $\tan x = \frac{1}{2} \sec x$, $\frac{\sin x}{\cos x} = \frac{1}{2 \cos x}$, $\sin x = \frac{1}{2}$; $\{\frac{\pi}{6}, \frac{5\pi}{6}\}$

Exercises 5-1 · Page 147

__A__   1. $\frac{\pi}{4}^R = \frac{180}{\pi}\left(\frac{\pi}{4}\right)^\circ = 45°$

2. $\frac{7\pi}{8}^R = \frac{180}{\pi}\left(\frac{7\pi}{8}\right)^\circ = 157\frac{1}{2}^\circ$

3. $\frac{2\pi}{3}^R = \frac{180}{\pi}\left(\frac{2\pi}{3}\right)^\circ = 120°$

4. $\frac{5\pi}{6}^R = \frac{180}{\pi}\left(\frac{5\pi}{6}\right)^\circ = 150°$

5. $4\pi^R = \frac{180}{\pi}(4\pi)° = 720°$

6. $-\frac{3\pi}{2}^R = \frac{180}{\pi}\left(-\frac{3\pi}{2}\right)^\circ = -270°$

7. $\frac{18\pi}{5}^R = \frac{180}{\pi}\left(\frac{18\pi}{5}\right)^\circ = 648°$

8. $-\frac{5\pi}{3}^R = \frac{180}{\pi}\left(-\frac{5\pi}{3}\right)^\circ = -300°$

9. $45° = \frac{\pi}{180}(45)^R = \frac{\pi}{4}^R$

10. $80° = \frac{\pi}{180}(80)^R = \frac{4\pi}{9}^R$

11. $120° = \frac{\pi}{180}(120)^R = \frac{2\pi}{3}^R$

12. $150° = \frac{\pi}{180}(150)^R = \frac{5\pi}{6}^R$

13. $210° = \frac{\pi}{180}(210)^R = \frac{7\pi}{6}^R$

14. $-30° = \frac{\pi}{180}(-30)^R = -\frac{\pi}{6}^R$

15. $-150° = \frac{\pi}{180}(-150)^R = -\frac{5\pi}{6}^R$

16. $420° = \frac{\pi}{180}(420)^R = \frac{7\pi}{3}^R$

__B__   17. $\frac{\pi}{2}^R = \frac{50}{\pi}\left(\frac{\pi}{2}\right)^c = 25^c$

18. $\frac{\pi}{5}^R = \frac{50}{\pi}\left(\frac{\pi}{5}\right)^c = 10^c$

19. $32° = \frac{5}{18}(32)^c = \frac{80^c}{9}$

20. $-298° = \frac{5}{18}(-298)^c = -\frac{745^c}{9}$

21. a. $45^c = \frac{18}{5}(45)° = 162°$

b. $45^c = \frac{\pi}{50}(45)^R = \frac{9\pi}{10}^R$

22. a. $130^c = \frac{18}{5}(130)° = 468°$

b. $130^c = \frac{\pi}{50}(130)^R = \frac{13\pi}{5}^R$

23. a. $260^c = \frac{18}{5}(260)° = 936°$

b. $260^c = \frac{\pi}{50}(260)^R = \frac{26\pi}{5}^R$

24. a. $-120^c = \frac{18}{5}(120)° = -432°$

b. $-120^c = -\frac{\pi}{50}(120) = -\frac{12\pi}{5}^R$

Exercises 5-2 · Page 151

<u>A</u>   <u>1.</u>

$m^R(\theta)$	$m°(\theta)$	$\cos \theta$	$\sin \theta$	$\tan \theta$
0	0	1	0	0
$\frac{\pi}{6}$	30	$\frac{\sqrt{3}}{2}$	$\frac{1}{2}$	$\frac{1}{\sqrt{3}}$
$\frac{\pi}{4}$	45	$\frac{\sqrt{2}}{2}$	$\frac{\sqrt{2}}{2}$	1
$\frac{\pi}{3}$	60	$\frac{1}{2}$	$\frac{\sqrt{3}}{2}$	$\sqrt{3}$
$\frac{\pi}{2}$	90	0	1	undef.
$\frac{2\pi}{3}$	120	$-\frac{1}{2}$	$\frac{\sqrt{3}}{2}$	$-\sqrt{3}$
$\frac{3\pi}{4}$	135	$-\frac{\sqrt{2}}{2}$	$\frac{\sqrt{2}}{2}$	-1
$\frac{5\pi}{6}$	150	$-\frac{\sqrt{3}}{2}$	$\frac{1}{2}$	$-\frac{1}{\sqrt{3}}$

<u>2.</u> $\cos 450° = \cos 90°$

<u>3.</u> $\tan 510° = \tan 150° = -\tan 30°$

<u>4.</u> $\sin 920° = \sin 200° = -\sin 20°$

<u>5.</u> $\cos 1000° = \cos (-80°) = \cos 80°$

<u>6.</u> $\tan 620° = \tan (260°) = \tan 80°$

<u>7.</u> $\sin 225° = -\sin 45°$

<u>8.</u> $\cos (-1200°) = \cos 1200° = \cos 120° = -\cos 60°$

<u>9.</u> $\tan (-420°) = -\tan 420° = -\tan 60°$

<u>10.</u> $\sin (-65°) = -\sin 65°$

<u>11.</u> $\cos (-120°) = -\cos 120° = -\cos 60°$

<u>12.</u> $\sin 390° = \sin 30° = 0.5$

<u>13.</u> $\cos 210° = -\cos 30° = -\frac{\sqrt{3}}{2}$

<u>14.</u> $\cos (-210°) = \cos 210° = -\cos 30° = -\frac{\sqrt{3}}{2}$

<u>15.</u> $\sin (-60°) = -\sin 60° = -\frac{\sqrt{3}}{2}$

<u>16.</u> $\tan 235° = \tan 55° \doteq 1.428$

<u>17.</u> $\tan (-150°) = -\tan 150° = \tan 30° = \frac{1}{\sqrt{3}}$

<u>18.</u> $\sin 330° = \sin (-30°) = -\sin 30° = -0.5$

<u>19.</u> $\cos 510° = \cos 150° = -\cos 30° = -\frac{\sqrt{3}}{2}$ <u>20.</u> $\cos (-300°) = \cos 60° = 0.5$

Exercises 5-3 · Pages 152-153

<u>A</u>   <u>1.</u> $\cot \theta = \dfrac{\cos \theta}{\sin \theta} = \dfrac{1}{\sin \theta} \cdot \dfrac{1}{\dfrac{1}{\cos \theta}} = \dfrac{\csc \theta}{\sec \theta}$

<u>2.</u> $\tan \theta = \dfrac{\sin \theta}{\cos \theta} = \dfrac{1}{\cos \theta} \cdot \dfrac{1}{\dfrac{1}{\sin \theta}} = \dfrac{\sec \theta}{\csc \theta}$

<u>3.</u> $\cos^2 \theta (1 + \tan^2 \theta) = \cos^2 \theta + \cos^2 \theta \cdot \dfrac{\sin^2 \theta}{\cos^2 \theta} = \cos^2 \theta + \sin^2 \theta = 1$

$\underline{4.}$ $\sin^2 \theta (1 + \cot^2 \theta) = \sin^2 \theta + \sin^2 \theta \cdot \dfrac{\cos^2 \theta}{\sin^2 \theta} = \sin^2 \theta + \cos^2 \theta = 1$

$\underline{5.}$ $(\sec \theta - \tan \theta)(\sec \theta + \tan \theta) = \sec^2 \theta - \tan^2 \theta = 1 + \tan^2 \theta - \tan^2 \theta = 1$

$\underline{6.}$ $(\csc \theta - \cot \theta)(\csc \theta + \cot \theta) = \csc^2 \theta - \cot^2 \theta = \cot^2 \theta + 1 - \cot^2 \theta = 1$

$\underline{7.}$ $\dfrac{\sin \theta \cos \theta}{1 - 2 \cos^2 \theta} = \dfrac{\sin \theta \cos \theta}{(1 - \cos^2 \theta) - \cos^2 \theta} = \dfrac{\dfrac{\sin \theta \cos \theta}{\sin \theta \cos \theta}}{\dfrac{\sin^2 \theta - \cos^2 \theta}{\sin \theta \cos \theta}} = \dfrac{1}{\dfrac{\sin \theta}{\cos \theta} - \dfrac{\cos \theta}{\sin \theta}} =$

$\dfrac{1}{\tan \theta - \cot \theta}$

$\underline{8.}$ $\dfrac{1 - \tan^2 \theta}{1 + \tan^2 \theta} = \dfrac{1 - \tan^2 \theta}{\sec^2 \theta} = \dfrac{1}{\sec^2 \theta} - \dfrac{\tan^2 \theta}{\sec^2 \theta} = \cos^2 \theta - \sin^2 \theta = 1 - 2 \sin^2 \theta$

$\underline{9.}$ $\dfrac{1 - \cos 2\theta}{\sin 2\theta} = \dfrac{1 - (1 - 2 \sin^2 \theta)}{2 \sin \theta \cos \theta} = \dfrac{\sin^2 \theta}{\sin \theta \cos \theta} = \tan \theta$

$\underline{10.}$ $\dfrac{\sin 2\theta}{1 - \cos 2\theta} = \dfrac{2 \sin \theta \cos \theta}{1 - (1 - 2 \sin^2 \theta)} = \dfrac{\cos \theta \sin \theta}{\sin^2 \theta} = \cot \theta$

$\underline{11.}$ $\csc^2 \dfrac{\theta}{2} = \dfrac{1}{\sin^2 \dfrac{\theta}{2}} = \dfrac{1}{\frac{1}{2}(1 - \cos \theta)} = \dfrac{2 \sec \theta}{\sec \theta (1 - \cos \theta)} = \dfrac{2 \sec \theta}{\sec \theta - 1}$

$\underline{12.}$ $2 \cos^2 \dfrac{\theta}{2} = 2(\frac{1}{2})(1 + \cos \theta) = \dfrac{\sec \theta (1 + \cos \theta)}{\sec \theta} = \dfrac{\sec \theta + 1}{\sec \theta}$

$\underline{13.}$ $\cos^3 \theta - \sin^3 \theta = (\cos \theta - \sin \theta)(\cos^2 \theta + \sin \theta \cos \theta + \sin^2 \theta) =$

$(\cos \theta - \sin \theta)(1 + \sin \theta \cos \theta) = (\cos \theta - \sin \theta)(1 + \frac{1}{2} \sin 2\theta)$

$\underline{14.}$ $\cot^2 \theta - \cos^2 \theta = \cos^2 \theta (\dfrac{1}{\sin^2 \theta} - 1) = \cos^2 \theta (\csc^2 \theta - 1) = \cos^2 \theta \cot^2 \theta$

$\underline{B}$ $\underline{15.}$ $\sin (\theta_2 + \theta_1) + \sin (\theta_2 - \theta_1) = \sin \theta_2 \cos \theta_1 + \cos \theta_2 \sin \theta_1 + \sin \theta_2 \cos \theta_1 -$

$\cos \theta_2 \sin \theta_1 = 2 \sin \theta_2 \cos \theta_1$

$\underline{16.}$ $\sin (\theta_2 + \theta_1) - \sin (\theta_2 - \theta_1) = \sin \theta_2 \cos \theta_1 + \cos \theta_2 \sin \theta_1 -$

$(\sin \theta_2 \cos \theta_1 - \cos \theta_2 \sin \theta_1) = 2 \cos \theta_2 \sin \theta_1$

$\underline{17.}$ $\cos (\theta_2 + \theta_1) + \cos (\theta_2 - \theta_1) = \cos \theta_2 \cos \theta_1 - \sin \theta_2 \sin \theta_1 + \cos \theta_2 \cos \theta_1 +$

$\sin \theta_2 \sin \theta_1 = 2 \cos \theta_2 \cos \theta_1$

$\underline{18.}$ $\cos (\theta_2 + \theta_1) - \cos (\theta_2 - \theta_1) = \cos \theta_2 \cos \theta_1 - \sin \theta_2 \sin \theta_1 -$

$(\cos \theta_2 \cos \theta_1 + \sin \theta_2 \sin \theta_1) = -2 \sin \theta_2 \sin \theta_1$

<u>19.</u> $\dfrac{\sin \theta_1 - \sin \theta_2}{\cos \theta_1 + \cos \theta_2} = \dfrac{2 \cos \left(\dfrac{\theta_1 + \theta_2}{2}\right) \sin \left(\dfrac{\theta_1 - \theta_2}{2}\right)}{2 \cos \left(\dfrac{\theta_1 + \theta_2}{2}\right) \cos \left(\dfrac{\theta_1 - \theta_2}{2}\right)} = \tan \left(\dfrac{\theta_1 - \theta_2}{2}\right)$

<u>20.</u> $\dfrac{\sin \theta_1 + \sin \theta_2}{\cos \theta_1 + \cos \theta_2} = \dfrac{2 \sin \left(\dfrac{\theta_1 + \theta_2}{2}\right) \cos \left(\dfrac{\theta_1 - \theta_2}{2}\right)}{2 \cos \left(\dfrac{\theta_1 + \theta_2}{2}\right) \cos \left(\dfrac{\theta_1 - \theta_2}{2}\right)} = \tan \left(\dfrac{\theta_1 + \theta_2}{2}\right)$

Exercises 5-4 · Pages 157-158

<u>A</u>   <u>1.</u> $\cos 58°40' \doteq 0.5200$ <u>2.</u> $\tan 62°50' \doteq 1.949$

<u>3.</u> $\sec 23°30' \doteq 1.090$ <u>4.</u> $\cot 84°10' \doteq 0.1022$

<u>5.</u> $\csc 74°30' \doteq 1.038$

<u>6.</u> $\sin 69°47' \doteq 0.9377 + 0.7(0.9387 - 0.9377) \doteq 0.9377 + 0.7(0.001) \doteq 0.9377 + 0.0007$

$= 0.9384$

<u>7.</u> $\cos 24°14' \doteq 0.9124 - 0.4(0.9124 - 0.9112) = 0.9124 - 0.4(0.0012) \doteq 0.9124 - 0.0005$

$= 0.9119$

<u>8.</u> $\sin 55°12' \doteq 0.8208 + 0.2(0.8225 - 0.8208) = 0.8208 + 0.2(0.0017) \doteq 0.8208 + 0.0003$

$= 0.8211$

<u>9.</u> $\cot 0.88^R \doteq 0.8267$ <u>10.</u> $\sin 1.14^R \doteq 0.9086$

<u>11.</u> $\tan 0.742^R \doteq 0.9131 + 0.2(0.9316 - 0.9131) = 0.9131 + 0.2(0.0185) \doteq$

$0.9131 + 0.0037 = 0.9168$

<u>12.</u> $\cos 1.333^R \doteq 0.2385 - 0.3(0.2385 - 0.2288) = 0.2385 - 0.3(0.0097) \doteq$

$0.2385 - 0.0029 = 0.2356$

<u>13.</u> $\sec 1.003^R \doteq 1.851 + 0.3(1.880 - 1.851) = 1.851 + 0.3(0.0029) \doteq$

$1.851 + 0.0009 = 1.860$

<u>14.</u> $\cos 1.502^R \doteq 0.0707 - 0.2(0.0707 - 0.0608) = 0.0707 - 0.2(0.0099) \doteq$

$0.0707 - 0.002 = 0.0687$

<u>15.</u> $\tan 1.323^R \doteq 3.903 + 0.3(4.072 - 3.903) = 3.903 + 0.3(0.169) \doteq$

$3.903 + 0.051 = 3.954$

<u>16.</u> $\sin 1.068^R \doteq 0.8724 + 0.8(0.8772 - 0.8724) = 0.8724 + 0.8(.0048) \doteq$

$0.8724 + 0.0038 = 0.8762$

<u>17.</u> $\theta = 36°20'$ <u>18.</u> $\theta = 18°10'$

19. $\theta \doteq 72°40' + 10'(\dfrac{3.224 - 3.204}{3.237 - 3.204}) = 72°40' + 10'(\dfrac{0.02}{0.033}) \doteq 72°40' + 10'(0.6) = 72°46'$

20. $\theta \doteq 15°50' + 10'(\dfrac{3.526 - 3.500}{3.526 - 3.487}) = 15°50' + 10'(\dfrac{0.026}{0.039}) \doteq 15°50' + 10'(0.7) = 15°57'$

21. $\theta \doteq 33°00' + 10'(\dfrac{1.836 - 1.832}{1.836 - 1.828}) = 33°00' + 10'(\dfrac{0.004}{0.008}) = 33°00' + 10'(0.5) = 33°5'$

22. $\theta \doteq 45°10' + 10'(\dfrac{1.420 - 1.418}{1.423 - 1.418}) = 45°10' + 10'(\dfrac{0.002}{0.005}) = 45°10' + 10'(0.4) = 45°14'$

23. $\theta \doteq 18°00' + 10'(\dfrac{0.30987 - 0.3090}{0.3118 - 0.3090}) = 18°00' + 10'(\dfrac{0.00087}{0.00280}) \doteq$
    $18°00' + 10'(0.3) = 18°3'$

24. $\theta \doteq 67°30' + 10'(\dfrac{0.3827 - 0.38088}{0.3827 - 0.3800}) = 67°30' + 10'(\dfrac{0.00182}{0.00270}) \doteq$
    $67°30' + 10'(0.7) = 67°37'$

25. $36°20' \doteq 0.63^R$ 26. $18°10' \doteq 0.32^R$ 27. $72°46' \doteq 1.27^R$

28. $15°57' \doteq 0.28^R$ 29. $33°5' \doteq 0.58^R$ 30. $45°14' \doteq 0.79^R$

31. $18°3' \doteq 0.31^R$ 32. $67°37' \doteq 1.18^R$

33. $\sin 195° = \sin (180° + 15°) = -\sin 15° \doteq -0.2588$

34. $\cos 275° = \cos (275° - 360°) = \cos (-85°) = \cos 85° \doteq 0.0872$

35. $\tan 572° = \tan (572° - 360°) = \tan 212° = \tan (180° + 32°) = \tan 32° = 0.6249$

36. $\cot 602° = \cot (602° - 360°) = \cot 242° = \cot (180° + 62°) = \cot 62° \doteq 0.5317$

37. $\sec (-123°) = \sec 123° = \sec (180° - 57°) = -\sec 57° = -1.836$

38. $\cos (-211°) = \cos 211° = \cos (180° - 31°) = -\cos 31° = -0.8572$

B 39. Expect the approximation to be less than $\sin \theta$ since the graph of $y = \sin \theta$ is concave downward and the line segment between any two points lies below the graph on this portion.

40. Expect the approximation to be greater than θ because the line segment between any two points in this portion lies to the right of the graph.

41. Equally good when θ is very close to either 0° or 90° (say within 5°).   Good near 0° because the graph of $\sin \theta$ is almost a straight line near 0° so the straight line joining two points lies very close to the graph.   Good near 90° because $\sin \theta$ changes very little when θ is close to 90° so the error must be very small.

$\underline{42}$. Expect the approximation to be greater than $\tan \theta$ since the graph of $y = \tan \theta$ is concave upward and the line segment between any two points lies above the graph on this portion.

$\underline{43}$. Expect the approximation to be less than θ because the line segment between any two points in this portion lies to the left of the graph.

$\underline{44}$. Expect a closer approximation when θ is close to $0°$. (See Ex. 41 for reasons.)

Exercises 5-5 · Page 161

$\underline{A}$ $\underline{1}$. $\cos \theta = \dfrac{-1}{\sqrt{(-1)^2 + (-1)^2}} = \dfrac{-1}{\sqrt{2}} = -\dfrac{\sqrt{2}}{2}$; $\sin \theta = \dfrac{1}{\sqrt{(-1)^2 + (-1)^2}} = \dfrac{1}{\sqrt{2}} = \dfrac{\sqrt{2}}{2}$;

$\tan \theta = \dfrac{-\frac{\sqrt{2}}{2}}{\frac{\sqrt{2}}{2}} = -1$; $\theta = 135°$

$\underline{2}$. $\cos \theta = \dfrac{-2}{\sqrt{(-2)^2 + (2\sqrt{3})^2}} = \dfrac{-2}{\sqrt{16}} = -\dfrac{1}{2}$; $\sin \theta = \dfrac{2\sqrt{3}}{\sqrt{(-2)^2 + (2\sqrt{3})^2}} = \dfrac{2\sqrt{3}}{\sqrt{16}} = \dfrac{\sqrt{3}}{2}$;

$\tan \theta = \dfrac{\frac{\sqrt{3}}{2}}{-\frac{1}{2}} = -\sqrt{3}$; $\theta = 120°$

$\underline{3}$. $\cos \theta = \dfrac{6}{\sqrt{6^2 + 8^2}} = \dfrac{6}{\sqrt{100}} = \dfrac{3}{5}$; $\sin \theta = \dfrac{8}{\sqrt{6^2 + 8^2}} = \dfrac{8}{\sqrt{100}} = \dfrac{4}{5}$; $\theta = 53°8'$

$\underline{4}$. $\cos \theta = \dfrac{8}{\sqrt{8^2 + (-15)^2}} = \dfrac{8}{\sqrt{289}} = \dfrac{8}{17}$; $\sin \theta = \dfrac{-15}{\sqrt{8^2 + (-15)^2}} = \dfrac{-15}{\sqrt{289}} = -\dfrac{15}{17}$;

$\tan \theta = \dfrac{-\frac{15}{17}}{\frac{8}{17}} = -\dfrac{15}{8}$; $\theta = 298°4'$

$\underline{5}$. $\cos \theta = \dfrac{2}{\sqrt{2^2 + 0^2}} = \dfrac{2}{\sqrt{4}} = 1$; $\sin \theta = \dfrac{0}{\sqrt{2^2 + 0^2}} = \dfrac{0}{\sqrt{4}} = 0$; $\tan \theta = \dfrac{0}{1} = 0$; $\theta = 0°$

$\underline{6}$. $\cos \theta = \dfrac{0}{\sqrt{0^2 + (-3)^2}} = \dfrac{0}{\sqrt{9}} = 0$; $\sin \theta = \dfrac{-3}{\sqrt{0^2 + (-3)^2}} = \dfrac{-3}{\sqrt{9}} = -1$; $\tan \theta = \dfrac{-1}{0}$,

not defined; $\theta = 270°$

$\underline{7}$. $\cos \theta = \dfrac{1}{\sqrt{1^2 + (-3)^2}} = \dfrac{1}{\sqrt{10}} = \dfrac{\sqrt{10}}{10}$; $\sin \theta = \dfrac{-3}{\sqrt{1^2 + (-3)^2}} = \dfrac{-3}{\sqrt{10}} = -\dfrac{3\sqrt{10}}{10}$;

$\tan \theta = \dfrac{-\frac{3\sqrt{10}}{10}}{\frac{\sqrt{10}}{10}} = -3$; $\theta = 288°26'$

8. $\cos \theta = \dfrac{2}{\sqrt{2^2 + 3^2}} = \dfrac{2}{\sqrt{13}} = \dfrac{2\sqrt{13}}{13}$; $\sin \theta = \dfrac{3}{\sqrt{2^2 + 3^2}} = \dfrac{3}{\sqrt{13}} = \dfrac{3\sqrt{13}}{13}$;

$\tan \theta = \dfrac{\frac{3\sqrt{13}}{13}}{\frac{2\sqrt{13}}{13}} = \dfrac{3}{2}$; $\theta = 56°19'$

9. $\cos \theta = \dfrac{1}{\sqrt{1^2 + (-\sqrt{3})^2}} = \dfrac{1}{\sqrt{4}} = \dfrac{1}{2}$; $\sin \theta = \dfrac{\sqrt{3}}{\sqrt{1^2 + (-\sqrt{3})^2}} = \dfrac{-\sqrt{3}}{\sqrt{4}} = -\dfrac{\sqrt{3}}{2}$;

$\tan \theta = \dfrac{-\frac{\sqrt{3}}{2}}{\frac{1}{2}} = -\sqrt{3}$; $\theta = 300°$

10. $x = r \cos \theta = 1 \cdot \cos 60° = 1 \cdot \dfrac{1}{2} = \dfrac{1}{2}$; $y = r \sin \theta = 1 \cdot \sin 60° = 1 \cdot \dfrac{\sqrt{3}}{2} = \dfrac{\sqrt{3}}{2}$;

$(\dfrac{1}{2}, \dfrac{\sqrt{3}}{2})$

11. $x = 2 \cdot \cos 120° = 2(-\dfrac{1}{2}) = -1$; $y = 2 \cdot \sin 120° = 2 \cdot \dfrac{\sqrt{3}}{2} = \sqrt{3}$; $(-1, \sqrt{3})$

12. $x = 4 \cdot \cos 45° = 4 \cdot \dfrac{\sqrt{2}}{2} = 2\sqrt{2}$; $y = 4 \sin 45° = 4 \cdot \dfrac{\sqrt{2}}{2} = 2\sqrt{2}$; $(2\sqrt{2}, 2\sqrt{2})$

13. $x = 4 \cos (-150°) = 4 \cos 150° = 4(-\dfrac{\sqrt{3}}{2}) = -2\sqrt{3}$; $y = 4 \sin (-150°) =$

$-4 \sin 150° = -4 \cdot \dfrac{1}{2} = -2$; $(-2\sqrt{3}, -2)$

14. $x = 10 \cos (-420°) = 10 \cos 420° = 10 \cos 60° = 10 \cdot \dfrac{1}{2} = 5$; $y = 10 \sin (-420°) =$

$-10 \sin 420° = -10 \sin 60° = -10 \cdot \dfrac{\sqrt{3}}{2} = -5\sqrt{3}$; $(5, -5\sqrt{3})$

15. $x = 5 \cos 810° = 5 \cos 90° = 5 \cdot 0 = 0$; $y = 5 \sin 810° = 5 \sin 90° = 5 \cdot 1 = 5$;

$(0, 5)$

16. $x = 10 \cos 22° \doteq 10(0.9272) = 9.272$; $y = 10 \sin 22° \doteq 10(0.3746) = 3.746$;

$(9.272, 3.746)$

17. $x = 3 \cos (-40°) = 3 \cos 40° \doteq 3(0.7660) = 2.298$; $y = 3 \sin (-40°) = -3 \sin 40° \doteq$

$-3(0.6428) = -1.9284$; $(2.298, -1.9284)$

B 18. Let $x = 1$, then $y = -1$ and $\cos \theta = \dfrac{1}{\sqrt{1^2 + (-1)^2}} = \dfrac{1}{\sqrt{2}} = \dfrac{\sqrt{2}}{2}$; $\sin \theta = \dfrac{-1}{\sqrt{2}} = -\dfrac{\sqrt{2}}{2}$

19. Let $x = -1$, then $y = -1$ and $\cos \theta = \dfrac{-1}{\sqrt{(-1)^2 + (-1)^2}} = -\dfrac{1}{\sqrt{2}} = -\dfrac{\sqrt{2}}{2}$;

$\sin \theta = \dfrac{-1}{\sqrt{2}} = -\dfrac{\sqrt{2}}{2}$

20. Let $y = -1$, then $x = -2$ and $\cos \theta = \dfrac{-2}{\sqrt{(-2)^2 + (-1)^2}} = -\dfrac{2}{\sqrt{5}}$; $\sin \theta = \dfrac{-1}{\sqrt{5}}$

<u>21.</u> Let $y = 1$, $x = 0$ and $\cos \theta = \dfrac{0}{\sqrt{0^2 + 1^2}} = 0$; $\sin \theta = \dfrac{1}{\sqrt{1}} = 1$

<u>22.</u> Let $x = -1$, $y = 0$ and $\cos \theta = \dfrac{-1}{\sqrt{0^2 + (-1)^2}} = \dfrac{-1}{1} = -1$; $\sin \theta = \dfrac{0}{1} = 0$

<u>23.</u> Let $x = 1$, then $y = \sqrt{3}$ and $\cos \theta = \dfrac{1}{\sqrt{1^2 + (\sqrt{3})^2}} = \dfrac{1}{2}$; $\sin \theta = \dfrac{\sqrt{3}}{2}$

Exercises 5-6 · Pages 163-164

<u>A</u>   <u>1.</u> Domain: $\{x: |x| \le 1, x \in R\}$; Range: $\{y: |y| \le \dfrac{\pi}{2}, y \in R\}$

<u>2.</u> Domain: $\{x: |x| \le 1, x \in R\}$; Range: $\{y: 0 \le y \le \pi, y \in R\}$

<u>3.</u> Domain: R; Range: $\{y: |y| < \dfrac{\pi}{2}, y \in R\}$

<u>4.</u> Domain: R; Range: $\{y: 0 < y < \pi, y \in R\}$

<u>5.</u> Domain: $\{x: |x| \ge 1, x \in R\}$; Range: $\{y: |y| < \dfrac{\pi}{2}, y \ne 0, y \in R\}$

<u>6.</u> Domain: $\{x: |x| \ge 1, x \in R\}$; Range: $\{y: 0 < y < \pi, y \ne \dfrac{\pi}{2}, y \in R\}$

<u>7.</u> $\text{Cot}^{-1} (\csc 90°) = \text{Cot}^{-1} 1 = 45°$ <u>8.</u> $\text{Sin}^{-1} (\cos 30°) = \text{Sin}^{-1} (\dfrac{\sqrt{3}}{2}) = 60°$

<u>9.</u> $\sin (\text{Tan}^{-1} \sqrt{3}) = \sin 60° = \dfrac{\sqrt{3}}{2}$ <u>10.</u> $\cos (\text{Sin}^{-1} 1) = \cos 90° = 0$

<u>11.</u> $\text{Sin}^{-1} (\tan 45°) = \text{Sin}^{-1} 1 = 90°$ <u>12.</u> $\text{Cos}^{-1} (\cot 45°) = \text{Cos}^{-1} 1 = 0°$

<u>13.</u> $\tan^2 \theta + \sqrt{3} \tan \theta = 0$, $\tan \theta (\tan \theta + \sqrt{3}) = 0$, $\tan \theta = 0$ or $\tan \theta = -\sqrt{3}$;
    $\{0°, 120°, 180°, 300°\}$

<u>14.</u> $\cos^2 \theta = \cos \theta$, $\cos \theta (\cos \theta - 1) = 0$, $\cos \theta = 0$ or $\cos \theta = 1$; $\{0°, 90°, 270°\}$

<u>15.</u> $\cot^2 \theta - 3 \csc \theta + 3 = 0$, $\csc^2 \theta - 1 - 3 \csc \theta + 3 = 0$, $\csc^2 \theta - 3 \csc \theta + 2 = 0$,
    $(\csc \theta - 2)(\csc \theta - 1) = 0$, $\csc \theta = 2$ or $\csc \theta = 1$; $\{30°, 90°, 150°\}$

<u>16.</u> $\tan 2\theta = \cot \theta$, $\dfrac{2 \tan \theta}{1 - \tan^2 \theta} = \dfrac{1}{\tan \theta}$, $2 \tan^2 \theta = 1 - \tan^2 \theta$, $3 \tan^2 \theta = 1$,
    $\tan \theta = \pm \dfrac{1}{\sqrt{3}}$; $\{30°, 150°, 210°, 330°\}$

<u>17.</u> $\cot \theta = \tan (2\theta - 270°)$, $\cot \theta = \dfrac{\sin (2\theta - 270°)}{\cos (2\theta - 270°)} = \dfrac{\cos 2\theta}{-\sin 2\theta} = -\cot 2\theta$, $\dfrac{1}{\tan \theta} =$
    $\dfrac{-1 + \tan^2 \theta}{2 \tan \theta}$, $2 = -1 + \tan^2 \theta$, $\tan^2 \theta = 3$, $\tan \theta = \pm\sqrt{3}$; $\{60°, 120°, 240°, 300°\}$

<u>18.</u> $\sin \theta = \sin (2\theta - 180°)$, $\sin \theta = -\sin 2\theta$, $\sin \theta = -2 \sin \theta \cos \theta$,
    $\sin \theta (1 + 2 \cos \theta) = 0$, $\sin \theta = 0$ or $\cos \theta = -\dfrac{1}{2}$; $\{0°, 120°, 180°, 240°\}$

B   19. $2 \cos^2 \theta \sin^2 \theta - \cos \theta \sin \theta = 0$, $\cos \theta \sin \theta (2 \cos \theta \sin \theta - 1) = 0$, $\cos \theta \sin \theta = 0$ or $\sin 2\theta = 1$, $\cos \theta = 0$, $\sin \theta = 0$, or $\sin 2\theta = 1$; $\{0°, 45°, 90°, 180°, 225°, 270°\}$

20. $\sec^2 2\theta + \tan 2\theta = 1$, $1 + \tan^2 2\theta + \tan 2\theta = 1$, $\tan 2\theta (\tan 2\theta + 1) = 0$, $\tan 2\theta = 0$ or $\tan 2\theta = -1$; $2\theta = 0°, 135°, 180°, 315°, 360°, 495°, 540°, 675°$; $\{0°, 67\frac{1}{2}°, 90°, 157\frac{1}{2}°, 180°, 247\frac{1}{2}°, 270°, 337\frac{1}{2}°\}$

21. $\cos (\text{Sec}^{-1} 2) = \cos (\text{Cos}^{-1} \frac{1}{2}) = \frac{1}{2}$; $\sin (\text{Sec}^{-1} 2) = \sqrt{1 - (\frac{1}{2})^2} = \frac{\sqrt{3}}{2}$; $\tan (\text{Sec}^{-1} 2) = \dfrac{\frac{\sqrt{3}}{2}}{\frac{1}{2}} = \sqrt{3}$

22. $\cos (\text{Cos}^{-1} \frac{2}{3}) = \frac{2}{3}$; $\sin (\text{Cos}^{-1} \frac{2}{3}) = \sqrt{1 - (\frac{2}{3})^2} = \frac{\sqrt{5}}{3}$; $\tan (\text{Cos}^{-1} \frac{2}{3}) = \dfrac{\frac{\sqrt{5}}{3}}{\frac{2}{3}} = \frac{\sqrt{5}}{2}$

23. $\cos (\text{Sin}^{-1} \frac{1}{2}) = \sqrt{1 - (\frac{1}{2})^2} = \frac{\sqrt{3}}{2}$; $\sin (\text{Sin}^{-1} \frac{1}{2}) = \frac{1}{2}$, $\tan (\text{Sin}^{-1} \frac{1}{2}) = \dfrac{\frac{1}{2}}{\frac{\sqrt{3}}{2}} = \frac{1}{\sqrt{3}}$

24. $\cos (\text{Sin}^{-1} (-\frac{2}{5})) = \sqrt{1 - (-\frac{2}{5})^2} = \frac{\sqrt{21}}{5}$; $\sin (\text{Sin}^{-1} (-\frac{2}{5})) = -\frac{2}{5}$; $\tan (\text{Sin}^{-1} (-\frac{2}{5})) = \dfrac{-\frac{2}{5}}{\frac{\sqrt{21}}{5}} = -\frac{2}{\sqrt{21}}$

C   25. $\tan (180° + \text{Arcsin} \frac{\sqrt{2}}{2}) = \tan (\text{Arcsin} \frac{\sqrt{2}}{2}) = \dfrac{\sin (\text{Arcsin} \frac{\sqrt{2}}{2})}{\cos (\text{Arcsin} \frac{\sqrt{2}}{2})} = \dfrac{\frac{\sqrt{2}}{2}}{1 - (\frac{\sqrt{2}}{2})^2} = \dfrac{\frac{\sqrt{2}}{2}}{\frac{\sqrt{2}}{2}} = 1$

26. $\cos [\text{Arcsin} (-\frac{1}{2}) + \text{Arccos} \frac{5}{13}] = \cos (\text{Arcsin} (-\frac{1}{2})) \cos (\text{Arccos} \frac{5}{13}) - \sin (\text{Arcsin} (-\frac{1}{2})) \sin (\text{Arccos} \frac{5}{13}) = (\sqrt{1 - (-\frac{1}{2})^2}) \frac{5}{13} - (-\frac{1}{2})(\sqrt{1 - (\frac{5}{13})^2}) = \frac{\sqrt{3}}{2} \cdot \frac{5}{13} - (-\frac{1}{2})(\frac{12}{13}) = \frac{5\sqrt{3} + 12}{26}$

27. $\cos (\frac{1}{2} \text{Arctan} \frac{3}{4}) = \sqrt{\frac{1}{2}(1 + \cos (\text{Arctan} \frac{3}{4}))} = \sqrt{\dfrac{1 + \frac{4}{5}}{2}} = \frac{3}{\sqrt{10}}$

28. $\sin (\frac{1}{2} \text{Arctan} \frac{3}{5}) = \sqrt{\frac{1}{2}(1 - \cos (\text{Arctan} \frac{3}{5}))} = \sqrt{\dfrac{1 - \frac{5}{\sqrt{34}}}{2}} = \sqrt{\dfrac{\sqrt{34} - 5}{2\sqrt{34}}}$

Exercises 5-7 · Pages 168-169

A   1. $a = \sqrt{13^2 - 12^2} = \sqrt{169 - 144} = \sqrt{25} = 5$; $\sin B = \frac{b}{c} = \frac{12}{13} \doteq 0.9231$, $B \doteq 67°20' + 10'(\frac{0.9231 - 0.9228}{0.9239 - 0.9228}) \doteq 67°23'$; $A = 90° - B \doteq 90° - 67°23' = 22°37'$

<u>2</u>. c = $\sqrt{9^2 + 12^2}$ = $\sqrt{81 + 144}$ = $\sqrt{225}$ = 15; sin A = $\frac{a}{c}$ = $\frac{9}{15}$ = 0.6000, A ≐ 36°50' +

10'($\frac{0.6000 - 0.5995}{0.6018 - 0.5995}$) ≐ 36°50' + 2' = 36°52'; B = 90° - A ≐ 90° - 36°52' = 53°8'

<u>3</u>. B = 90° - A = 90° - 36° = 54°; sin 36° = $\frac{a}{13}$, a ≐ 13 (sin 36°) ≐ 13 (0.5878) ≐

7.64; sin 54° = $\frac{b}{13}$, b ≐ 13 (sin 54°) ≐ 13(0.8090) = 10.52

<u>4</u>. B = 90° - A = 90° - 53°10' = 36°50'; tan 53°10' = $\frac{a}{6}$, a ≐ 6(1.335) ≐ 8.01;

sec 53°10' = $\frac{c}{6}$, c = 6(1.668) ≐ 10.01

<u>5</u>. A = 90° - B = 90° - 64°15' = 25°45'; cos B = cos 64°15' = $\frac{a}{20}$, a ≐ 20(0.4344) ≐

8.69; sin B = sin 64°15' = $\frac{b}{20}$, b ≐ 20(0.9007) ≐ 18.01

<u>6</u>. A = 90° - B = 90° - 80°25' = 9°35'; csc B = csc 80°25' = $\frac{c}{10}$, c ≐ 10(1.014) =

10.14; cot B = cot 80°25' = $\frac{a}{10}$, a ≐ 10(0.1688) ≐ 1.69

<u>7</u>. b = $\sqrt{20^2 - 10^2}$ = $\sqrt{400 - 100}$ = $\sqrt{300}$ ≐ 17.32; sin A = $\frac{a}{c}$ = $\frac{10}{20}$ = 0.5, A = 30°;

B = 90° - A = 90° - 30° = 60°

<u>8</u>. a = $\sqrt{20^2 - 15^2}$ = $\sqrt{400 - 225}$ = $\sqrt{175}$ ≐ 13.23; sin B = $\frac{b}{c}$ = $\frac{15}{20}$ = 0.75, B ≐

48°30' + 10'($\frac{0.7500 - 0.7490}{0.7509 - 0.7490}$) ≐ 48°30' + 5' = 48°35'; A = 90° - B ≐

90° - 48°35' = 41°25'

<u>9</u>. sin θ = $\frac{50}{70}$ ≐ 0.7143, θ ≐ 46°

<u>10</u>. x = $\sqrt{30^2 - 5^2}$ = $\sqrt{900 - 25}$ = $\sqrt{875}$ ≐ 30 ft.

<u>11</u>. csc 22$\frac{1}{2}°$ = $\frac{x}{4}$, x = 4(2.613) ≐ 10 in.

<u>12</u>. The hypotenuse of the rt. Δ is a diameter, so sin A = $\frac{8}{20}$ = 0.4000, A ≐ 24° and

B ≐ 90° - 24° = 66°

<u>13</u>. By the Pythagorean Theorem, a² + b² = 1².

sin θ = $\frac{a}{1}$ and cos θ = $\frac{b}{1}$, so

a² + b² = sin² θ + cos² θ = 1² = 1.

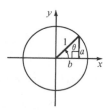

<u>14.</u> See art for Ex. 13.   By the Pythagorean Theorem, $a^2 + b^2 = 1^2$. $a^2 = 1 - b^2$, but

$\sin \theta = \frac{a}{1} = a$ and $\cos \theta = \frac{b}{1} = b$, so $\sin^2 \theta = 1 - \cos^2 \theta$ by substitution.

<u>15.</u> By the Pythagorean Theorem, $1^2 + a^2 = b^2$.

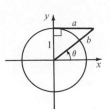

$\cot \theta = \frac{a}{1} = a$ and $\csc \theta = \frac{b}{1} = b$ so

$1^2 + a^2 = b^2$ yields $1 + \cot^2 \theta = \csc^2 \theta$

by substitution.

<u>16.</u> By the Pythagorean Theorem, $1 + a^2 = b^2$.

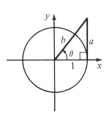

$\tan \theta = \frac{a}{1} = a$ and $\sec \theta = \frac{b}{1} = b$, so

$1^2 + a^2 = b^2$ yields $1 + \tan^2 \theta = \sec^2 \theta$

by substitution.

<u>B</u>   <u>17.</u> Since $x^2 + y^2 = 1$, $y = \sqrt{1 - x^2}$ and $\csc \theta = \frac{1}{y} = \frac{1}{\sqrt{1 - x^2}}$.

Ex. 17   Ex. 18   Ex. 19

<u>18.</u> Since $1 + y^2 = x^2$, $y = \sqrt{x^2 - 1}$ and $\sin \theta = \frac{y}{x} = \frac{\sqrt{x^2 - 1}}{x}$.

<u>19.</u> Since $y^2 + x^2 = 1$, $y = \sqrt{1 - x^2}$ and $\cot \theta = \frac{y}{x} = \frac{\sqrt{1 - x^2}}{x}$.

<u>20.</u> Since $x^2 + y^2 = 1$, $y = \sqrt{1 - x^2}$ and $\tan \theta = \frac{y}{x} = \frac{\sqrt{1 - x^2}}{x}$.

    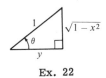

Ex. 20   Ex. 21   Ex. 22

<u>21.</u> Since $1 + (\sqrt{1 - x^2})^2 = y$, $y = \sqrt{2 - x^2}$ and $\sin \theta = \frac{1}{y} = \frac{1}{\sqrt{2 - x^2}}$.

<u>22.</u> Since $y + (\sqrt{1 - x^2})^2 = 1$, $y = x$ and $\tan \theta = \frac{\sqrt{1 - x^2}}{y} = \frac{\sqrt{1 - x^2}}{x}$.

<u>C</u>   <u>23.</u> If x is the height of the building, $\tan 67° = \frac{x}{125}$, $x \doteq 125(2.356) = 294$ ft.; If y is

the height of the flagpole, $\tan 70° = \frac{y + 294}{125}$, $y \doteq 125(2.747) - 294 = 343 - 294 =$

49 ft.

24. Let x be the height of the building and y be the distance of the point from the building.

Then $\tan 45° = 1 = \frac{x}{y}$, so $x = y$. $\tan 47° = \frac{25 + x}{y} = \frac{25 + x}{x}$, $1.072x \doteq 25 + x$,

$0.072x \doteq 25$, $x \doteq \frac{25}{0.072} \doteq 347$ ft.

25. If x is the distance from nearer ship to observer and y the distance from farther ship

to observer, $\cot 18° = \frac{x}{3000}$ and $\cot 13° = \frac{y}{3000}$. $y - x = 3000 (\cot 13°) -$

$3000 (\cot 18°) = 3000 (\cot 13° - \cot 18°) \doteq 3000 (4.331 - 3.078) =$

$3000(1.253) \doteq 3760$ ft.

26. Let x be the height of B. $x = 100 \tan 38° + 100 \tan 50° \doteq 100(0.7813 + 1.192) =$

$100(1.9733) \doteq 197$ ft.

Exercises 5-8 · Pages 172-173

A   1. $c^2 = 4^2 + 2^2 - 2 \cdot 4 \cdot 2 \cos 30° = 16 + 4 - 16 \frac{\sqrt{3}}{2} = 20 - 8\sqrt{3} \doteq 20 - 13.86 =$

6.14, $c \doteq 2.5$

2. $c^2 = 1^2 + 3^2 - 2 \cdot 1 \cdot 3 \cos 30° = 1 + 9 - 6 \frac{\sqrt{3}}{2} = 10 - 3\sqrt{3} \doteq 10 - 5.20 = 4.80,$

$c \doteq 2.2$

3. $a^2 = 6^2 + 4^2 - 2 \cdot 6 \cdot 4 \cos 60° = 36 + 16 - 48 \cdot \frac{1}{2} = 52 - 24 = 28$, $a \doteq 5.3$

4. $a^2 = 3^2 + 4^2 - 2 \cdot 3 \cdot 4 \cos 30° = 9 + 16 - 24 \frac{\sqrt{3}}{2} = 25 - 12\sqrt{3} \doteq 25 - 20.78 =$

4.22, $a \doteq 2.1$

5. $a^2 = 7^2 + 2^2 - 2 \cdot 7 \cdot 2 \cos 135° = 49 + 4 - 28 (-\frac{\sqrt{2}}{2}) = 53 + 14\sqrt{2} \doteq 53 + 19.8 =$

72.8, $a \doteq 8.5$

6. $3^2 = 5^2 + 6^2 - 2 \cdot 5 \cdot 6 \cos A$, $9 = 25 + 36 - 60 \cos A$, $60 \cos A = 52$,

$\cos A = \frac{52}{60} \doteq 0.8667$, $A \doteq 30°$

7. $8^2 = 5^2 + 6^2 - 2 \cdot 5 \cdot 6 \cos C$, $64 = 25 + 36 - 60 \cos C$, $60 \cos C = -3$, $\cos C =$

$-\frac{3}{60} = -0.0500$, $C \doteq 180° - 87°10' = 92°50'$

8. $9^2 = 5^2 + 5^2 - 2 \cdot 5 \cdot 5 \cos C$, $81 = 25 + 25 - 50 \cos C$, $50 \cos C = -31$,

$\cos C = -\frac{31}{50} = -0.6200$, $C \doteq 180° - 51°40' = 128°20'$

9. $c^2 = 6^2 + 8^2 - 2 \cdot 6 \cdot 8 \cos 53° \doteq 36 + 64 - 96(0.6018) \doteq 100 - 57.77 = 42.2,$

$c \doteq 6.5$; $6^2 = 8^2 + (\sqrt{42.2})^2 - 2 \cdot 8 \cdot 6.5 \cos A$, $36 = 64 + 42.2 - 104 \cos A$,

$104 \cos A = 70.2$, $\cos A = \frac{70.2}{104} = 0.675$, $A \doteq 47°30'$

10. $c^2 = 7^2 + 8^2 - 2 \cdot 7 \cdot 8 \cos 10° \doteq 49 + 64 - 112(0.7848) \doteq 113 - 110.3 = 2.7$,
$c \doteq 1.6$; $7^2 = 8^2 + (\sqrt{2.7})^2 - 2(8)(1.6) \cos A$, $49 = 64 + 2.7 - 25.6 \cos A$,
$25.6 \cos A = 17.7$, $\cos A = \frac{17.7}{25.6} \doteq 0.6914$, $A \doteq 46°20'$

11. $x^2 = [2(30)]^2 + [2(40)]^2 - 2[2(30)][2(40)] \cos 40 = 3600 + 6400 - 9600(0.7660) =$
$10000 - 7353.6 = 2646.4$, $x \doteq 51.4$ mi.

12. $x^2 = 200^2 + 50^2 - 2 \cdot 50 \cdot 200 \cos 20° \doteq 40,000 + 2500 - 20,000(0.9397) =$
$42,500 - 18,794 = 23706$, $x \doteq 154.0$ mi.; $200^2 = 50^2 + 23706 - 2(50)(154) \cos C$,
$40,000 = 2500 + 23706 - 15400 \cos C$, $\cos C = -\frac{13794}{15400} \doteq -0.896$, $C =$
$\text{Cos}^{-1}(-0.896) = 180° - \text{Cos}^{-1}(0.896)$; If θ is the angle of correction,
$\theta = 180° - (180° - \text{Cos}^{-1}(0.896)) = \text{Cos}^{-1}(0.896) \doteq 26°20'$

13. $x^2 = 11^2 + 8^2 - 2 \cdot 11 \cdot 8 \cos 138° = 121 + 64 + 176 \cos 42° \doteq 185 + 130.8 =$
315.8, $x \doteq 17.8$ cm.; $y^2 = 11^2 + 8^2 - 2 \cdot 11 \cdot 8 \cos 42° = 121 + 64 - 176 \cos 42°$
$\doteq 185 - 130.8 = 54.2$, $y \doteq 7.3$ cm.

14. $c^2 = 86^2 + 61^2 - 2 \cdot 86 \cdot 61 \cos 76° = 7396 + 3721 - 10492 \cos 76° \doteq$
$11117 - 10492(0.2419) \doteq 11,117 - 2538.0 = 8579.0$, $c \doteq 92.6$ m.

15. $x^2 = 310^2 + 250^2 - 2(310)(250) \cos 52° = 96100 + 62500 - 155,000(0.6157) \doteq$
$158600 - 95433.5 = 63166.5$, $x \doteq 251.3$; $P = 250 + 310 + 251.3 = 811.3$ m.

16. $22^2 = 43^2 + 52^2 - 2 \cdot 43 \cdot 52 \cos C$, $484 = 1849 + 2704 - 4472 \cos C$,
$\cos C = \frac{4069}{4472} \doteq 0.9099$, $C \doteq 24°31'$

B 17. $a^2 = b^2 + c^2 - 2bc \cos A$, $\cos A = \frac{b^2 + c^2 - a^2}{2bc}$, $1 = \frac{2bc}{2bc}$, $1 + \cos A =$
$\frac{b^2 + c^2 - a^2 + 2bc}{2bc} = \frac{(b + c)^2 - a^2}{2bc} = \frac{(b + c + a)(b + c - a)}{2bc}$

18. $a^2 = b^2 + c^2 - 2bc \cos A$, $-\cos A = \frac{a^2 - b^2 - c^2}{2bc}$, $1 = \frac{2bc}{2bc}$, $1 - \cos A =$
$\frac{a^2 - b^2 - c^2 + 2bc}{2bc} = \frac{a^2 - (b - c)^2}{2bc} = \frac{(a - b + c)(a + b - c)}{2bc}$

19. $d = \sqrt{a^2 + b^2 - 2ab \cos B}$, $m°(A) = 180° - m°(B)$, $\cos A = -\cos B$, $2ab \cos A =$
$-2ab \cos B$, and $d = \sqrt{a^2 + b^2 + 2ab \cos A}$

20. $a^2 = b^2 + c^2 - 2bc \cos A$ so $2bc \cos A = b^2 + c^2 - a^2$; similarly $2ac \cos B =$
$a^2 + c^2 - b^2$ and $2ab \cos C = a^2 + b^2 - c^2$; adding corres. members of all
3 equations yields $2(bc \cos A + ac \cos B + ab \cos C) = a^2 + b^2 + c^2$

C   21. $\tan \frac{A}{2} = \dfrac{\sin \frac{A}{2}}{\cos \frac{A}{2}} = \sqrt{\dfrac{\frac{1}{2}(1 - \cos A)}{\frac{1}{2}(1 + \cos A)}} = \sqrt{\dfrac{\dfrac{(a - b + c)(a + b - c)}{2bc}}{\dfrac{(b + c + a)(b + c - a)}{2bc}}} =$

$\sqrt{\dfrac{(a - b + c)(a + b - c)}{(b + c + a)(b + c - a)}}$

22. $s - a = \dfrac{a + b + c}{2} - \dfrac{2a}{2} = \dfrac{b + c - a}{2}, \quad s - b = \dfrac{a + b + c}{2} - \dfrac{2b}{2} = \dfrac{a + c - b}{2},$

$s - c = \dfrac{a + b + c}{2} - \dfrac{2c}{2} = \dfrac{a + b - c}{2}; \quad \tan \dfrac{A}{2} = \sqrt{\dfrac{2(s - b)\,2(s - c)}{(2s)\,2(s - a)}} =$

$\sqrt{\dfrac{(s - b)(s - c)}{s(s - a)}} = \dfrac{1}{s - a}\sqrt{\dfrac{(s - a)(s - b)(s - c)}{s}} = \dfrac{r}{s - a}$

Exercises 5-9 · Pages 175-176

A   1. Area $= \frac{1}{2}ab \sin C = \frac{1}{2}(12)(20) \sin 30° = 120(\frac{1}{2}) = 60$

2. Area $= \frac{1}{2}ab \sin C = \frac{1}{2}(6)(15) \sin 60° = 45(\frac{\sqrt{3}}{2}) \doteq 39.0$

3. Area $= \frac{1}{2}bc \sin A = \frac{1}{2}(18)(40) \sin 28° = 360(0.4695) \doteq 169.0$

4. Area $= \frac{1}{2}ac \sin B = \frac{1}{2}(25)(30) \sin 42° \doteq 375(0.6991) \doteq 262.2$

5. Area $= \frac{1}{2}ab \sin C = \frac{1}{2}(35)(25) \sin 150° = \frac{875}{2}(\frac{1}{2}) \doteq 218.8$

6. Area $= \frac{1}{2}ab \sin C = \frac{1}{2}(40)(46) \sin 135° = 920(\frac{\sqrt{2}}{2}) \doteq 650.4$

7. $b = \dfrac{a \sin B}{\sin A} = \dfrac{6 \sin 45°}{\sin 60°} = \dfrac{6(\frac{\sqrt{2}}{2})}{\frac{\sqrt{3}}{2}} = 2\sqrt{6} \doteq 4.9$

8. $a = \dfrac{b \sin A}{\sin B} = \dfrac{30 \sin 45°}{\sin 30°} = \dfrac{30(\frac{\sqrt{2}}{2})}{\frac{1}{2}} = 30\sqrt{2} \doteq 42.4$

9. $\sin B = \dfrac{b \sin A}{a} = \dfrac{20 \sin 30°}{25} = \dfrac{20(\frac{1}{2})}{25} = 0.4000, \quad B \doteq 23°30'$

10. $\sin A = \dfrac{a \sin C}{c} = \dfrac{8 \sin 135°}{16} = \dfrac{8(\frac{\sqrt{2}}{2})}{16} = \dfrac{\sqrt{2}}{4} \doteq 0.3535, \quad C \doteq 20°40'$

11. $\sin A = \dfrac{a \sin B}{b} = \dfrac{12(\frac{1}{4})}{28} = \dfrac{3}{28} \doteq 0.1071, \quad A \doteq 6°10'$

12. $\sin B = \dfrac{b \sin A}{a} = \dfrac{6(\frac{4}{5})}{18} = \dfrac{4}{15} \doteq 0.2667, \quad B \doteq 15°30'$

13. $\dfrac{\overline{DF}}{\sin 23°50'} = \dfrac{\overline{EF}}{\sin 70°30'} = \dfrac{12.0}{\sin 85°40'}; \quad \overline{DF} = \dfrac{12.0(\sin 23°50')}{\sin 85°40'} =$

12.0($\sin 23°50'$)($\csc 85°40'$) $\doteq$ (12.0)(0.4041)(1.003) $\doteq$ 4.9 mi.; $\overline{EF} =$

12.0($\sin 70°30'$)($\csc 85°40'$) $\doteq$ 12.0(0.9426)(1.003) $\doteq$ 11.3 mi.

14. Let θ be a base angle. Then $2\theta + 54° = 180°$ and $\theta = 63°$. Let the equal legs have

length x, then $x = \dfrac{24 \sin \theta}{\sin 54°} = 24 \sin 63° \csc 54° \doteq 24(0.8910)(1.236) \doteq 26.4$,

$2x + 24 \doteq 52.8 + 24 = 76.8$

15. $\dfrac{42.0}{\sin 70°} = \dfrac{x}{\sin 12°}$, $x = 42.0 \sin 12° \csc 70° \doteq 42.0(0.2079)(1.064) \doteq 9.3$ ft.

16. $\dfrac{x}{\sin 42°} = \dfrac{300}{\sin 11°}$, $x = 300 \sin 42° \csc 11° \doteq 300(0.6691)(5.241) \doteq 1052.0$ m.

B   17. Area $= \frac{1}{2}bc \sin A$; Since $\dfrac{\sin C}{c} = \dfrac{\sin B}{b}$, $c = \dfrac{b \sin C}{\sin B}$, Area $=$

$\frac{1}{2}b(\dfrac{b \sin C}{\sin B}) \sin A = \dfrac{b^2 \sin A \sin C}{2 \sin B}$

18. $\dfrac{a}{b} = \dfrac{\sin A}{\sin B}$, so $\dfrac{a}{b} - 1 = \dfrac{\sin A}{\sin B} - 1$ and $\dfrac{a}{b} - \dfrac{b}{b} = \dfrac{\sin A}{\sin B} - \dfrac{\sin B}{\sin B}$ or $\dfrac{a - b}{b} =$

$\dfrac{\sin A - \sin B}{\sin B}$

19. $\dfrac{a}{b} = \dfrac{\sin A}{\sin B}$, so $\dfrac{a}{b} + 1 = \dfrac{\sin A}{\sin B} + 1$ and $\dfrac{a + b}{b} = \dfrac{\sin A + \sin B}{\sin B}$

20. By Exercises 18 and 19, $\dfrac{\frac{a - b}{b}}{\frac{a + b}{b}} = \dfrac{\frac{\sin A - \sin B}{\sin B}}{\frac{\sin A + \sin B}{\sin B}}$ or $\dfrac{a - b}{a + b} = \dfrac{\sin A - \sin B}{\sin A + \sin B}$

21. $\dfrac{a - b}{a + b} = \dfrac{\sin A - \sin B}{\sin A + \sin B} = \dfrac{\frac{\sin A - \sin B}{\cos A + \cos B}}{\frac{\sin A + \sin B}{\cos A + \cos B}} = \dfrac{\tan (\frac{A - B}{2})}{\tan (\frac{A + B}{2})}$

C   22. Area $= \frac{1}{2}bc \sin A = \frac{1}{2}bc(2 \sin \frac{A}{2} \cos \frac{A}{2}) = bc \sin \frac{A}{2} \cos \frac{A}{2}$

23. $(\text{Area})^2 = (bc \sin \frac{A}{2} \cos \frac{A}{2})^2 = b^2c^2 \sin^2 \frac{A}{2} \cos^2 \frac{A}{2} = b^2c^2(\dfrac{1 - \cos A}{2})(\dfrac{1 + \cos A}{2})$

$= b^2c^2\left[\dfrac{(a - b + c)(a + b - c)}{2(2bc)}\right]\left[\dfrac{(b + c + a)(b + c - a)}{2(2bc)}\right] =$

$\dfrac{(a + b + c)(b + c - a)(a + b - c)(a - b + c)}{16}$

24. Area $= \sqrt{\dfrac{(a + b + c)(b + c - a)(a + b - c)(a - b + c)}{16}} =$

$\sqrt{\dfrac{2s\,[2(s - a)][2(s - c)][2(s - b)]}{16}} = \sqrt{s(s - a)(s - b)(s - c)}$

Exercises 5-10 · Page 179

<u>A</u> 1. $\dfrac{\sin B}{7} = \dfrac{\sin 42°}{8}$, $\sin B = \dfrac{7 \sin 42°}{8} \doteq \dfrac{7(0.6691)}{8} \doteq 0.5854$, B $\doteq$ 35°50';

A = 180° - B - C = 180° - 35°50' - 42° = 102°10'; $a = \dfrac{c \sin A}{\sin C} =$

$\dfrac{8 \sin 102°10'}{\sin 42°}$ = 8 sin 77°50' csc 42° $\doteq$ 8(0.9775)(1.494) $\doteq$ 11.68

2. Since a $\le$ b and $\angle$ A is obtuse, no triangle exists.

<u>3.</u> $\dfrac{\sin C}{9} = \dfrac{\sin 35°}{7}$, $\sin C = \dfrac{9 \sin 35°}{7} \doteq \dfrac{9(0.5736)}{7} = 0.7375$, C = 47°30' +

10'$(\dfrac{0.7375 - 0.7373}{0.7392 - 0.7373})$ = 47°30' + 10'$(\dfrac{2}{19})$ $\doteq$ 47°30' + 1' = 47°31';

A = 180° - 35° - 47°31' = 97°29'; $a = \dfrac{b \sin A}{\sin B} = \dfrac{7 \sin 97°29'}{\sin 35°} =$

7 sin 82°31' csc 35° = 7[0.9914 + 0.1(0.9918 - 0.9914)](1.743) =

7[0.9914 + 0.1(0.0004)](1.743) = 7(0.9914)(1.743) $\doteq$ 12.10; or C = 180 - 47°31' =

132°29'; A = 180° - 132°29' - 35° = 12°31'; $a = \dfrac{b \sin A}{\sin B} = \dfrac{7 \sin 12°31'}{\sin 35°} =$

7 sin 12°31' csc 35° = 7[0.2164 + 0.1(0.2193 - 0.2164)](1.743) =

7[0.2164 + 0.1(0.0029)](1.743) = 7(0.2167)(1.743) $\doteq$ 2.64

4. Since C > 90° and c < b, we can see no triangle will exist.

<u>5.</u> $\dfrac{\sin B}{13} = \dfrac{\sin 87°}{15}$, $\sin B = \dfrac{13 \sin 87°}{15} \doteq \dfrac{13(0.9986)}{15} \doteq 0.8655$, B $\doteq$ 59°50' +

10'$(\dfrac{0.8655 - 0.8646}{0.8660 - 0.8646})$ = 59°50' + 10'$(\dfrac{9}{14})$ $\doteq$ 59°50' + 6' = 59°56'; C =

180° - 87° - 59°56' = 33°4'; $c = \dfrac{a \sin C}{\sin A} = \dfrac{15 \sin 33°4'}{\sin 87°}$ = 15 sin 33°4' csc 87° $\doteq$

15[0.5446 + 0.4(0.5471 - 0.5446)](1.001) = 15[0.5446 + 0.4(0.0025)](1.001) =

15(0.5446 + 0.001)(1.001) = 15(0.5456)(1.001) $\doteq$ 8.19

<u>6.</u> $\dfrac{\sin A}{16} = \dfrac{\sin 127°}{26}$, $\sin A = \dfrac{16 \sin 127°}{26} = \dfrac{8 \sin 53°}{13} \doteq \dfrac{8(0.7986)}{13} \doteq 0.4914$,

A $\doteq$ 29°20' + 10'$(\dfrac{0.4914 - 0.4899}{0.4924 - 0.4899})$ = 29°20' + 10'$(\dfrac{15}{25})$ = 29°20' + 6' = 29°26';

B $\doteq$ 180° - 29°26' - 127° = 23°34'; $b = \dfrac{c \sin B}{\sin C} = \dfrac{26 \sin 23°34'}{\sin 127°} =$

26 sin 23°34' csc 53° $\doteq$ 26[0.3987 + 0.4(0.4014 - 0.3987)](1.252) =

26[0.3987 + 0.4(0.0027)](1.252) $\doteq$ 26(0.3987 + 0.0011)(1.252) = 26(0.3998)(1.252) $\doteq$

13.01

7. $\dfrac{6.7 \times 10^7}{\sin 28^\circ} = \dfrac{9.3 \times 10^7}{\sin SVE} = \dfrac{x}{\sin VSE}$; $\sin SVE = \dfrac{9.3 \times 10^7 \sin 28^\circ}{6.7 \times 10^7} \doteq \dfrac{9.3(0.4695)}{6.7} \doteq$

0.6517; $\angle SVE \doteq 40^\circ 40'$ or $180^\circ - 40^\circ 40' = 139^\circ 20'$; $\angle VSE \doteq 180^\circ - 28^\circ - 40^\circ 40' =$

$111^\circ 20'$ or $180^\circ - 28^\circ - 139^\circ 20' = 12^\circ 40'$; $x \doteq \dfrac{6.7 \times 10^7 \sin VSE}{\sin 28^\circ} =$

$6.7 \times 10^7 \sin 12^\circ 40' \csc 28^\circ \doteq 6.7 \times 10^7 (0.2193)(2.130) \doteq 3.13 \times 10^7$ mi. or

$x \doteq 6.7 \times 10^7 \sin 111^\circ 20' \csc 28^\circ = 6.7 \times 10^7 \sin 68^\circ 40' \csc 28^\circ \doteq$

$6.7 \times 10^7 (0.9315)(2.130) \doteq 1.33 \times 10^8$ mi.

8. $\dfrac{152}{\sin V} = \dfrac{120}{\sin 47^\circ}$, $\sin V = \dfrac{152 \sin 47^\circ}{120} \doteq \dfrac{19(0.7314)}{15} \doteq 0.9264$; $V \doteq 67^\circ 50'$;

$\dfrac{x}{\sin (180^\circ - 2V)} = \dfrac{120}{\sin 67^\circ 50'}$, $x = \dfrac{120 \sin (180^\circ - 135^\circ 40')}{\sin 67^\circ 50'} =$

$120 \sin 44^\circ 20' \csc 67^\circ 50' \doteq 120(0.6988)(1.084) \doteq 91$ mi.

B 9. If $a < b \sin A$, then $\dfrac{a}{\sin A} < b$ as $\sin A > 0$. $\dfrac{\sin A}{a} > \dfrac{1}{b}$ and, since $\sin B \leq 1$,

$\dfrac{1}{6} \geq \dfrac{\sin B}{b}$ so that $\dfrac{\sin A}{a} > \dfrac{\sin B}{b}$. This contradicts $\dfrac{\sin A}{a} = \dfrac{\sin B}{b}$.

10. Using $\sin B = \dfrac{b \sin A}{a}$, there are two solutions, B and $180^\circ - B$. Since $a \geq b$ and

$a = \dfrac{b \sin A}{\sin B}$, we have $b \leq \dfrac{b \sin A}{\sin B}$ or $\sin B \leq \sin A$. Arbitrarily select

$B < 180^\circ - B$; then $0^\circ < B \leq A$ and $180^\circ > 180^\circ - B \geq 180^\circ - A$, so $C =$

$180^\circ - A - B$ or $C = 180^\circ - A - (180^\circ - B) = B - A$. But $B - A \leq 0$ and

$C > 0$; $\therefore$ at most one solution exists. Also, $A + B \leq A + A < 90^\circ + 90^\circ = 180^\circ$,

so $C = 180^\circ - A - B > 0$ and at least one solution exists.

Chapter Test · Page 180

1. a. $432^\circ = \dfrac{\pi}{180}(432)^R = \dfrac{12\pi}{5}^R$ b. $(-\dfrac{5}{36}\pi) = \dfrac{180}{\pi}(-\dfrac{5}{36}\pi)^\circ = -25^\circ$

2. a. $\sin 172^\circ = \sin (180^\circ - 8^\circ) = \sin 8^\circ$ b. $\cos 305^\circ = \cos (360 - 55^\circ) = \cos 55^\circ$

3. a. $\sin 750^\circ = \sin (750^\circ - 720^\circ) = \sin 30^\circ = \dfrac{1}{2}$

   b. $\csc (-1020^\circ) = \csc (-1020^\circ + 3 \cdot 360^\circ) = \csc 60^\circ = \dfrac{2}{\sqrt{3}}$

4. $\csc \theta - \cos \theta \cot \theta = \dfrac{1}{\sin \theta} - \dfrac{\cos \theta \cos \theta}{\sin \theta} = \dfrac{1 - \cos^2 \theta}{\sin \theta} = \dfrac{\sin^2 \theta}{\sin \theta} = \sin \theta$

5. $\cos 32^\circ 17' \doteq 0.8465 + 0.7(0.8450 - 0.8465) = 0.8465 + (0.7)(-0.0015) \doteq$

   $0.8465 - 0.0011 = 0.8454$

6. $\cos \theta = \dfrac{3}{\sqrt{(3)^2 + (-7)^2}} = \dfrac{3}{\sqrt{58}} = \dfrac{3\sqrt{58}}{58}$; $\sin \theta = -\dfrac{7}{\sqrt{58}} = \dfrac{-7\sqrt{58}}{58}$

7. $\cos 2\theta = 1 - \sin^2 \theta$, $1 - 2\sin^2 \theta = 1 - \sin^2 \theta$, $\sin^2 \theta = 0$, $\{0, 180°\}$

8. $\cos 35° = \dfrac{x}{9}$, $x = 9 \cos 35°$ (inches)

9. $a^2 = 7^2 + 8^2 - 2(7)(8)\cos 120° = 49 + 64 - 112(-\frac{1}{2}) = 113 + 56 = 169$, $a = 13$.

10. $\dfrac{\sin 30°}{6} = \dfrac{\sin 45°}{b}$, $b = \dfrac{6 \sin 45°}{\sin 30°} = \dfrac{6 \frac{\sqrt{2}}{2}}{\frac{1}{2}} = 6\sqrt{2} \doteq 8$ feet.

11. a. none   b. two   c. none or one.

CHAPTER 6. Vectors

Exercises 6-1 · Pages 190-191

A 1. $\|\vec{u}\| = \sqrt{1^2 + 3^2} = \sqrt{10}$; $\|\vec{v}\| = \sqrt{6^2 + 2^2} = \sqrt{40} = 2\sqrt{10}$

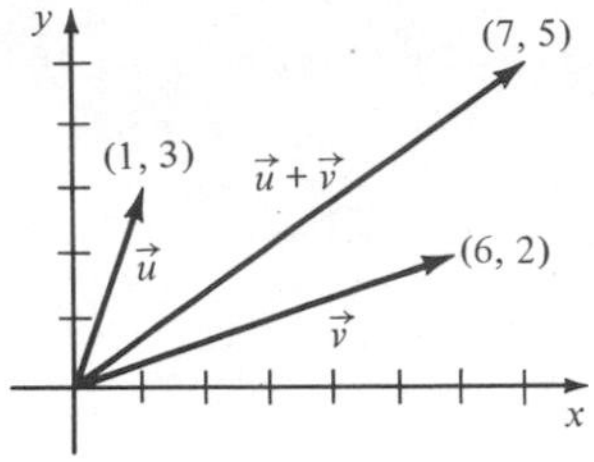

Ex. 1

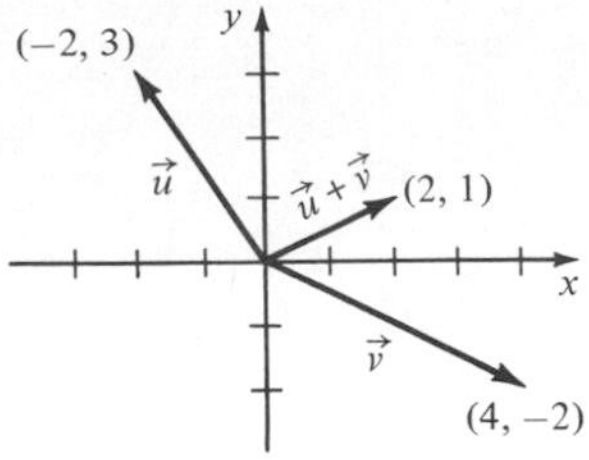

Ex. 2

2. $\|\vec{u}\| = \sqrt{(-2)^2 + 3^2} = \sqrt{13}$; $\|\vec{v}\| = \sqrt{4^2 + (-2)^2} = \sqrt{20} = 2\sqrt{5}$

3. $\|\vec{u}\| = \sqrt{6^2 + 8^2} = \sqrt{100} = 10$; $\|\vec{v}\| = \sqrt{3^2 + (-4)^2} = \sqrt{25} = 5$

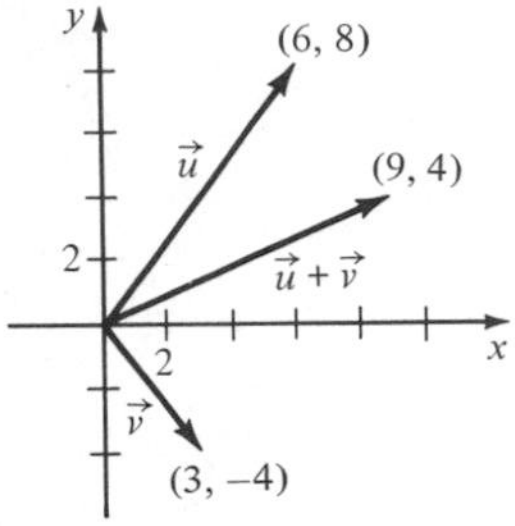

Ex. 3

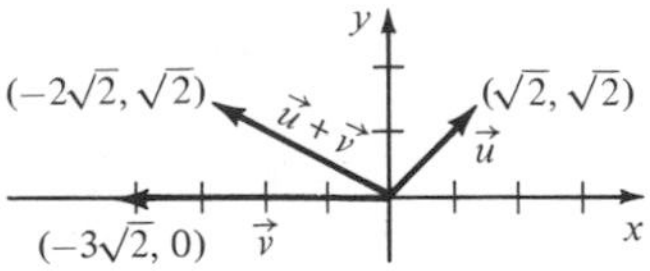

Ex. 4

4. $\|\vec{u}\| = \sqrt{(\sqrt{2})^2 + (\sqrt{2})^2} = \sqrt{4} = 2$; $\|\vec{v}\| = \sqrt{(-3\sqrt{2})^2 + 0^2} = \sqrt{18} = 3\sqrt{2}$

5. $\|\vec{u}\| = \sqrt{(-4)^2 + 1^2} = \sqrt{17}$; $\|\vec{v}\| = \sqrt{1^2 + 4^2} = \sqrt{17}$

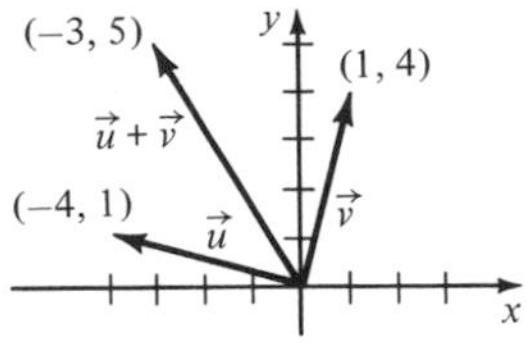

Ex. 5

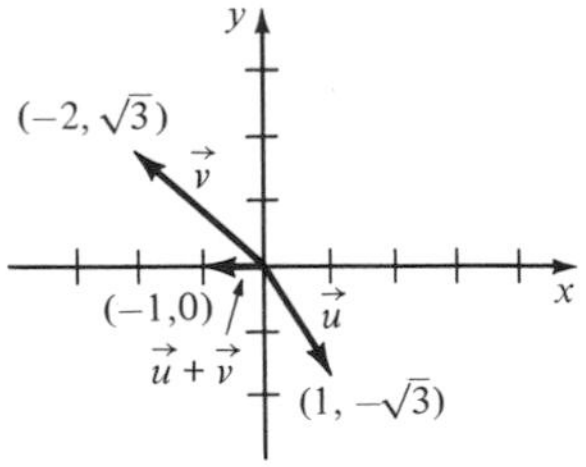

Ex. 6

6. $\|\vec{u}\| = \sqrt{1^2 + (-\sqrt{3})^2} = \sqrt{4} = 2$; $\|\vec{v}\| = \sqrt{(-2)^2 + (\sqrt{3})^2} = \sqrt{7}$

<u>7</u>. $\cos 45° = \dfrac{x}{2\sqrt{2}}$, $x = 2\sqrt{2} \cos 45°$, $x = 2\sqrt{2}(\dfrac{\sqrt{2}}{2}) = 2$; $\sin 45° = \dfrac{y}{2\sqrt{2}}$,

$y = 2\sqrt{2} \sin 45°$, $y = 2\sqrt{2}(\dfrac{\sqrt{2}}{2}) = 2$; the required vector is (2, 2).

<u>8</u>. $\cos 30° = \dfrac{x}{4}$, $x = 4 \cos 30°$, $x = 4(\dfrac{\sqrt{3}}{2}) = 2\sqrt{3}$; $\sin 30° = \dfrac{y}{4}$, $y = 4 \sin 30°$,

$y = 4(\dfrac{1}{2}) = 2$; the required vector is $(2\sqrt{3},\ 2)$.

<u>9</u>. $\cos 180° = \dfrac{x}{3}$, $x = 3 \cos 180°$, $x = 3(-1)$, $x = -3$; $\sin 180° = \dfrac{y}{3}$, $y = 3 \sin 180°$,

$y = 3(0)$, $y = 0$; the required vector is (-3, 0).

<u>10</u>. $\cos 270° = \dfrac{x}{2}$, $x = 2 \cos 270°$, $x = 2(0)$, $x = 0$; $\sin 270° = \dfrac{y}{2}$, $y = 2 \sin 270°$,

$y = 2(-1)$, $y = -2$; the required vector is (0, -2).

<u>11</u>. $\cos 225° = \dfrac{x}{8}$, $x = 8 \cos 225°$, $x = 8(-\dfrac{\sqrt{2}}{2})$, $x = -4\sqrt{2}$; $\sin 225° = \dfrac{y}{8}$,

$y = 8 \sin 225°$, $y = 8(-\dfrac{\sqrt{2}}{2})$, $y = -4\sqrt{2}$; the required vector is $(-4\sqrt{2},\ -4\sqrt{2})$.

<u>12</u>. $\cos 120° = \dfrac{x}{4}$, $x = 4 \cos 120°$, $x = 4(-\dfrac{1}{2})$, $x = -2$; $\sin 120° = \dfrac{y}{4}$, $y = 4 \sin 120°$,

$y = 4(\dfrac{\sqrt{3}}{2})$, $y = 2\sqrt{3}$; the required vector is $(-2,\ 2\sqrt{3})$.

<u>13</u>. $2\vec{u} = 2(-1,\ 5) = (-2,\ 10)$

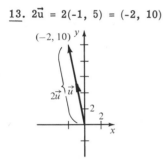

<u>14</u>. $-3\vec{v} = -3(3,\ -2) = (-9,\ 6)$

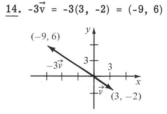

<u>15</u>. $2\vec{u} + (-3)\vec{v} = 2(-1,\ 5) + (-3)(3,\ -2) =$

(-2, 10) + (-9, 6) = (-11, 16)

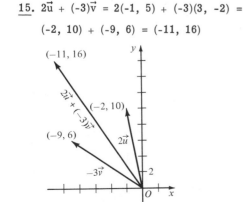

<u>16</u>. $4\vec{u} + 4\vec{v} = 4(-1,\ 5) + 4(3,\ -2) =$

(-4, 20) + (12, -8) = (8, 12)

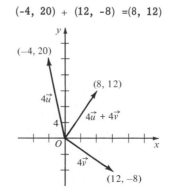

17. $4(\vec{u} + \vec{v}) = 4[(-1, 5) + (3, -2)] =$

$4(2, 3) = (8, 12)$

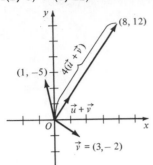

18. $\vec{v} + (-1)\vec{u} = (3, -2) + (-1)(-1, 5) =$

$(3, -2) + (1, -5) = (4, -7)$

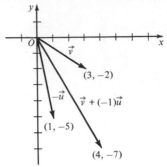

19. $\vec{u} - \vec{v} = \vec{u} + (-\vec{v}) = (-1, 5) + (-3, 2) =$

$(-4, 7)$

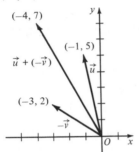

20. $\frac{1}{2}\vec{u} + \frac{2}{3}\vec{v} = \frac{1}{2}(-1, 5) + \frac{2}{3}(3, -2) =$

$(-\frac{1}{2}, \frac{5}{2}) + (2, -\frac{4}{3}) = (\frac{3}{2}, \frac{7}{6})$

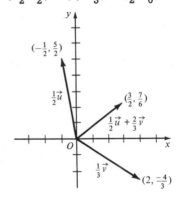

21. $(5, 8) + (x, y) = (0, 0)$, $(5 + x, 8 + y) = (0, 0)$, $5 + x = 0$ and $8 + y = 0$, $x = -5$ and $y = -8$. The required vector is $(-5, -8)$.

22. $(7, 4) + (x, y) = (10, -5)$, $(7 + x, 4 + y) = (10, -5)$, $7 + x = 10$ and $4 + y = -5$, $x = 3$ and $y = -9$. The required vector is $(3, -9)$.

23. $-2(5, -1) + 3(x, y) = (2, -1)$, $(-10, 2) + (3x, 3y) = (2, -1)$, $-10 + 3x = 2$ and $2 + 3y = -1$; $x = 4$ and $y = -1$. The required vector is $(4, -1)$.

24. $-\frac{1}{2}(x, y) = (6, 4)$, $(-\frac{1}{2}x, -\frac{1}{2}y) = (6, 4)$, $-\frac{1}{2}x = 6$ and $-\frac{1}{2}y = 4$, $x = -12$ and $y = -8$. The required vector is $(-12, -8)$.

25. $4[(3, 7) + (x, y)] = (20, 16)$, $4(3 + x, 7 + y) = (20, 16)$, $(12 + 4x, 28 + 4y) = (20, 16)$, $12 + 4x = 20$ and $28 + 4y = 16$, $4x = 8$ and $4y = -12$, $x = 2$ and $y = -3$. The required vector is $(2, -3)$.

<u>26.</u> $3(x, y) + (-1)(x, y) = (6, 10), (3x, 3y) + (-x, -y) = (6, 10), (2x, 2y) = (6, 10),$

$2x = 6$ and $2y = 10$, $x = 3$ and $y = 5$.   The required vector is $(3, 5)$.

<u>27.</u> $5(x, y) + (2, -8) = 3(x, y), (5x, 5y) + (2, -8) = (3x, 3y), (5x + 2, 5y - 8) =$

$(3x, 3y)$, $5x + 2 = 3x$ and $5y - 8 = 3y$, $x = -1$ and $y = 4$.   The required vector

is $(-1, 4)$.

B   <u>28.</u> $\vec{u} + \vec{v} = (a_1, b_1) + (a_2, b_2) = (a_1 + a_2, b_1 + b_2)$.   Since $a_1, a_2, b_1, b_2 \in \mathcal{R}$,

we know $a_1 + a_2 \in \mathcal{R}$ and $b_1 + b_2 \in \mathcal{R}$, so $\vec{u} + \vec{v} \in V$.

<u>29.</u> $(\vec{u} + \vec{v}) + \vec{w} = [(a_1, b_1) + (a_2, b_2)] + (a_3, b_3) = (a_1 + a_2, b_1 + b_2) + (a_3, b_3) =$

$([a_1 + a_2] + a_3, [b_1 + b_2] + b_3) = (a_1 + [a_2 + a_3], b_1 + [b_2 + b_3])$ [by associative

property of addition in $\mathcal{R}$] $= (a_1, b_1) + (a_2 + a_3, b_2 + b_3) = (a_1, b_1) +$

$[(a_2, b_2) + (a_3, b_3)] = \vec{u} + (\vec{v} + \vec{w})$.

<u>30.</u> $\vec{u} + \vec{0} = (a_1, b_1) + (0, 0) = (a_1 + 0, b_1 + 0) = (0 + a_1, 0 + b_1)$ [by commutative

property of addition in $\mathcal{R}$] $= (0, 0) + (a_1, b_1) = \vec{0} + \vec{u} = (0, 0) + (a_1, b_1) =$

$(0 + a_1, 0 + b_1) = (a_1, b_1)$ [by zero property of addition in $\mathcal{R}$] $= \vec{u}$.

<u>31.</u> $u + (-u) = (a_1, b_1) + (-a_1, -b_1) = (a_1 - a_1, b_1 - b_1) = (0, 0)$ [by additive inverse

property in $\mathcal{R}$]. $(-u) + u = (-a_1, -b_1) + (a_1, b_1) = (-a_1 + a_1, -b_1 + b_1) = (0, 0)$.

<u>32.</u> $\vec{u} + \vec{v} = (a_1, b_1) + (a_2, b_2) = (a_1 + a_2, b_1 + b_2) = (a_2 + a_1, b_2 + b_1)$ [by

commutative property of addition in $\mathcal{R}$] $= (a_2, b_2) + (a_1, b_1) = \vec{v} + \vec{u}$.

<u>33.</u> $r\vec{u} = r(a_1, b_1) = (ra_1, rb_1)$.   Since r, a_1, and $b_1 \in \mathcal{R}$, $ra_1, rb_1 \in \mathcal{R}$, so $r\vec{u} \in V$.

<u>34.</u> $r(s\vec{u}) = r[s(a_1, b_1)] = r(sa_1, sb_1) = (r(sa_1), r(sb_1)) = ((rs)a_1, (rs)b_1)$ [by

associative property of multiplication in $\mathcal{R}$] $= (rs)(a_1, b_1) = (rs)\vec{u}$

<u>35.</u> $(r + s)\vec{u} = (r + s)(a_1, b_1) = ((r + s)a_1, (r + s)b_1) = (ra_1 + sa_1, rb_1 + sb_1)$ [by

the distributive property in $\mathcal{R}$] $= (ra_1, rb_1) + (sa_1, sb_1) = r(a_1, b_1) + s(a_1, b_1) =$

$r\vec{u} + s\vec{u}$

<u>36.</u> $r(\vec{u} + \vec{v}) = r((a_1, b_1) + (a_2, b_2)) = r(a_1 + a_2, b_1 + b_2) = (r(a_1 + a_2), r(b_1 + b_2))$

$(ra_1 + ra_2, rb_1 + rb_2)$ [by the distributive property in $\mathcal{R}$] $= (ra_1, rb_1) + (ra_2, rb_2) =$

$r(a_1, b_1) + r(a_2, b_2) = r\vec{u} + r\vec{v}$.

37. $1\vec{u} = 1(a_1, b_1) = (1 \cdot a_1, 1 \cdot b_1) = (a_1, b_1) = \vec{u}$.

38. $-1\vec{u} = -1(a_1, b_1) = (-1 \cdot a_1, -1 \cdot b_1) = (-a_1, -b_1) = -\vec{u}$.

39. $0\vec{u} = 0(a_1, b_1) = (0 \cdot a_1, 0 \cdot b_1) = (0, 0) = \vec{0}$.

40. $r\vec{0} = r(0, 0) = (r \cdot 0, r \cdot 0) = (0, 0) = \vec{0}$.

C   41. $\|\vec{u} + \vec{v}\| = \|(a_1, b_1) + (a_2, b_2)\| = \|(a_1 + a_2, b_1 + b_2)\| = \sqrt{(a_1 + a_2)^2 + (b_1 + b_2)^2} =$

$\sqrt{a_1^2 + b_1^2 + a_2^2 + b_2^2 + 2(a_1 a_2 + b_1 b_2)}$.   We know that $2(a_1 a_2 + b_1 b_2) \le$

$2\sqrt{(a_1 a_2 + b_1 b_2)^2} \le 2\sqrt{(a_1 a_2 + b_1 b_2)^2 + (a_1 b_2 - a_2 b_1)^2} =$

$2\sqrt{a_1^2 a_2^2 + b_1^2 b_2^2 + a_1^2 b_2^2 + a_2^2 b_1^2}$.   Then, $\|u + v\| =$

$\sqrt{a_1^2 + b_1^2 + 2(a_1 a_2 + b_1 b_2) + a_2^2 + b_2^2} \le$

$\sqrt{(a_1^2 + b_1^2) + 2\sqrt{a_1^2 a_2^2 + b_1^2 b_2^2 + a_1^2 b_2^2 + a_2^2 b_1^2} + (a_2^2 + b_2^2)} =$

$\sqrt{(\sqrt{a_1^2 + b_1^2} + \sqrt{a_2^2 + b_2^2})^2} = \sqrt{a_1^2 + b_1^2} + \sqrt{a_2^2 + b_2^2} = \|\vec{u}\| + \|\vec{v}\|$.

Therefore, we have $\|\vec{u} + \vec{v}\| \le \|\vec{u}\| + \|\vec{v}\|$.

42. Let $\vec{v} = (x, y)$.   Since $\|\vec{v}\| \ne 0$, we have $\dfrac{1}{\|\vec{v}\|} = \dfrac{1}{\sqrt{x^2 + y^2}}$, and then

$\dfrac{1}{\|\vec{v}\|} \vec{v} = \dfrac{1}{\sqrt{x^2 + y^2}}(x, y) = (\dfrac{x}{\sqrt{x^2 + y^2}}, \dfrac{y}{\sqrt{x^2 + y}})$.   The norm of $\dfrac{1}{\|\vec{v}\|}\vec{v}$ is

$\left\|(\dfrac{x}{\sqrt{x^2 + y^2}}, \dfrac{y}{\sqrt{x^2 + y^2}})\right\| = \sqrt{\dfrac{x^2 + y^2}{x^2 + y^2}} = \sqrt{1} = 1$.   The direction angle θ of $\vec{v}$ is

given by $\sin \theta = \dfrac{y}{\sqrt{x^2 + y^2}}$ and $\cos \theta = \dfrac{x}{\sqrt{x^2 + y^2}}$.   The direction angle α for

$\dfrac{1}{\|\vec{v}\|}\vec{v}$ is given by $\sin \alpha = \dfrac{\frac{y}{\sqrt{x^2 + y^2}}}{1} = \dfrac{y}{\sqrt{x^2 + y^2}} = \sin \theta$, and $\cos \alpha =$

$\dfrac{\frac{x}{\sqrt{x^2 + y^2}}}{1} = \dfrac{x}{\sqrt{x^2 + y^2}} = \cos \theta$; therefore $\theta = \alpha$.

43. If $\vec{u}$ or $\vec{v} = (0, 0)$, then $x_1 y_2 - x_2 y_1 = 0$ and $\vec{u}$ and $\vec{v}$ can be considered collinear

since $\vec{0}$ may be assigned any convenient direction.   Suppose then that $\vec{u}, \vec{v} \ne \vec{0}$.

If $\vec{u}$ and $\vec{v}$ are collinear, then $\vec{u} = r\vec{v}$, $r \in \mathcal{R}$.   Hence $x_1 = rx_2$, $y_1 = ry_2$, and

$x_1 y_2 - x_2 y_1 = (rx_2)y_2 - x_2(ry_2) = 0$.   Conversely, if $x_1 y_2 - x_2 y_1 = 0$, then

$x_1 y_2 = x_2 y_1$, $\dfrac{x_1}{x_2} = \dfrac{y_1}{y_2}$, and $\dfrac{x_1}{x_2} = \dfrac{y_1}{y_2} = r$ for some real number r.   Then $x_1 = rx_2$,

$y_1 = ry_2$, and $\vec{u} = r\vec{v}$.   Hence $\vec{u}$ and $\vec{v}$ are collinear.

Exercises 6-2 · Pages 196-197

<u>A</u>   1. <u>a</u>. $\vec{u} + \vec{v} = (-4, 1) + (5, 2) = (1, 3) = 1\vec{i} + 3\vec{j}$

   <u>b</u>. $2\vec{u} + (-3)\vec{v} = 2(-4, 1) + (-3)(5, 2) = (-8, 2) + (-15, -6) = (-23, -4) =$
   $-23\vec{i} + (-4)\vec{j}$

2. <u>a</u>. $\vec{u} + \vec{v} = (-6, 4) + (1, 7) = (-5, 11) = -5\vec{i} + 11\vec{j}$

   <u>b</u>. $2\vec{u} + (-3)\vec{v} = 2(-6, 4) + (-3)(1, 7) = (-12, 8) + (-3, -21) = (-15, -13) =$
   $-15\vec{i} + (-13)\vec{j}$

3. <u>a</u>. $\vec{u} + \vec{v} = (0, 8) + (4, 0) = (4, 8) = 4\vec{i} + 8\vec{j}$

   <u>b</u>. $2\vec{u} + (-3)\vec{v} = 2(0, 8) + (-3)(4, 0) = (0, 16) + (-12, 0) = (-12, 16) = -12\vec{i} + 16\vec{j}$

4. <u>a</u>. $\vec{u} + \vec{v} = (-3, 3) + (1, -1) = (-2, 2) = -2\vec{i} + 2\vec{j}$

   <u>b</u>. $2\vec{u} + (-3)\vec{v} = 2(-3, 3) + (-3)(1, -1) = (-6, 6) + (-3, 3) = (-9, 9) = -9\vec{i} + 9\vec{j}$

5. <u>a</u>. $\vec{u} + \vec{v} = (10, 2) + (\frac{1}{3}, -\frac{2}{3}) = (10\frac{1}{3}, 1\frac{1}{3}) = 10\frac{1}{3}\vec{i} + 1\frac{1}{3}\vec{j}$

   <u>b</u>. $2\vec{u} + (-3)\vec{v} = 2(10, 2) + (-3)(\frac{1}{3}, -\frac{2}{3}) = (20, 4) + (-1, 2) = (19, 6) = 19\vec{i} + 6\vec{j}$

6. <u>a</u>. $\vec{v} = \sqrt{3}\,\vec{i} - \vec{j} = (\sqrt{3}, -1)$; $\|\vec{v}\| = \sqrt{(\sqrt{3})^2 + (-1)^2} = 2$; $\vec{u} = \dfrac{1}{\|\vec{v}\|}\vec{v} = \frac{1}{2}(\sqrt{3}, -1) =$
   $(\frac{\sqrt{3}}{2}, -\frac{1}{2})$, so $\vec{v} = 2(\frac{\sqrt{3}}{2}, -\frac{1}{2})$;

   <u>b</u>. $\cos\theta = \frac{\sqrt{3}}{2}$ and $\sin\theta = -\frac{1}{2}$; therefore the direction angle $\theta = 330°$.

7. <u>a</u>. $\vec{v} = -5\vec{i} + 5\vec{j} = (-5, 5)$; $\|\vec{v}\| = \sqrt{(-5)^2 + 5^2} = 5\sqrt{2}$; $\vec{u} = \dfrac{1}{\|\vec{v}\|}\vec{v} = \dfrac{1}{5\sqrt{2}}(-5, 5) =$
   $(\dfrac{-5}{5\sqrt{2}}, \dfrac{5}{5\sqrt{2}}) = (-\dfrac{1}{\sqrt{2}}, \dfrac{1}{\sqrt{2}})$, so $\vec{v} = 5\sqrt{2}(-\dfrac{1}{\sqrt{2}}, \dfrac{1}{\sqrt{2}})$;

   <u>b</u>. $\cos\theta = -\dfrac{1}{\sqrt{2}}$ and $\sin\theta = \dfrac{1}{\sqrt{2}}$; therefore the direction angle $\theta = 135°$.

8. <u>a</u>. $\vec{v} = -3\vec{i} - 4\vec{j} = (-3, -4)$; $\|\vec{v}\| = \sqrt{(-3)^2 + (-4)^2} = 5$; $\vec{u} = \dfrac{1}{\|\vec{v}\|}\vec{v} = \frac{1}{5}(-3, -4) =$
   $(-\frac{3}{5}, -\frac{4}{5})$, so $\vec{v} = 5(-\frac{3}{5}, -\frac{4}{5})$

   <u>b</u>. $\cos\theta = -\frac{3}{5}$ and $\sin\theta = -\frac{4}{5}$; therefore the direction angle $\theta \doteq 233°$.

9. <u>a</u>. $\vec{v} = 7\vec{i} + \vec{j} = (7, 1)$; $\|\vec{v}\| = \sqrt{7^2 + 1^2} = \sqrt{50} = 5\sqrt{2}$; $\vec{u} = \dfrac{1}{\|\vec{v}\|}\vec{v} = \dfrac{1}{5\sqrt{2}}(7, 1) =$
   $(\dfrac{7}{5\sqrt{2}}, \dfrac{1}{5\sqrt{2}})$, so $\vec{v} = 5\sqrt{2}(\dfrac{7}{5\sqrt{2}}, \dfrac{1}{5\sqrt{2}})$

   <u>b</u>. $\cos\theta = \dfrac{7}{5\sqrt{2}} \doteq .9901$ and $\sin\theta = \dfrac{1}{5\sqrt{2}} \doteq .1414$; therefore the direction angle
   $\theta \doteq 8°$.

10. **a.** $\vec{v} = 6\vec{i} + 0\vec{j} = (6, 0)$; $\|\vec{v}\| = \sqrt{6^2 + 0^2} = 6$; $\vec{u} = \dfrac{1}{\|\vec{v}\|}\vec{v} = \dfrac{1}{6}(6, 0) = (1, 0)$, so

$\vec{v} = 6(1, 0)$

b. $\cos\theta = 1$ and $\sin\theta = 0$; therefore the direction angle $\theta = 0°$.

11. **a.** $\vec{v} = \sqrt{3}\,i + \sqrt{3}\,j = (\sqrt{3}, \sqrt{3})$; $\|\vec{v}\| = \sqrt{(\sqrt{3})^2 + (\sqrt{3})^2} = \sqrt{6}$; $\vec{u} = \dfrac{1}{\|\vec{v}\|}\vec{v} =$

$\dfrac{1}{\sqrt{6}}(\sqrt{3}, \sqrt{3}) = (\dfrac{1}{\sqrt{2}}, \dfrac{1}{\sqrt{2}})$, so $\vec{v} = \sqrt{6}(\dfrac{1}{\sqrt{2}}, \dfrac{1}{\sqrt{2}})$

b. $\cos\theta = \dfrac{1}{\sqrt{2}}$ and $\sin\theta = \dfrac{1}{\sqrt{2}}$; therefore the direction angle $\theta = 45°$.

12. Let $\vec{v} = \vec{i} + \vec{j} = (1, 1)$; $\|\vec{v}\| = \sqrt{1^2 + 1^2} = \sqrt{2}$; $\vec{u} = \dfrac{1}{\|\vec{v}\|}\vec{v} =$

$\dfrac{1}{\sqrt{2}}(1, 1) = (\dfrac{1}{\sqrt{2}}, \dfrac{1}{\sqrt{2}})$

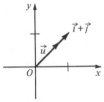   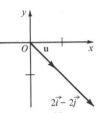

Ex. 12   Ex. 13

13. Let $\vec{v} = 2\vec{i} - 2\vec{j} = (2, -2)$; $\|\vec{v}\| = \sqrt{2^2 + (-2)^2} = 2\sqrt{2}$; $\vec{u} = \dfrac{1}{\|\vec{v}\|}\vec{v} = \dfrac{1}{2\sqrt{2}}(2, -2) =$

$(\dfrac{1}{\sqrt{2}}, -\dfrac{1}{\sqrt{2}})$

14. Let $\vec{v} = -3\vec{i} + 9\vec{j} = (-3, 9)$; $\|\vec{v}\| = \sqrt{(-3)^2 + 9^2} = 3\sqrt{10}$; $\vec{u} = \dfrac{1}{\|\vec{v}\|}\vec{v} = \dfrac{1}{3\sqrt{10}}(-3, 9)$

$= (-\dfrac{1}{\sqrt{10}}, \dfrac{3}{\sqrt{10}})$

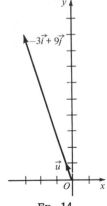   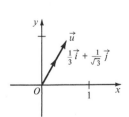

Ex. 14   Ex. 15

15. Let $\vec{v} = \dfrac{1}{3}\vec{i} + \dfrac{1}{\sqrt{3}}\vec{j} = (\dfrac{1}{3}, \dfrac{1}{\sqrt{3}})$; $\|\vec{v}\| = \sqrt{(\dfrac{1}{3})^2 + (\dfrac{1}{\sqrt{3}})^2} = \dfrac{2}{3}$; $\vec{u} = \dfrac{1}{\|\vec{v}\|}\vec{v} =$

$\dfrac{3}{2}(\dfrac{1}{3}, \dfrac{1}{\sqrt{3}}) = (\dfrac{1}{2}, \dfrac{\sqrt{3}}{2\sqrt{3}})$

16. Let $\vec{v} = \|\vec{v}\|$ (cos θ, sin θ) = 3(cos 60°, sin 60°); so $\vec{u}$ = (cos 60°, sin 60°) =

$(\frac{1}{2}, \frac{\sqrt{3}}{2})$.

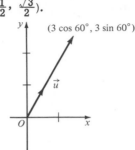

Ex. 16

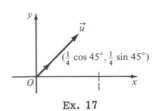

Ex. 17

17. Let $\vec{v} = \|\vec{v}\|$ (cos θ, sin θ) = $\frac{1}{4}$(cos 45°, sin 45°); $\vec{u}$ = (cos 45°, sin 45°) =

$(\frac{\sqrt{2}}{2}, \frac{\sqrt{2}}{2})$

18. cos θ = $\frac{x}{\|\vec{v}\|}$; cos 60° = $\frac{x}{3}$, x = 3 cos 60° = $\frac{3}{2}$; sin θ = $\frac{y}{\|\vec{v}\|}$, sin 60° = $\frac{y}{3}$,

y = 3 sin 60° = $\frac{3\sqrt{3}}{2}$; $\vec{v} = x\vec{i} + y\vec{j} = \frac{3}{2}\vec{i} + \frac{3\sqrt{3}}{2}\vec{j}$.

19. cos 90° = $\frac{x}{7}$, x = 7 cos 90° = 0; sin 90° = $\frac{y}{7}$, y = 7 sin 90° = 7;

$\vec{v} = x\vec{i} + 0\vec{i} = 7\vec{j}$.

20. cos 150° = $\frac{x}{\frac{1}{2}}$; x = $\frac{1}{2}$ cos 150°, x = $-\frac{\sqrt{3}}{4}$; sin 150° = $\frac{y}{\frac{1}{2}}$; y = $\frac{1}{2}$ sin 150°,

y = $\frac{1}{4}$; $\vec{v} = x\vec{i} + y\vec{j} = -\frac{\sqrt{3}}{4}i + \frac{1}{4}j$

21. cos 160° = $\frac{x}{10}$, x = 10 cos 160°, x = -9.397; sin 160° = $\frac{y}{10}$, y = 10 sin 160° =

3.420; $\vec{v} = x\vec{i} + y\vec{j}$ = -9.397i + 3.420j

22. cos 112° = $\frac{x}{\frac{3}{4}}$, x = $\frac{3}{4}$ cos 112° = $\frac{3}{4}$(-.3746) = -0.28095; sin 112° = $\frac{y}{\frac{3}{4}}$,

y = $\frac{3}{4}$ sin 112° = $\frac{3}{4}$(.9272) = 0.6954; $\vec{v} = x\vec{i} + y\vec{j}$ = -0.28095$\vec{i}$ + 0.6954$\vec{j}$

B 23. $\vec{k}$ = (cos 30°)$\vec{i}$ + (sin 30°)$\vec{j}$ = $\frac{\sqrt{3}}{2}\vec{i} + \frac{1}{2}\vec{j}$; $\vec{m}$ = (cos 120°)$\vec{i}$ + (sin 120°)$\vec{j}$ = $-\frac{1}{2}\vec{i} + \frac{\sqrt{3}}{2}\vec{j}$;

then r = $\dfrac{(1)(\frac{\sqrt{3}}{2}) - (-\frac{1}{2})(-1)}{(\frac{\sqrt{3}}{2})(\frac{\sqrt{3}}{2}) - (-\frac{1}{2})(\frac{1}{2})}$ = $\dfrac{\frac{\sqrt{3}}{2} - \frac{1}{2}}{1}$ = $\dfrac{\sqrt{3} - 1}{2}$; and s =

$\dfrac{(\frac{\sqrt{3}}{2})(-1) - (1)(\frac{1}{2})}{1}$ = $\dfrac{-\sqrt{3} - 1}{2}$. Thus $\vec{i} - \vec{j} = r\vec{k} + s\vec{m} = (\frac{\sqrt{3} - 1}{2})\vec{k} - (\frac{\sqrt{3} + 1}{2})\vec{m}$

24. As shown in Exercise 23, $\vec{k} = \frac{\sqrt{3}}{2}\vec{i} + \frac{1}{2}\vec{j}$; $\vec{m} = -\frac{1}{2}\vec{i} + \frac{\sqrt{3}}{2}\vec{j}$.   Then,

$$r = \frac{(2)(\frac{\sqrt{3}}{2}) - (-\frac{1}{2})(2)}{(\frac{\sqrt{3}}{2})(\frac{\sqrt{3}}{2}) - (-\frac{1}{2})(\frac{1}{2})} = \frac{\sqrt{3} + 1}{1}; \quad \text{and } s = \frac{(\frac{\sqrt{3}}{2})(2) - (2)(\frac{1}{2})}{1} = \frac{\sqrt{3} - 1}{1}.$$

Thus $2\vec{i} + 2\vec{j} = r\vec{k} + s\vec{m} = (\sqrt{3} + 1)\vec{k} + (\sqrt{3} - 1)\vec{m}$.

25. $\vec{k} = (\cos 60°)\vec{i} + (\sin 60°)\vec{j} = \frac{1}{2}\vec{i} + \frac{\sqrt{3}}{2}\vec{j}$; $\vec{m} = (\cos 150°)\vec{i} + (\sin 150°)\vec{j} =$

$-\frac{\sqrt{3}}{2}\vec{i} + \frac{1}{2}\vec{j}$; then $r = \frac{(3)(\frac{1}{2}) - (-\frac{\sqrt{3}}{2})(1)}{(\frac{1}{2})(\frac{1}{2}) - (-\frac{\sqrt{3}}{2})(\frac{\sqrt{3}}{2})} = \frac{\frac{3}{2} + \frac{\sqrt{3}}{2}}{1} = \frac{3 + \sqrt{3}}{2}$; and

$s = \frac{(\frac{1}{2})(1) - (3)(\frac{\sqrt{3}}{2})}{1} = \frac{1 - 3\sqrt{3}}{2}$.   Thus, $3\vec{i} + \vec{j} = r\vec{k} + s\vec{m} =$

$(\frac{3 + \sqrt{3}}{2})\vec{k} + (\frac{1 - 3\sqrt{3}}{2})\vec{m}$.

26. As shown in Exercise 25, $\vec{k} = \frac{1}{2}\vec{i} + \frac{\sqrt{3}}{2}\vec{j}$; $\vec{m} = -\frac{\sqrt{3}}{2}\vec{i} + \frac{1}{2}\vec{j}$.   Then

$$r = \frac{(4)(\frac{1}{2}) - (-\frac{\sqrt{3}}{2})(-2)}{1} = 2 - \sqrt{3}; \text{ and } s = \frac{(\frac{1}{2})(-2) - (4)(\frac{\sqrt{3}}{2})}{1} = -1 - 2\sqrt{3};$$

$4\vec{i} - 2\vec{j} = r\vec{k} + s\vec{m} = (2 - \sqrt{3})\vec{k} - (1 + 2\sqrt{3})\vec{m}$.

27. $\vec{k} = (\cos 90°)\vec{i} + (\sin 90°)\vec{j} = 0\vec{i} + \vec{j} = \vec{j}$; $\vec{m} = (\cos 180°)\vec{i} + (\sin 180°)\vec{j} =$

$-\vec{i} + 0\vec{j} = -\vec{i}$; and so $r = \frac{(1)(0) - (-1)(1)}{(0)(0) - (-1)(1)} = \frac{0 + 1}{1} = 1$; and $s = \frac{(0)(1) - (1)(1)}{1} =$

$-\frac{1}{1} = -1$.   Thus $\vec{i} + \vec{j} = r\vec{k} + s\vec{m} = \vec{k} - \vec{m}$.

28. As shown in Exercise 27, $\vec{k} = \vec{j}$, and $\vec{m} = -\vec{i}$.   Then $r = \frac{5(0) - (-1)(-2)}{1} = -2$;

and $s = \frac{0(-2) - 5(1)}{1} = -5$.   Thus $5\vec{i} - 2\vec{j} = r\vec{k} + s\vec{m} = -2\vec{k} - 5\vec{m}$.

C   29. Let $\vec{v} = (x, y)$, $r \in R$, and $r \geq 1$.   Then if θ is the direction angle of $\vec{v}$,

$\sin \theta = \frac{y}{\sqrt{x^2 + y^2}}$ and $\cos \theta = \frac{x}{\sqrt{x^2 + y^2}}$.   Let α be the direction angle of

$\vec{v} + r\vec{v} = (x, y) + r(x, y) = ((1 + r)x, (1 + r)y)$.   Then $\sin \alpha =$

$\frac{(1 + r)y}{\sqrt{(1 + r)^2x^2 + (1 + r)^2y^2}} = \frac{(1 + r)y}{\sqrt{(1 + r)^2(x^2 + y^2)}} = \frac{y}{\sqrt{x^2 + y^2}} = \sin \theta$;

and $\cos \alpha = \frac{(1 + r)x}{\sqrt{(1 + r)^2x^2 + (1 + r^2)y^2}} = \frac{(1 + r)x}{\sqrt{(1 + r)^2(x^2 + y^2)}} = \frac{x}{\sqrt{x^2 + y^2}} =$

$\cos \theta$.   Since $\sin \alpha = \sin \theta$ and $\cos \alpha = \theta$, we know $\alpha = \theta$, and so $\vec{v} + r\vec{v}$

has the same direction as $\vec{v}$.

30. Let $\vec{u} = x_1\vec{i} + y_1\vec{j}$ and $\vec{v} = x_2\vec{i} + y_2\vec{j}$.   Let θ_1 and θ_2 be the direction angles of $\vec{u}$ and v, respectively.   Then $\cos\theta_1 = \dfrac{x_1}{\sqrt{x_1^2 + y_1^2}}$ and $\sin\theta_1 = \dfrac{y_1}{\sqrt{x_1^2 + y_1^2}}$, $\cos\theta_2 = \dfrac{x_2}{\sqrt{x_2^2 + y_2^2}}$, and $\sin\theta_2 = \dfrac{y_2}{\sqrt{x_2^2 + y_2^2}}$.   We know $\theta_2 = \theta_1 \pm 90°$, whenever both $\cos\theta_2 = \mp \sin\theta_1$ and $\sin\theta_2 = \pm \cos\theta_1$ (reading either the top sign or bottom sign in both cases).   Then $\cos^2\theta_2 = \sin^2\theta_1$, $\dfrac{x_2^2}{x_2^2 + y_2^2} = \dfrac{y_1^2}{x_1^2 + y_1^2}$, or $x_2^2(x_1^2 + y_1^2) = y_1^2(x_2^2 + y_2^2)$, $x_1^2 x_2^2 = y_1^2 y_2^2$, $x_1 x_2 = \pm y_1 y_2$.

We know that $x_1 x_2 \neq y_1 y_2$, since we want $\vec{u}$ and $\vec{v}$ to be orthogonal.   Thus we have $x_1 x_2 = -y_1 y_2$, or $x_1 x_2 + y_1 y_2 = 0$; so whenever $x_1 x_2 + y_1 y_2 = 0$, $\vec{u}$ and $\vec{v}$ are orthogonal.

Exercises 6-3 · Page 203

A   1. $\vec{u} \cdot \vec{v} = (0, 1) \cdot (0, 1) = (0)(0) + (1)(1) = 1$

2. $\vec{u} \cdot \vec{v} = (1, 0) \cdot (1, 0) = (1)(1) + (0)(0) = 1$

3. $\vec{u} \cdot \vec{v} = (1, 1) \cdot (1, -1) = (1)(1) + (1)(-1) = 0$

4. $\vec{u} \cdot \vec{v} = (1, 0) \cdot (0, -1) = (1)(0) + (0)(-1) = 0$

5. $\vec{u} \cdot \vec{v} = (5, 2) \cdot (7, -3) = (5)(7) + (2)(-3) = 29$

6. $\vec{u} \cdot \vec{v} = (2, -3) \cdot (5, 7) = (2)(5) + (-3)(7) = -11$

7. $\vec{u} \cdot \vec{v} = (2, 3) \cdot (4, 6) = (2)(4) + (3)(6) = 26$

8. $\vec{u} \cdot \vec{v} = (k_1, k_1) \cdot (k_2, -k_2) \cdot (k_1)(k_2) + (k_1)(-k_2) = 0$

9. $\vec{u} \cdot \vec{v} = (\sqrt{2}, -\sqrt{2}) \cdot (\frac{\sqrt{3}}{2}, \frac{1}{2}) = \dfrac{\sqrt{6} - \sqrt{2}}{2}$; $\|\vec{u}\| = \sqrt{(\sqrt{2})^2 + (-\sqrt{2})^2} = 2$; $\|v\| = \sqrt{(\frac{\sqrt{3}}{2})^2 + (\frac{1}{2})^2} = 1$; $\cos\alpha = \dfrac{\vec{u} \cdot \vec{v}}{\|\vec{u}\|\,\|\vec{v}\|} = \dfrac{\sqrt{6} - \sqrt{2}}{2(2)} = \dfrac{\sqrt{2}(\sqrt{3} - 1)}{4}$.

Since $(\sqrt{3} - 1)^2 = 4 - 2\sqrt{3}$, we can rewrite the last expression as $\sqrt{\dfrac{2(4 - 2\sqrt{3})}{16}} = \sqrt{\dfrac{1 - \frac{\sqrt{3}}{2}}{2}} = \sqrt{\dfrac{1 + \cos 150°}{2}} = \cos 75°$; $\alpha = 75°$

10. $\vec{u} \cdot \vec{v} = (\frac{1}{2}, \frac{\sqrt{3}}{2}) \cdot (\frac{1}{2}, \frac{1}{2}) = \dfrac{1 + \sqrt{3}}{4}$; $\|\vec{u}\| = \sqrt{(\frac{1}{2})^2 + (\frac{\sqrt{3}}{2})^2} = 1$; $\|\vec{v}\| = \sqrt{(\frac{1}{2})^2 + (\frac{1}{2})^2} = \dfrac{1}{\sqrt{2}}$; so $\cos\alpha = \dfrac{\frac{1 + \sqrt{3}}{4}}{\frac{1}{\sqrt{2}}} = \dfrac{\sqrt{2}(1 + \sqrt{3})}{4}$.   Since $(1 + \sqrt{3})^2 =$

$4 + 2\sqrt{3}$, we can rewrite the last expression as $\sqrt{\dfrac{2(4 + 2\sqrt{3})}{16}} = \sqrt{\dfrac{1 + \frac{\sqrt{3}}{2}}{2}} =$

$\sqrt{\dfrac{1 + \cos 30°}{2}} = \cos 15°; \ \alpha = 15°.$

11. $\vec{u} \cdot \vec{v} = (-1, 0) \cdot (-2\sqrt{2}, 2\sqrt{2}) = 2\sqrt{2}; \ \|\vec{u}\| = \sqrt{(-1)^2 + 0^2} = 1; \ \|\vec{v}\| =$

$\sqrt{(-2\sqrt{2})^2 + (2\sqrt{2})^2} = 4; \ \cos \alpha = \dfrac{2\sqrt{2}}{4} = \dfrac{\sqrt{2}}{2}; \ \alpha = 45°$

12. $\vec{u} \cdot \vec{v} = (3, 2) \cdot (-1, 4) = 5; \ \|\vec{u}\| = \sqrt{3^2 + 2^2} = \sqrt{13}; \ \|\vec{v}\| = \sqrt{(-1)^2 + (4)^2} = \sqrt{17};$

$\cos \alpha = \dfrac{5}{\sqrt{13}\,\sqrt{17}} = \dfrac{5}{\sqrt{221}} = \dfrac{5}{14.9} \doteq 0.335; \ \alpha \doteq 70°20'$

13. $\vec{u} \cdot \vec{v} = (0, 1) \cdot (\tfrac{1}{2}, -\tfrac{\sqrt{3}}{2}) = -\dfrac{\sqrt{3}}{2}; \ \|\vec{u}\| = \sqrt{0^2 + 1^2} = 1; \ \|\vec{v}\| = \sqrt{(\tfrac{1}{2})^2 + (-\tfrac{\sqrt{3}}{2})^2}$

$= 1; \ \cos \alpha = \dfrac{-\frac{\sqrt{3}}{2}}{1} = -\dfrac{\sqrt{3}}{2}; \ \alpha = 150°$

14. $\vec{u} \cdot \vec{v} = (-3, -2) \cdot (2, -3) = 0, \ \|\vec{u}\| = \sqrt{(-3)^2 + (-2)^2} = \sqrt{13}; \ \|\vec{v}\| = \sqrt{2^2 + (-3)^2} =$

$\sqrt{13}, \ \cos \alpha = \dfrac{0}{\sqrt{13}\,\sqrt{13}} = 0; \ \alpha = 90°$

15. Since $\vec{u} \cdot \vec{v} = (\tfrac{\sqrt{3}}{2}, \tfrac{1}{2}) \cdot (-\tfrac{1}{2}, \tfrac{\sqrt{3}}{2}) = 0$, the vectors $\vec{u}$ and $\vec{v}$ are orthogonal.

Since $\|\vec{u}\|^2 = (\tfrac{\sqrt{3}}{2})^2 + (\tfrac{1}{2})^2 = 1$ and $\|\vec{v}\|^2 = (-\tfrac{1}{2})^2 + (\tfrac{\sqrt{3}}{2})^2 = 1, \ \vec{u}$ and $\vec{v}$ are also

unit vectors. Let $\vec{a} = 2\vec{i} + 0\vec{j}$; using the equation $\vec{a} = (\vec{u} \cdot \vec{a})\vec{u} + (\vec{v} \cdot \vec{a})\vec{v}$, the

result is $\vec{a} = (\tfrac{\sqrt{3}}{2}, \tfrac{1}{2}) \cdot (2, 0)\vec{u} + (-\tfrac{1}{2}, \tfrac{\sqrt{3}}{2}) \cdot (2, 0)\vec{v} = \sqrt{3}\,\vec{u} - \vec{v}$

16. From Exercise 15, we know that u and v are orthogonal, unit vectors. Letting

$\vec{a} = 0 - 2\vec{j}$ and using the equation $\vec{a} = (\vec{u} \cdot \vec{a})\vec{u} + (\vec{v} \cdot \vec{a})\vec{v} = (\tfrac{\sqrt{3}}{2}, \tfrac{1}{2}) \cdot (0, -2)\vec{u} +$

$(-\tfrac{1}{2}, \tfrac{\sqrt{3}}{2}) \cdot (0, -2)\vec{v} = -\vec{u} - \sqrt{3}\,\vec{v}.$

In Exercises 17-20: From Exercise 15, we know that $\vec{u}$ and $\vec{v}$ are orthogonal, unit vectors.

17. Let $\vec{a} = \tfrac{1}{2}\vec{i} + \tfrac{\sqrt{3}}{2}\vec{j} = (\tfrac{\sqrt{3}}{2}, \tfrac{1}{2}) \cdot (\tfrac{1}{2}, \tfrac{\sqrt{3}}{2})\vec{u} + (-\tfrac{1}{2}, \tfrac{\sqrt{3}}{2}) \cdot (\tfrac{1}{2}, \tfrac{\sqrt{3}}{2})\vec{v} = \tfrac{\sqrt{3}}{2}\vec{u} + \tfrac{1}{2}\vec{v}$

18. Let $\vec{a} = \tfrac{1}{2}\vec{i} - \tfrac{\sqrt{3}}{2}\vec{j} = (\tfrac{\sqrt{3}}{2}, \tfrac{1}{2}) \cdot (\tfrac{1}{2}, -\tfrac{\sqrt{3}}{2})\vec{u} + (-\tfrac{1}{2}, \tfrac{\sqrt{3}}{2}) \cdot (\tfrac{1}{2}, -\tfrac{\sqrt{3}}{2})\vec{v} = 0\vec{u} - \vec{v} = -\vec{v}.$

19. Let $\vec{a} = -\vec{i} + \vec{j} = (\tfrac{\sqrt{3}}{2}, \tfrac{1}{2}) \cdot (-1, 1)\vec{u} + (-\tfrac{1}{2}, \tfrac{\sqrt{3}}{2}) \cdot (-1, 1)\vec{v} = (\dfrac{1 - \sqrt{3}}{2})\vec{u} +$

$(\dfrac{1 + \sqrt{3}}{2})\vec{v}.$

20. Let $\vec{a} = \vec{i} - \vec{j} = (\frac{\sqrt{3}}{2}, \frac{1}{2}) \cdot (1, -1)\vec{u} + (-\frac{1}{2}, \frac{\sqrt{3}}{2})(1, -1)\vec{v} = (\frac{\sqrt{3} - 1}{2})\vec{u} - (\frac{1 + \sqrt{3}}{2})\vec{v}.$

B 21. $\vec{u} \cdot \vec{v} = (a_1, a_2) \cdot (b_1, b_2) = a_1 b_1 + a_2 b_2 = b_1 a_1 + b_2 a_2$ [by the commutative

property of multiplication in $\mathcal{R}$] $= (b_1, b_2) \cdot (a_1, a_2) = \vec{v} \cdot \vec{u}$

22. $r(\vec{u} \cdot \vec{v}) = r[(a_1, a_2) \cdot (b_1, b_2)] = r(a_1 b_1 + a_2 b_2) = r(a_1 b_1) + r(a_2 b_2) =$

$(ra_1)b_1 + (ra_2)b$ [by the associative property of multiplication in $\mathcal{R}$] $=$

$(ra_1, ra_2) \cdot (b_1, b_2) = (r(a_1, a_2)) \cdot (b_1, b_2) = (r\vec{u}) \cdot \vec{v}$

23. $(r\vec{u}) \cdot (s\vec{v}) = (ra_1, ra_2) \cdot (sb_1, sb_2) = (ra_1)(sb_1) + (ra_2)(sb_2) = r(a_1 s)b_1 + r(a_2 s)b_2$

[by the associative property of multiplication in $\mathcal{R}$] $= r(sa_1)b_1 + r(sa_2)b_2$ [by the

commutative property of multiplication in $\mathcal{R}$] $= (rs)a_1 b_1 + (rs)a_2 b_2 =$

$((rs)a_1)b_1 + ((rs)a_2)b_2 = ((rs)a_1, (rs)a_2) \cdot (b_1, b_2) = (rs)(a_1, a_2) \cdot (b_1, b_2) =$

$(rs)\vec{u} \cdot \vec{v}.$

24. $\vec{u} \cdot (\vec{v} + \vec{t}) = (a_1, a_2) \cdot [(b_1, b_2) + (c_1, c_2)] = (a_1, a_2) \cdot (b_1 + c_1, b_2 + c_2) =$

$a_1(b_1 + c_1) + a_2(b_2 + c_2) = a_1 b_1 + a_1 c_1 + a_2 b_2 + a_2 c_2$ [by the distributive property

in $\mathcal{R}$] $= a_1 b_1 + a_2 b_2 + a_1 c_1 + a_2 c_2 = (a_1 b_1 + a_2 b_2) + (a_1 c_1 + a_2 c_2) =$

$(a_1, a_2) \cdot (b_1, b_2) + (a_1, a_2) \cdot (c_1, c_2) = \vec{u} \cdot \vec{v} + \vec{u} \cdot \vec{t}$

25. $(\vec{u} + \vec{v}) \cdot \vec{t} = [(a_1, a_2) + (b_1, b_2)] \cdot (c_1, c_2) = (a_1 + b_1, a_2 + b_2) \cdot (c_1, c_2) =$

$(a_1 + b_1)c_1 + (a_2 + b_2)c_2 = a_1 c_1 + b_1 c_1 + a_2 c_2 + b_2 c_2$ [by the distributive property

in $\mathcal{R}$] $= (a_1 c_1 + a_2 c_2) + (b_1 c_1 + b_2 c_2) = (a_1, a_2) \cdot (c_1, c_2) + (b_1, b_2) \cdot (c_1, c_2) =$

$\vec{u} \cdot \vec{t} + \vec{v} \cdot \vec{t}.$

26. $(\vec{u} + \vec{v}) \cdot (\vec{u} + \vec{v}) = [(a_1, a_2) + (b_1, b_2)] \cdot [(a_1, a_2) + (b_1, b_2)] =$

$(a_1 + b_1, a_2 + b_2) \cdot (a_1 + b_1, a_2 + b_2) = (a_1 + b_1)^2 + (a_2 + b_2)^2 =$

$a_1^2 + 2a_1 b_1 + b_1^2 + a_2^2 + 2a_2 b_2 + b_2^2 = (a_1^2 + a_2^2) + 2(a_1 b_1 + a_2 b_2) + (b_1^2 + b_2^2)$

$= (a_1, a_2) \cdot (a_1, a_2) + 2(a_1, a_2) \cdot (b_1, b_2) + (b_1, b_2) \cdot (b_1, b_2) = \vec{u} \cdot \vec{v} +$

$2\vec{u} \cdot \vec{v} + \vec{v} \cdot \vec{v}$

27. $(\vec{u} + \vec{v}) \cdot (\vec{u} - \vec{v}) = (a_1 + b_1, a_2 + b_2) \cdot (a_1 - b_1, a_2 - b_2) = (a_1 + b_1)(a_1 - b_1) +$

$(a_2 + b_2)(a_2 - b_2) = a_1^2 - b_1^2 + a_2^2 - b_2^2 = a_1^2 + a_2^2 - (b_1^2 + b_2^2) =$

$(a_1, a_2) \cdot (a_1, a_2) - (b_1, b_2) \cdot (b_1, b_2) = \vec{u} \cdot \vec{u} - \vec{v} \cdot \vec{v}.$

28. $\|\vec{u} + \vec{v}\|^2 = \|(a_1 + b_1, a_2 + b_2)\|^2 = (a_1 + b_1)^2 + (a_2 + b_2)^2 =$

$(a_1 + b_1)(a_1 + b_1) + (a_2 + b_2)(a_2 + b_2) = (a_1 + b_1, a_2 + b_2) \cdot (a_1 + b_1, a_2 + b_2) =$

$(\vec{u} + \vec{v}) \cdot (\vec{u} + \vec{v}).$

29. $\|\vec{u} + \vec{v}\|^2 = (\vec{u} + \vec{v}) \cdot (\vec{u} + \vec{v})$ by Exercise 28, and $(\vec{u} + \vec{v}) \cdot (\vec{u} + \vec{v}) =$

$\vec{u} \cdot \vec{u} + 2\vec{u} \cdot \vec{v} + \vec{v} \cdot \vec{v}$ (by Exercise 26) $= \|\vec{u}\|^2 + 2\vec{u} \cdot \vec{v} + \|\vec{v}\|^2.$

30. $(\vec{u} \cdot \vec{v})^2 = (\|\vec{u}\| \|\vec{v}\| \cos \alpha)^2 = \|\vec{u}\|^2 \|\vec{v}\|^2 \cos^2 \alpha = (\vec{u} \cdot \vec{u})(\vec{v} \cdot \vec{v}) \cos^2 \alpha \le$

$(\vec{u} \cdot \vec{u})(\vec{v} \cdot \vec{v}).$

Exercises 6-4 · Pages 209-211

1. x $\doteq$ 136 miles, $\theta \doteq 8°$, bearing $\doteq 98°$

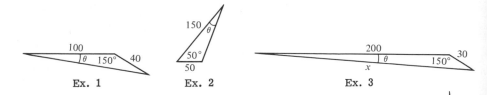

Ex. 1   Ex. 2   Ex. 3

2. x $\doteq$ 124 miles, $\theta \doteq 18°$, bearing $\doteq 202°$

3. Ground speed = x $\doteq$ 226 miles per hour, $\theta \doteq 4°$, true course $\doteq 94°$

4. Ground speed = x $\doteq$ 266 miles per hour, $\theta \doteq 6°$, true course $\doteq 174°$

5. Air speed = x $\doteq$ 204 miles per hour, $\theta \doteq 11°$, heading $\doteq 79°$

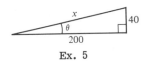
Ex. 5

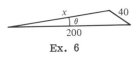
Ex. 6

Ex. 4

6. Air speed = x $\doteq$ 169 miles per hour, $\theta = 8°$, heading $\doteq 82°$

7. $\dfrac{x}{\sin 15°} = \dfrac{20}{\sin 135°}$, x = 20 sin 15° csc 135° =

20 sin 15° csc 45° $\doteq$ 20(0.2588)(1.414) $\doteq$ 7 knots

8. $\dfrac{x}{\sin 30°} = \dfrac{300}{\sin 95°}$, x = 300 sin 30° csc 95° = 300 sin 30° csc 85° $\doteq$

300(0.500)(1.004) = 151 miles per hour

<u>9.</u> $x^2 = 40^2 + 48^2 - 2(40)(48) \cos 60° = 1600 + 2304 - 1920 = 1984$, $x = 8\sqrt{31}$,

$x \doteq 45$ nautical miles; $\gamma + \theta = 80°$. $\dfrac{\sin \gamma}{48} = \dfrac{\sin 60°}{\sqrt{1984}}$, $\sin \gamma = 48(\frac{\sqrt{3}}{2}) \cdot \dfrac{1}{8\sqrt{31}} =$

$\dfrac{3\sqrt{3}}{\sqrt{31}} \doteq \dfrac{3(1.73)}{5.56} \doteq 0.933$, $\gamma \doteq 69°$, $\theta \doteq 80° - 69° = 11°$, bearing $\doteq 180° + 11° = 191°$

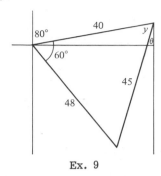

Ex. 9

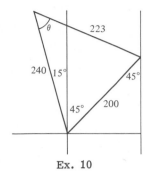

Ex. 10

<u>10.</u> $x^2 = 240^2 + 200^2 - 2(240)(200) \cos 60° = 57{,}600 + 40{,}000 - 48{,}000 = 49{,}600$,

$x \doteq 223$ miles; $\dfrac{\sin \theta}{200} \doteq \dfrac{\sin 60°}{223}$, $\sin \theta = \dfrac{200 \sin 60°}{223} = \dfrac{200(0.866)}{223} \doteq 0.7767$,

$\theta \doteq 51°$, bearing $\doteq 180° - 15° - 51° = 114°$

<u>11.</u> $\dfrac{320}{\sin 80°} = \dfrac{30}{\sin \theta}$, $\sin \theta = \dfrac{30 \sin 80°}{320} \doteq \dfrac{30(0.9848)}{320} = 0.0923$, $\theta \doteq 5°$, heading $\doteq$

$60° + 5° = 65°$; $x^2 = 30^2 + 320^2 - 2(30)(320) \cos (180° - 80° - 5°) =$

$900 + 102{,}400 - 19{,}200 \cos 95° = 103{,}300 + 19{,}200 \cos 85° \doteq 103{,}300 +$

$19{,}200(0.0872) \doteq 103{,}300 + 1674 = 104{,}974$, ground speed $= x \doteq 324$; time $\doteq$

$\dfrac{400}{324} \doteq 1$ hour 14 minutes, ETA $\doteq$ 1:14 p.m.

<u>12.</u> $\dfrac{40}{\sin \theta} = \dfrac{320}{\sin 30°}$, $\sin \theta = \dfrac{40 \sin 30°}{320} = \dfrac{1}{16} = 0.0625$, $\theta \doteq 4°$, bearing $= 240° - 4° =$

$236°$; $\dfrac{x}{\sin (180° - 30° - 4°)} = \dfrac{320}{\sin 30°}$, $\dfrac{x}{\sin 34°} = 640$, $x = 640 \sin 34° \doteq$

$640(0.5592) = 358$; $t = \dfrac{400}{358} \doteq 1$ hour 7 minutes, ETA $\doteq$ 7:07 p.m.

Exercises 6-5 · Pages 215-216

<u>1.</u> $500 \cos 30° = 500(\frac{\sqrt{3}}{2}) = 250\sqrt{3} \doteq 433$ lb.; $500 \sin 30° = 500(\frac{1}{2}) = 250$ lb.

<u>2.</u> $800 \cos 16° \doteq 800(0.9613) \doteq 769$ lb.; $800 \sin 16° \doteq 800(0.2756) \doteq 220$ lb.

<u>3.</u> $55 - 120 \sin 20° \doteq 55 - 120(0.3420) \doteq 55 - 41 = 14$ lb.

<u>4.</u> $400 \sin 18° \doteq 400(0.3090) \doteq 124$ lb.

<u>5.</u> weight $= 60 \cot 35° \doteq 60(1.428) \doteq 86$ lb.; tension $= 60 \csc 35° \doteq 60(1.743) \doteq$

105 lb.

6. $\tan \theta = \frac{80}{40} = 2.000$, $\theta = 63°$

7. A + C = 15 lb. at 210°.   Let α = angle of -(A + B + C) and let $\theta = \alpha - 300°$.

Then $\tan \theta = \frac{15}{30} = 0.5000$, $\theta \doteq 27°$, $\alpha \doteq 327°$.   Magnitude of -(A + B + C) =

$\sqrt{15^2 + 30^2} = \sqrt{225 + 900} = \sqrt{1125} \doteq 34$ lb.

8. $\vec{v}_x = -30 \cos 60° - 50 \cos 90° - 70 \cos 240° - 80 \cos 330° = -30(\frac{1}{2}) - 50(0) -$

$70(-\frac{1}{2}) - 80(\frac{\sqrt{3}}{2}) = -15 + 35 - 40\sqrt{3} = 20 - 40\sqrt{3}$; $\vec{v}_y = -30 \sin 60° - 50 \sin 90° -$

$70 \sin 240° - 80 \sin 330° = -30(\frac{\sqrt{3}}{2}) - 50(1) - 70(-\frac{\sqrt{3}}{2}) - 80(-\frac{1}{2}) = -15\sqrt{3} -$

$50 + 35\sqrt{3} + 40 = -10 + 20\sqrt{3}$; $\|\vec{v}\| = \sqrt{(20 - 40\sqrt{3})^2 + (-10 + 20\sqrt{3})^2} =$

$10\sqrt{4(1 - 2\sqrt{3})^2 + (-1 + 2\sqrt{3})^2} = 10\sqrt{5(1 - 2\sqrt{3})^2} = 10(2\sqrt{3} - 1)\sqrt{5} \doteq$

$10(3.5 - 1)(2.2) = 10(2.5)(2.2) = 55$ lb. $\tan \theta = \frac{v_y}{v_x} = \frac{-10 + 20\sqrt{3}}{20 - 40\sqrt{3}} = \frac{-1 + 2\sqrt{3}}{2 - 4\sqrt{3}} =$

$\frac{-1 + 2\sqrt{3}}{-2(-1 + 2\sqrt{3})} = \frac{1}{-2} = -0.500$, $\theta \doteq 180° - 27° = 153°$

9. $T_1 \sin 23° = T_2 \sin 20°$, $T_2 = T_1 \sin 23° \csc 20° \doteq$

1.148 T_1; 1000 = $T_1 \cos 23° + T_2 \cos 20° =$

0.9205 T_1 + (1.148 T_1)(9.397) = 0.9205 T_1 + 1.0787 T_1

= 1.992 T_1, T_1 = 502 lb.; T_2 = (1.148)(502) =

576 lb.

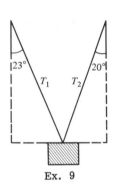

10. $T_1 = 50\sqrt{2} \csc 105° \doteq 50(1.414)(1.04) \doteq 74$ lb.;

$T_2 = 50 \csc 105° \doteq 50(1.04) \doteq 52$ lb.

Ex. 9

Exercises 6-6 · Pages 223-224

1. x = 3 cos 180° = -3, y = 3 sin 180° = 0, (-3, 0)

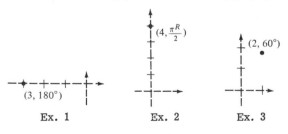

Ex. 1   Ex. 2   Ex. 3   Ex. 4

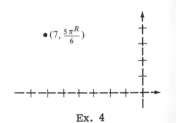

2. x = 4 cos $\frac{\pi^R}{2}$ = 0, y = 4 sin $\frac{\pi^R}{2}$ = 4, (0, 4)

3. x = 2 cos 60° = 1, y = 2 sin 60° = $\sqrt{3}$, (1, $\sqrt{3}$)

4. x = 7 cos $\frac{5\pi}{6}^R$ = $-\frac{7\sqrt{3}}{2}$, y = 7 sin $\frac{5\pi}{6}^R$ = $\frac{7}{2}$, $(-\frac{7\sqrt{3}}{2}, \frac{7}{2})$

<u>5</u>. x = -2 cos 210° = $\sqrt{3}$, y = -2 sin 210° = 1, ($\sqrt{3}$, 1)

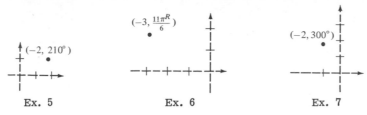

Ex. 5   Ex. 6   Ex. 7

<u>6</u>. x = -3 cos $\frac{11\pi^R}{6}$ = $-\frac{3\sqrt{3}}{2}$, y = -3 sin $\frac{11\pi^R}{6}$ = $\frac{3}{2}$, $(-\frac{3\sqrt{3}}{2}, \frac{3}{2})$

<u>7</u>. x = -2 cos 300° = -1, y = -2 sin 300° = $\sqrt{3}$; (-1, $\sqrt{3}$)

<u>8</u>. x = -3 cos $(-\frac{9\pi^R}{4})$ = $-\frac{3\sqrt{2}}{2}$, y = -3 sin $(-\frac{9\pi^R}{4})$ = $\frac{3\sqrt{2}}{2}$, $(-\frac{3\sqrt{2}}{2}, \frac{3\sqrt{2}}{2})$

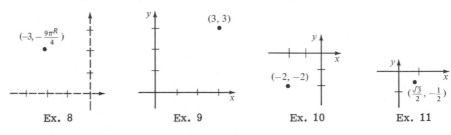

Ex. 8   Ex. 9   Ex. 10   Ex. 11

<u>9</u>. ρ = $\sqrt{3^2 + 3^2}$ = $3\sqrt{2}$, θ = Tan^{-1} $\frac{3}{3}$ = $\frac{\pi^R}{4}$, $(3\sqrt{2}, \frac{\pi^R}{4})$

<u>10</u>. ρ = $\sqrt{(-2)^2 + (-2)^2}$ = $2\sqrt{2}$, θ = Tan^{-1} $\frac{-2}{-2}$ + π^R = $\frac{5\pi^R}{4}$, $(2\sqrt{2}, \frac{5\pi^R}{4})$

<u>11</u>. ρ = $\sqrt{(\frac{\sqrt{3}}{2})^2 + (-\frac{1}{2})^2}$ = 1, θ = Tan^{-1} $\frac{-\frac{1}{2}}{\frac{\sqrt{3}}{2}}$ + π^R = $\frac{5\pi^R}{6}$ + π^R = $\frac{11\pi^R}{6}$, $(1, \frac{11\pi^R}{6})$

<u>12</u>. ρ = $\sqrt{(-\frac{3\sqrt{3}}{2})^2 + (\frac{3}{2})^2}$ = 3, θ = Tan^{-1} $\frac{\frac{3}{2}}{-\frac{3\sqrt{3}}{2}}$ = $\frac{5\pi^R}{6}$, $(3, \frac{5\pi^R}{6})$

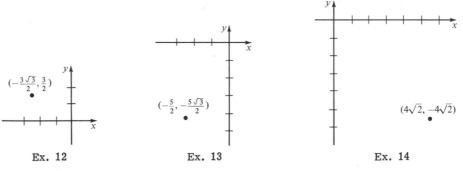

Ex. 12   Ex. 13   Ex. 14

<u>13</u>. ρ = $\sqrt{(-\frac{5}{2})^2 + (-\frac{5\sqrt{3}}{2})^2}$ = 5, θ = Tan^{-1} $\frac{-\frac{5\sqrt{3}}{2}}{-\frac{5}{2}}$ + π^R = $\frac{\pi^R}{3}$ + π^R = $\frac{4\pi^R}{3}$, $(5, \frac{4\pi^R}{3})$

<u>14</u>. ρ = $\sqrt{(4\sqrt{2})^2 + (-4\sqrt{2})^2}$ = 8, θ = Tan$^{-1}\frac{-4\sqrt{2}}{4\sqrt{2}}$ + π^R = $\frac{3\pi^R}{4}$ + π^R = $\frac{7\pi^R}{4}$, $(8, \frac{7\pi^R}{4})$

<u>15.</u> $\rho = \sqrt{3^2 + 4^2} = 5$, $\theta = \text{Tan}^{-1} \frac{4}{3}$, $(5, 53°10')$

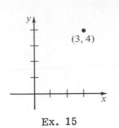

Ex. 15

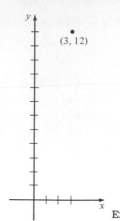

Ex. 16

<u>16.</u> $\rho = \sqrt{3^2 + 12^2} = \sqrt{153} = 3\sqrt{17}$, $\theta = \text{Tan}^{-1} \frac{12}{3} =$

$\text{Tan}^{-1} 4$, $(3\sqrt{17}, 76°)$

<u>17.</u> $x = \rho \cos \theta$; $4 = \rho \cos \theta$

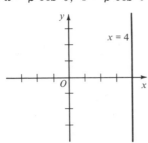

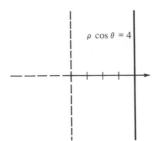

<u>18.</u> $x = \rho \cos \theta$; $-3 = \rho \cos \theta$

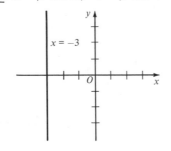

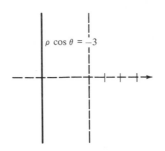

<u>19.</u> $y = \rho \sin \theta$; $2 = \rho \sin \theta$

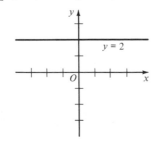

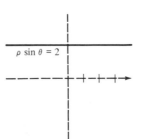

<u>20.</u> $y = \rho \sin \theta$; $-1 = \rho \sin \theta$

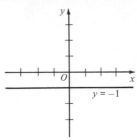

$y = -1$

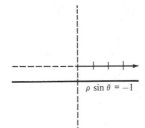

$\rho \sin \theta = -1$

<u>21.</u> $\rho \sin \theta + \rho \cos \theta = 0$, $\rho(\sin \theta + \cos \theta) = 0$; $\rho = 0$ or $\sin \theta + \cos \theta = 0$,
$\rho = 0$ or $\tan \theta = -1$, $\theta = 135°$.

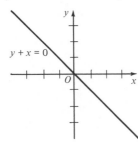

$y + x = 0$

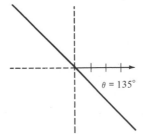

$\theta = 135°$

<u>22.</u> $\rho \sin \theta - \rho \cos \theta = 1$; $\rho(\sin \theta - \cos \theta) = 1$

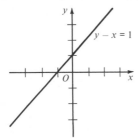

$y - x = 1$

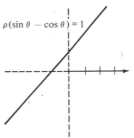

$\rho(\sin \theta - \cos \theta) = 1$

<u>23.</u> $(\rho \cos \theta)^2 + (\rho \sin \theta)^2 = 16$, $\rho^2 \cos^2 \theta + \rho^2 \sin^2 \theta = 16$, $\rho^2(\cos^2 \theta + \sin^2 \theta) = 16$,
$\rho^2 = 16$, $\rho = \pm 4$.

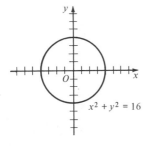

$x^2 + y^2 = 16$

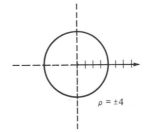

$\rho = \pm 4$

<u>24.</u> ρ cos θ + ρ sin θ = 4, ρ(cos θ + sin θ) = 4

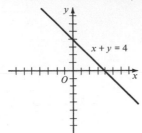

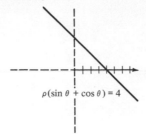

<u>25.</u>

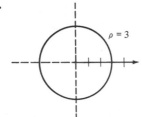

<u>26.</u>

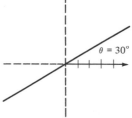

<u>27.</u>

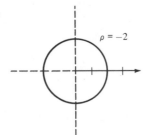

<u>28.</u>

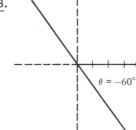

<u>29.</u>

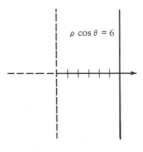

<u>30.</u>

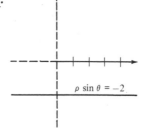

<u>31.</u> $\rho = \sqrt{x^2 + y^2},\; 6 = \sqrt{x^2 + y^2},\; x^2 + y^2 = 36$

<u>32.</u> $\sqrt{x^2 + y^2} = 2,\; x^2 + y^2 = 4$

<u>33.</u> $\theta = \tan^{-1}\frac{y}{x},\; \frac{3\pi^R}{4} = \tan^{-1}\frac{y}{x},\; \tan\frac{3\pi^R}{4} = \frac{y}{x},\; -1 = \frac{y}{x},\; -x = y.$

<u>34.</u> $\rho = 3\cos\theta,\; \rho^2 = 3\rho\cos\theta;\;$ so $x^2 + y^2 = 3x,\; x^2 + y^2 - 3x = 0.$

<u>35.</u> $\rho^2 = 4\sin\theta;\; \rho^3 = 4\rho\sin\theta,\; (\sqrt{x^2 + y^2})^3 = 4y,\;$ so $[(\sqrt{x^2 + y^2})^3]^2 = (4y)^2,$
$(x^2 + y^2)^3 = 16y^2.$

36. $\rho \sin \theta = \rho^2 \cos \theta$, $(\rho \sin \theta)^2 = (\rho^2 \cos \theta)^2$, $(\rho \sin \theta)^2 = \rho^4 \cos^2 \theta =$

$\rho^2 (\rho \cos \theta)^2$, so $y^2 = (x^2 + y^2)x^2$, $y^2 = x^4 + x^2 y^2$, $x^4 + x^2 y^2 - y^2 = 0$.

B 37. $\rho = 4$ and $\rho = 4 \sin \theta$, $4 = 4 \sin \theta$, $\sin \theta = 1$, $\theta = 90°$. The solution set is

$\{(4, 90°)\}$

38. $\rho = 2$ and $\rho = 4 \sin 2\theta$, $2 = 4 \sin 2\theta$, $\sin 2\theta = \frac{1}{2}$. As θ goes from $0°$ to $360°$,

2θ goes from $0°$ to $720°$. In the interval $0°$ to $720°$, $\sin 2\theta = \frac{1}{2}$ at $2\theta = 30°$, $150°$,

$390°$, and $510°$; so $\theta = 15°$, $75°$, $195°$, and $255°$. Although the graphs of the two

sets intersect at eight points, only these four points satisfy both equations. The

solution set is $\{(2, 15°), (2, 75°), (2, 195°), \text{and } (2, 255°)\}$.

39. $\rho = 3$ and $\theta = 120°$; the point of intersection is $(3, 120°)$. Although the graphs of

the two sets intersect at two points, only $(3, 120°)$ satisfies both equations, and hence

the solution set is $\{(3, 120°)\}$.

40. $\rho = 2 \sin \theta$, $\rho^2 = 2\rho \sin \theta$, so

$x^2 + y^2 = 2y$, $x^2 + y^2 + 2y = 0$,

$x^2 + y^2 + 2y + 1 = 0 + 1$, $x^2 + (y + 1)^2 = 1$.

This is Cartesian equation of a circle with

center $(0, 1)$ and radius $= 1$. At $x = 0$,

$y = 0$, we have $0^2 + (0 + 1)^2 = 1$, so the

point $(0, 0)$, the origin, lies on the circle.

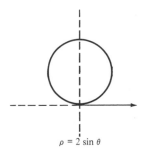

$\rho = 2 \sin \theta$

C 41. Let $A(\rho_1, \theta_1)$ and $B(\rho_2, \theta_2)$ be two points in the plane. The points A and B have

Cartesian coordinates (x_1, y_1) and (x_2, y_2). Then $x_1 = \rho_1 \cos \theta_1$, $y_1 = \rho_1 \sin \theta_1$,

$x_2 = \rho_2 \cos \theta_2$, and $y_2 = \rho_2 \sin \theta_2$. We know that $d(AB) = \sqrt{(x_1 - x_2)^2 + (y_1 - y_2)^2}$

$= \sqrt{(\rho_1 \cos \theta_1 - \rho_2 \cos \theta_2)^2 + (\rho_1 \sin \theta_1 - \rho_2 \sin \theta_2)^2} =$

$\sqrt{(\rho_1^2 \cos^2 \theta_1 - 2\rho_1 \rho_2 \cos \theta_1 \cos \theta_2 + \rho_2^2 \cos^2 \theta_2) + }$

$\sqrt{(\rho_1^2 \sin^2 \theta_1 - 2\rho_1 \rho_2 \sin \theta_1 \sin \theta_2 + \rho_2^2 \sin^2 \theta_2)} =$

$\sqrt{\rho_1^2 (\cos^2 \theta_1 + \sin^2 \theta_1) - 2\rho_1 \rho_2 (\cos \theta_1 \cos \theta_2 + \sin \theta_1 \sin \theta_2) + }$

$\sqrt{\rho_2^2 (\cos^2 \theta_2 + \sin^2 \theta_2)} = \sqrt{\rho_1^2 - 2\rho_1 \rho_2 \cos (\theta_1 - \theta_2) + \rho_2^2}$.

42. Let circle have radius = a and center at (ρ_1, θ_1). Let (ρ, θ) be any point on circle. Then distance between (ρ, θ) and (ρ_1, θ_1) is equal to a; a = $\sqrt{\rho_1{}^2 + \rho^2 - 2\rho_1\rho \cos (\theta_1 - \theta)}$ by results of Exercise 41; $a^2 = \rho_1{}^2 + \rho^2 - 2\rho_1\rho \cos (\theta_1 - \theta)$, or $a^2 = \rho^2 + \rho_1{}^2 - 2\rho\rho_1 \cos (\theta - \theta_1)$.

Exercises 6-7 · Pages 229-230

A   1.

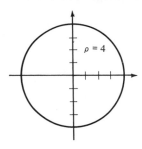

$\rho = 4$

2.

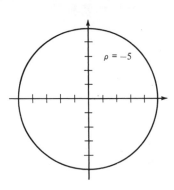

$\rho = -5$

3.

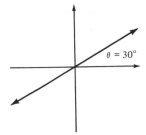

$\theta = 30°$

4.

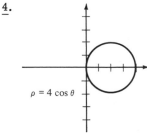

$\rho = 4 \cos \theta$

5.

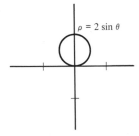

$\rho = 2 \sin \theta$

6.

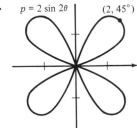

$p = 2 \sin 2\theta$ $(2, 45°)$

7.

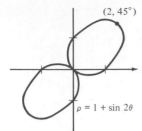

$(2, 45°)$

$\rho = 1 + \sin 2\theta$

8.

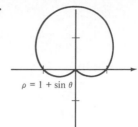

$\rho = 1 + \sin \theta$

9.

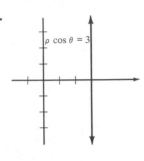

$\rho \cos \theta = 3$

10.

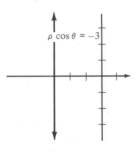

$\rho \cos \theta = -3$

11.

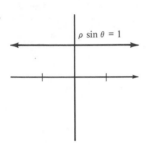

$\rho \sin \theta = 1$

12.

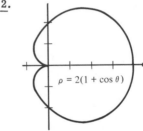

$\rho = 2(1 + \cos \theta)$

13.

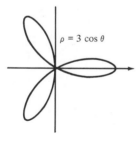

$\rho = 3 \cos \theta$

14.

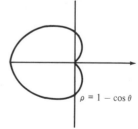

$\rho = 1 - \cos \theta$

15.

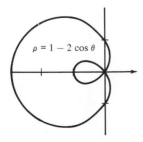

$\rho = 1 - 2 \cos \theta$

16.

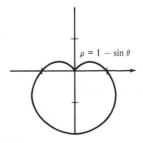

$\rho = 1 - \sin \theta$

17.
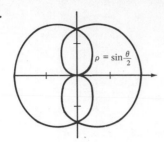
$\rho = \sin\frac{\theta}{2}$

18.
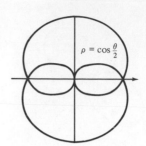
$\rho = \cos\frac{\theta}{2}$

19.
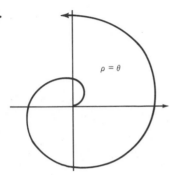
$\rho = \theta$

20.
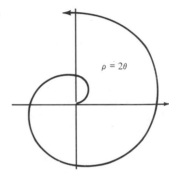
$\rho = 2\theta$

21.
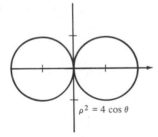
$\rho^2 = 4\cos\theta$

22.
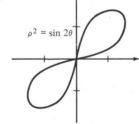
$\rho^2 = \sin 2\theta$

B   23.
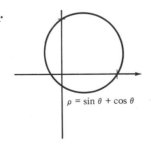
$\rho = \sin\theta + \cos\theta$

24.
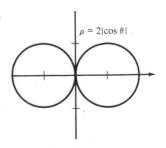
$\rho = 2|\cos\theta|$

<u>25.</u>

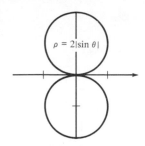

$\rho = 2|\sin\theta|$

<u>26.</u>

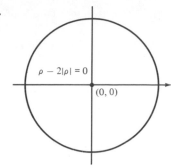

$\rho - 2|\rho| = 0$

$(0, 0)$

<u>27.</u>

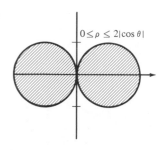

$0 \le \rho \le 2|\cos\theta|$

28. Circle whose center is at the origin is
    in the form of $x^2 + y^2 = r^2$,

    $(\sqrt{x^2 + y^2})^2 = r^2$, therefore the

    desired equation is $\rho^2 = 36$, $\rho = 6$

    or $\rho = -6$.

Chapter Test · Page 232

<u>1.</u> <u>a.</u> $\|\vec{u} + \vec{v}\| = \|(2, -6) + (3, 5)\| = \|(5, -1)\| = \sqrt{5^2 + (-1)^2} = \sqrt{26}$

   <u>b.</u> $4\vec{u} + (-1)\vec{v} = 4(2, -6) + (-1)(3, 5) = (8, -24) + (3, -5) = (5, -29)$

<u>2.</u> $\cos 120° = \frac{x}{5}$, $x = 5\cos 120° = 5(-\frac{1}{2}) = -\frac{5}{2}$, $\sin 120° = \frac{y}{5}$, $y = 5\sin 120° =$

   $5(\frac{\sqrt{3}}{2}) = \frac{5\sqrt{3}}{2}$, $(-\frac{5}{2}, \frac{5\sqrt{3}}{2})$.

<u>3.</u> <u>a.</u> $\vec{v} = (-2, 6)$, so $\|\vec{v}\| = \sqrt{(-2)^2 + 6^2} = \sqrt{40} = 2\sqrt{10}$; and $\vec{u} = \frac{1}{\|\vec{v}\|}\vec{v} =$

   $\frac{1}{2\sqrt{10}}(-2, 6) = (-\frac{1}{\sqrt{10}}, \frac{3}{\sqrt{10}})$.   So $\vec{v} = 2\sqrt{10}(-\frac{1}{\sqrt{10}}, \frac{3}{\sqrt{10}})$

   <u>b.</u> $v = \|v\|(\cos\theta, \sin\theta)$, so $\cos\theta = -\frac{1}{\sqrt{10}}$ and $\sin\theta = \frac{3}{\sqrt{10}}$.

<u>4.</u> $\vec{u} \cdot \vec{v} = 0$, $(2, 4) \cdot (x, -3) = 0$, $2x - 12 = 0$, $2x = 12$, $x = 6$

<u>5.</u> $v_a = 20\sin 60° + 50\sin 180° + 40\sin 90° = 20(\frac{\sqrt{3}}{2}) + 50(0) + 40(1) = 10\sqrt{3} + 40$;

   $v_b = 20\cos 60° + 50\cos 180° + 40\cos 90° = 20(\frac{1}{2}) + 50(-1) + 40(0) = 10 - 50 = -40$;

   $\|v\| = \sqrt{(10\sqrt{3} + 40)^2 + (-40)^2} = \sqrt{100(3) + 800\sqrt{3} + 1600 + 1600} \doteq \sqrt{4885.6} \doteq$

   69.9 miles; $\tan\theta = \frac{v_b}{v_a} = \frac{-40}{10\sqrt{3} + 40} = \frac{-4}{4 + \sqrt{3}}$.

<u>6.</u> $\sin 30° = \frac{y}{60}$, $y = 60 \sin 30°$, $y = 60(\frac{1}{2})$, $y = 30$ lbs.

<u>7.</u> Three possible answers: $(-4, 255°)$, $(4, 435°)$, and $(-4, 615°)$

<u>8.</u>
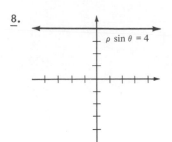

<u>9.</u>
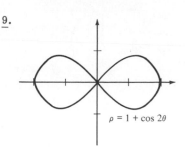

Cumulative Review: Chapters 4-6 · Pages 232-233

Chapter 4

<u>1.</u> $\frac{4\pi}{3}$, $\frac{5\pi}{3}$ <u>2.</u> $\frac{\pi}{3}$, $\frac{5\pi}{3}$ <u>3.</u> $\frac{2\pi}{3}$, $\frac{5\pi}{3}$ <u>4.</u> $\frac{\pi}{3}$, $\frac{5\pi}{3}$ <u>5.</u> $-\frac{\pi}{2}$

<u>6.</u> $\frac{2\pi}{3}$ <u>7.</u> $\frac{4}{5}$ <u>8.</u> $\frac{1}{4}$

<u>9.</u> $\cot^2 x + \cot x = 0$, $\cot x (\cot x + 1) = 0$, $\cot x = 0$ or $\cot x + 1 = 0$, $\cot x = 0$

     or $\cot x = -1$. $x = \frac{\pi}{2}$, $\frac{3\pi}{2}$ or $x = \frac{3\pi}{4}$, $\frac{7\pi}{4}$

<u>10.</u> $\sin x - \cos x = 1$, $\sin^2 x - 2 \sin x \cos x + \cos^2 x = 1$, $-2 \sin x \cos x = 0$,

     $\sin x \cos x = 0$, so either $\sin x = 0$ or $\cos x = 0$, $x = 0$, π or $x = \frac{\pi}{2}$, $\frac{3\pi}{2}$; 0 and

     $\frac{3\pi}{2}$ are not solutions; the general solution is $\{x: x = \frac{\pi}{2} + 2k\pi\} \cup \{x: x = (2k + 1)\pi\}$

Chapter 5

<u>11.</u> $\frac{7}{6}(180°) = 210°$ <u>12.</u> $\cos 141° = -\cos (180 - 141) = -\cos 39°$

<u>13.</u> <u>a.</u> $\tan 1095° = \tan 15° \doteq .2679$ <u>b.</u> $\sin (-690°) = \sin 30° = .5000$

<u>14.</u> $\dfrac{\sin 2x}{1 + \cos 2x} = \dfrac{2 \sin x \cos x}{1 + (2 \cos^2 x - 1)} = \dfrac{2 \sin x \cos x}{2 \cos^2 x} = \dfrac{\sin x}{\cos x} = \tan x.$

<u>15.</u> $\sin \theta = \dfrac{-5}{\sqrt{29}}$ and $\cos \theta = \dfrac{2}{\sqrt{29}}$

16. $2 \cos^2 2\theta + \cos 2\theta - 1 = 0$, $(2 \cos 2\theta - 1)(\cos 2\theta + 1) = 0$, $2 \cos 2\theta - 1 = 0$ or

$\cos 2\theta + 1 = 0$, $2 \cos 2\theta = 1$ or $\cos 2\theta = -1$, $\cos 2\theta = \frac{1}{2}$ or $\cos 2\theta = -1$, for

$0 \le m^\circ(2\theta) < 720^\circ$, we have $2\theta = 60^\circ$, 300°, 420°, 660°, or $2\theta = 180^\circ$, 540°; this

means $\theta = 30^\circ$, 150°, 210°, 330°, or $\theta = 90^\circ$, 270°.

17. $A = 90^\circ - 54^\circ = 36^\circ$. $\cot 54^\circ = \frac{a}{120}$, $a = 120 \cot 54^\circ$, $a = 120(.7265)$, $a \doteq 87$.

$c^2 = a^2 + b^2$, $c = \sqrt{a^2 + b^2}$, $c = \sqrt{120^2 + 87^2}$, $c = \sqrt{14400 + 7569}$, $c = \sqrt{21969}$,

$c \doteq 148$.

18. $c^2 = a^2 + b^2 - 2ab \cos C$, $c^2 = 4^2 + 10^2 - 2(4)(10) \cos 30^\circ$, $c^2 = 16 + 100 - 40\sqrt{3}$,

$c^2 = 116 - 69.28$, $c^2 = 46.72$, $c = \sqrt{46.72}$, $c = 7$.

19. $\frac{a}{\sin A} = \frac{b}{\sin B}$, $b = \frac{a \sin B}{\sin A}$, $b = \frac{20 \sin 45^\circ}{\sin 60^\circ}$, $b = \frac{20(\frac{\sqrt{2}}{2})}{\frac{\sqrt{3}}{2}}$, $b = \frac{20\sqrt{6}}{3}$,

$b = \frac{20(2.449)}{3}$, $b = 16.67$

20. 2.

Chapter 6

21. a. $\| \vec{u} - \vec{v} \| = \|(-3, 7) - (4, 7)\| = \|(-7, 0)\| = \sqrt{(-7)^2 + 0^2} = 7$

b. $3\vec{u} + 2\vec{v} = 3(-3, 7) + 2(4, 7) = (-9, 21) + (8, 14) = (-1, 35)$.

22. $\cos 210^\circ = \frac{x}{10}$, $x = 10 \cos 210^\circ = 10(-\frac{\sqrt{3}}{2}) = -5\sqrt{3}$; $\sin 210^\circ = \frac{y}{10}$,

$y = 10 \sin 210^\circ = 10(-\frac{1}{2}) = -5$; $(-5\sqrt{3}, -5)$

23. $\vec{v} = (4, -3)$, $\| \vec{v} \| = \sqrt{4^2 + (-3)^2} = 5$, $\vec{u} = (\frac{4}{5}, -\frac{3}{5})$, so $\vec{v} = 5(\frac{4}{5}, -\frac{3}{5})$.

24. $\| v \| = \|(4, -5)\| = \sqrt{4^2 + (-5)^2} = \sqrt{41} \doteq 6.4$; the direction angle θ has $\cos \theta = \frac{4}{\sqrt{41}}$

and $\sin \theta = -\frac{5}{\sqrt{41}}$, $\cos \theta \doteq 0.6250$ and $\sin < 0$, $\theta \doteq -51^\circ$; so the direction

angle is 309°.

25. $\vec{u} \cdot \vec{v} = 0$, $(-1, 3) \cdot (x, -4) = 0$, $-x - 12 = 0$, $x = -12$

26. $\vec{u} \cdot \vec{v} = (-\frac{1}{2})(\frac{\sqrt{3}}{2}) + (\frac{\sqrt{3}}{2})(\frac{1}{2}) = 0$, so $\vec{u}$ and $\vec{v}$ are orthogonal; $\| \vec{u} \|^2 =$

$(-\frac{1}{2})^2 + (\frac{\sqrt{3}}{2})^2 = 1$, $\| \vec{v} \|^2 = (\frac{\sqrt{3}}{2})^2 + (\frac{1}{2})^2 = 1$, so $\vec{u}$ and $\vec{v}$ are unit vectors;

$\vec{a} = (1, 1) = (\vec{u} \cdot \vec{a})\vec{u} + (\vec{v} \cdot \vec{a})\vec{v} = [(-\frac{1}{2})(1) + (\frac{\sqrt{3}}{2})(1)]\vec{u} + [(\frac{\sqrt{3}}{2})(1) + (\frac{1}{2})(1)]\vec{v} = (\frac{\sqrt{3} - 1}{2})\vec{u} + (\frac{\sqrt{3} + 1}{2})\vec{v}.$

<u>27</u>. $\cos \alpha = \dfrac{\vec{u} \cdot \vec{v}}{\|\vec{u}\| \|\vec{v}\|} = \dfrac{(2, 2\sqrt{3}) \cdot (-\sqrt{2}, \sqrt{2})}{\sqrt{2^2 + (2\sqrt{3})^2} \sqrt{(-\sqrt{2})^2 + (\sqrt{2})^2}} = \dfrac{-2\sqrt{2} + 2\sqrt{6}}{(4)(2)} =$

$\sqrt{2 \left(\dfrac{\sqrt{3} - 1}{4}\right)} = \sqrt{\dfrac{2(\sqrt{3} - 1)^2}{16}} = \sqrt{\dfrac{4 - 2\sqrt{3}}{8}} = \sqrt{\dfrac{1}{2}\left(1 - \dfrac{\sqrt{3}}{2}\right)} = \sqrt{\dfrac{1}{2}(1 + \cos 150)} =$

$\cos 75°$.   So $\alpha = 75°$

<u>28</u>. $x^2 = 400^2 + 300^2 - 2(400)(300) \cos 60° = 160,000 + 90,000 - 120,000 = 130,000,$

$x^2 = 130,000, \ x = \sqrt{130,000} \doteq 361$ miles

<u>29</u>. $y = 160 \sin 30°, \ y = 160(\frac{1}{2}) = 80$ lbs.; $x = 160 \cos 30°, \ x = 160(\frac{\sqrt{3}}{2}), \ x = 80\sqrt{3},$

$x \doteq 138$ lbs.

<u>30</u>. $y^2 = 80^2 - 40^2 = 4800 = 40\sqrt{3}$ lbs., $\theta = 90° + 60° = 150°.$

CHAPTER 7. Complex Numbers

A 1. a. $z_1 + z_2 = (3, 2) + (1, 6) = (4, 8)$

　　b. $z_1 z_2 = (3, 2)(1, 6) = (3 - 12, 18 + 2) = (-9, 20)$

　2. a. $z_1 + z_2 = (3, -2) + (-1, 6) = (2, 4)$

　　b. $z_1 z_2 = (3, -2)(-1, 6) = (-3 + 12, 18 + 2) = (9, 20)$

　3. a. $z_1 + z_2 = (4, 8) + (5, 1) = (9, 9)$

　　b. $z_1 z_2 = (4, 8)(5, 1) = (20 - 8, 4 + 40) = (12, 44)$

　4. a. $z_1 + z_2 = (4, 8) + (-5, -1) = (-1, 7)$

　　b. $z_1 z_2 = (4, 8)(-5, -1) = (-20 + 8, -4 - 40) = (-12, -44)$

　5. a. $z_1 + z_2 = (\frac{1}{3}, 2) + (3, \frac{1}{2}) = (3\frac{1}{3}, 2\frac{1}{2})$

　　b. $z_1 z_2 = (\frac{1}{3}, 2)(3, \frac{1}{2}) = (1 - 1, \frac{1}{6} + 6) = (0, 6\frac{1}{6})$

　6. a. $z_1 + z_2 = (\frac{1}{4}, \frac{1}{2}) + (\frac{1}{8}, \frac{1}{4}) = (\frac{3}{8}, \frac{3}{4})$

　　b. $z_1 z_2 = (\frac{1}{4}, \frac{1}{2})(\frac{1}{8}, \frac{1}{4}) = (\frac{1}{32} - \frac{1}{8}, \frac{1}{16} + \frac{1}{16}) = (-\frac{3}{32}, \frac{1}{8})$

　7. a. $z^2 = (1, 1)(1, 1) = (1 - 1, 1 + 1) = (0, 2)$

　　b. $z^3 = (0, 2)(1, 1) = (0 - 2, 0 + 2) = (-2, 2)$

　8. a. $z^2 = (2, 3)(2, 3) = (4 - 9, 6 + 6) = (-5, 12)$

　　b. $z^3 = (-5, 12)(2, 3) = (-10 - 36, -15 + 24) = (-46, 9)$

　9. a. $z^2 = (0, 1)(0, 1) = (0 - 1, 0 + 0) = (-1, 0)$

　　b. $z^3 = (-1, 0)(0, 1) = (0 - 0, -1 + 0) = (0, -1)$

　10. a. $z^2 = (5, 0)(5, 0) = (25 - 0, 0 + 0) = (25, 0)$

　　b. $z^3 = (25, 0)(5, 0) = (125 - 0, 0 + 0) = (125, 0)$

　11. a. $z^2 = (\sqrt{3}, -\sqrt{2})(\sqrt{3}, -\sqrt{2}) = (3 - 2, -\sqrt{6} - \sqrt{6}) = (1, -2\sqrt{6})$

　　b. $z^3 = (1, -2\sqrt{6})(\sqrt{3}, -\sqrt{2}) = (\sqrt{3} - 4\sqrt{3}, -\sqrt{2} - 6\sqrt{2}) = (-3\sqrt{3}, -7\sqrt{2})$

　12. a. $z^2 = (0, \sqrt{5})(0, \sqrt{5}) = (0 - 5, 0 + 0) = (-5, 0)$

　　b. $z^3 = (-5, 0)(0, \sqrt{5}) = (0 - 0, -5\sqrt{5} + 0) = (0, -5\sqrt{5})$

B 13. $z_1 + z_2 = (x_1, y_1) + (x_2, y_2) = (x_1 + x_2, y_1 + y_2) = (x_2 + x_1, y_2 + y_1) =$
　　$(x_2, y_2) + (x_1, y_1) = z_2 + z_1$

14. $(z_1 + z_2) + z_3 = [(x_1, y_1) + (x_2, y_2)] + (x_3, y_3)$

$= (x_1 + x_2, y_1 + y_2) + (x_3, y_3)$

$= ((x_1 + x_2) + x_3, (y_1 + y_2) + y_3)$

$= (x_1 + (x_2 + x_3), y_1 + (y_2 + y_3))$

$= (x_1, y_1) + (x_2 + x_3, y_2 + y_3)$

$= (x_1, y_1) + [(x_2, y_3) + (x_2, y_3)]$

$= z_1 + (z_2 + z_3)$

15. $z_1 z_2 = (x_1, y_1)(x_2, y_2) = (x_1 x_2 - y_1 y_2, x_1 y_2 + x_2 y_1)$

$x_1 x_2 - y_1 y_2$ and $x_1 y_2 + x_2 y_1 \in \mathcal{R}$

$\therefore (x_1 x_2 - y_1 y_2, x_1 y_2 + x_2 y_1) \in C$

16. $(z_1 z_2)z_3 = [(x_1, y_1)(x_2, y_2)](x_3 y_3)$

$= (x_1 x_2 - y_1 y_2, x_1 y_2 + x_2 y_1)(x_3, y_3)$

$= ((x_1 x_2)x_3 - (y_1 y_2)x_3 - (x_1 y_2)y_3 - (x_2 y_1)y_3, (x_1 x_2)y_3 - y_1 y_2 (y_3)$

$+ (x_1 y_2)x_3 + (x_2 y_1)x_3)$

$= (x_1(x_2 x_3) - y_1(y_2 x_3) - x_1(y_2 y_3) - y_1(x_2 y_3), x_1(x_2 y_3) - y_1(y_2 y_3)$

$+ x_1(y_2 x_3) + y_1(x_2 x_3)$

$= (x_1(x_2 x_3) - x_1(y_2 y_3) - y_1(x_2 y_3) - y_1(x_3 y_2), x_1(x_2 y_3) + x_1(x_3 y_2)$

$+ y_1(x_2 x_3) - y_1(y_2 y_3))$

$= (\{x_1(x_2 x_3 - y_2 y_3) - y_1(x_2 y_3 + x_3 y_2)\}, \{x_1(x_2 y_3 + y_3 y_2) + y_1(x_2 x_3 - y_2 y_3)\})$

$= (x_1, y_1)(x_2 x_3 - y_2 y_3, x_2 y_3 + x_3 y_2)$

$= (x_1, y_1)[(x_2, y_2)(x_3, y_3)]$

$= z_1(z_2 z_3)$

17. $z_1 z_2 = (x_1, y_1)(x_2, y_2)$

$= (x_1 x_2 - y_1 y_2, x_1 y_2 + x_2 y_3)$

$= (x_2 x_1 - y_2 y_1, y_2 x_1 + y_3 x_2)$

$= (x_2, y_2)(x_1, y_1)$

$= z_2 z_1$

18. $z_1(z_2 + z_3) = (x_1, y_1)[(x_2, y_2) + (x_3, y_3)]$

$= (x_1, y_1)(x_2 + x_3, y_2 + y_3)$

$= (x_1 x_2 + x_1 x_3 - y_1 y_2 - y_1 y_3, x_1 y_2 + x_1 y_3 + x_2 y_1 + x_3 y_1)$

$= ((x_1 x_2 - y_1 y_2) + (x_1 x_3 - y_1 y_3), (x_1 y_2 + x_2 y_1) + (x_1 y_3 + x_3 y_1))$

$= (x_1 x_2 - y_1 y_2, x_1 y_2 + x_2 y_1) + (x_1 x_3 - y_1 y_3, x_1 y_3 + x_3 y_1)$

$= [(x_1, y_1)(x_2, y_2) + (x_1, y_1)(x_3, y_3)]$

$= z_1 z_2 + z_1 z_3$

C 19. $(x, y) + (a, b) = (a, b) + (x, y) = (a + x, b + y)$, if $(a + b) + (x, y) = (x, y)$ then

$(a + x, b + y) = (x, y)$ so $a + x = x$ and $b + y = y$, $a = 0$ and $b = 0$, therefore

$(a, b) = (0, 0)$

20. $(a, b) + (x, y) = (x, y) + (a, b) = (x + a, y + b) = (0, 0)$; $x + a = 0$ and $y + b = 0$,

$a = -x$ and $b = -y$, $(a, b) = (-x, -y)$

21. $z_1 z_2 = (r_1 \cos \theta_1, r_1 \sin \theta_1)(r_2 \cos \theta_2, r_2 \sin \theta_2)$

$= (r_1 r_2 \cos \theta_1 \cos \theta_2 - r_1 r_2 \sin \theta_1 \sin \theta_2, r_1 r_2 \cos \theta_1 \sin \theta_2$

$+ r_1 r_2 \cos \theta_2 \sin \theta_1)$

$= (r_1 r_2(\cos \theta_1 \cos \theta_2 - \sin \theta_1 \sin \theta_2), r_1 r_2(\sin \theta_1 \cos \theta_2 + \sin \theta_2 \cos \theta_1))$

$= [r_1 r_2 \cos (\theta_1 + \theta_2), r_1 r_2 \sin (\theta_1 + \theta_2)]$

Exercises 7-2 · Page 245

A 1. $(0, 4)(0, 1) = (0 - 4, 0 + 0) = (-4, 0)$; $\frac{(0, 4)}{(0, 1)} = \frac{(0, 4)(0, -1)}{(0, 1)(0, -1)} = \frac{(4, 0)}{(1, 0)} =$

$(\frac{4}{1}, \frac{0}{1}) = (4, 0)$

2. $(0, 4)(0, -1) = (0 + 4, 0 + 0) = (4, 0)$; $\frac{(0, 4)}{(0, -1)} = \frac{(0, 4)(0, 1)}{(0, -1)(0, 1)} = \frac{(-4, 0)}{(1, 0)} =$

$(-\frac{4}{1}, \frac{0}{1}) = (-4, 0)$

3. $(0, 4)(0, 3) = (-12, 0)$; $\frac{(0, 4)}{(0, 3)} = \frac{(0, 4)(0, -3)}{(0, 3)(0, -3)} = \frac{(12, 0)}{(9, 0)} = (\frac{12}{9}, \frac{0}{9}) = (\frac{4}{3}, 0)$

4. $(0, 4)(0, -3) = (12, 0)$; $\frac{(0, 4)}{(0, -3)} = \frac{(0, 4)(0, 3)}{(0, -3)(0, 3)} = \frac{(-12, 0)}{(9, 0)} = (-\frac{12}{9}, \frac{0}{9}) = (-\frac{4}{3}, 0)$

5. $(0, 4)(2, 0) = (0, 8)$; $\frac{(0, 4)}{(2, 0)} = \frac{(0, 4)(2, 0)}{(2, 0)(2, 0)} = \frac{(0, 8)}{(4, 0)} = (\frac{0}{4}, \frac{8}{4}) = (0, 2)$

6. $(0, 4)(-2, 0) = (0, -8)$; $\frac{(0, 4)}{(-2, 0)} = \frac{(0, 4)(-2, 0)}{(-2, 0)(-2, 0)} = \frac{(0, -8)}{(4, 0)} = (\frac{0}{4}, -\frac{8}{4}) = (0, -2)$

7. a. $z_1 - z_2 = z_1 + (-z_2) = (1, 2) + (-4, -3) = (-3, -1)$

b. $z_2^{-1} = (\frac{x_2}{x_2^2 + y_2^2}, \frac{-y_2}{x_2^2 + y_2^2}) = (\frac{4}{16 + 9}, \frac{-3}{16 + 9}) = (\frac{4}{25}, -\frac{3}{25})$

c. $z_2 = (x_2, -y_2) = (4, -3)$

d. $\frac{z_1}{z_2} = z_1 z_2^{-1} = (1, 2)(\frac{4}{25}, -\frac{3}{25}) = (\frac{4}{25} + \frac{6}{25}, -\frac{3}{25} + \frac{8}{25}) = (\frac{10}{25}, \frac{5}{25}) = (\frac{2}{5}, \frac{1}{5})$

8. a. $z_1 - z_2 = z_1 + (-z_2) = (6, -2) + (-2, 0) = (4, -2)$

b. $z_2^{-1} = (\frac{x_2}{x_2^2 + y_2^2}, \frac{-y_2}{x_2^2 + y_2^2}) = (\frac{2}{4 + 0}, \frac{-0}{4 + 0}) = (\frac{1}{2}, 0)$

c. $\overline{z}_2 = (x_2, -y_2) = (2, 0)$

d. $\dfrac{z_1}{z_2} = z_1 z_2^{-1} = (6, -2)(\tfrac{1}{2}, 0) = (3 - 0, 0 - 1) = (3, -1)$

9. a. $z_1 - z_2 = z_1 + (-z_2) = (-2, 0) + (-1, -4) = (-3, -4)$

b. $z_2^{-1} = (\dfrac{x_2}{x_2^2 + y_2^2}, \dfrac{-y_2}{x_2^2 + y_2^2}) = (\dfrac{1}{1 + 16}, \dfrac{-4}{1 + 16}) = (\tfrac{1}{17}, -\tfrac{4}{17})$

c. $\overline{z}_2 = (x_2, -y_2) = (1, -4)$

d. $\dfrac{z_1}{z_2} = z_1 z_2^{-1} = (-2, 0)(\tfrac{1}{17}, -\tfrac{4}{17}) = (-\tfrac{2}{17}, \tfrac{8}{17})$

10. a. $z_1 - z_2 = z_1 + (-z_2) = (3, 5) + (-5, -3) = (-2, 2)$

b. $z_2^{-1} = (\dfrac{x_2}{x_2^2 + y_2^2}, \dfrac{-y_2}{x_2^2 + y_2^2}) = (\dfrac{5}{25 + 9}, \dfrac{-3}{25 + 9}) = (\tfrac{5}{34}, -\tfrac{3}{34})$

c. $\overline{z}_2 = (x_2, -y_2) = (5, -3)$

d. $\dfrac{z_1}{z_2} = z_1 z_2^{-1} = (3, 5)(\tfrac{5}{34}, -\tfrac{3}{34}) = (\tfrac{15}{34} + \tfrac{15}{34}, -\tfrac{9}{34} + \tfrac{25}{34}) = (\tfrac{30}{34}, \tfrac{16}{34}) = (\tfrac{15}{17}, \tfrac{8}{17})$

11. a. $z_1 - z_2 = z_1 + (-z_2) = (12, 0) + (-4, 0) = (8, 0)$

b. $z_2^{-1} = (\dfrac{x_2}{x_2^2 + y_2^2}, \dfrac{-y_2}{x_2^2 + y_2^2}) = (\dfrac{4}{16 + 0}, \dfrac{-0}{16 + 0}) = (\tfrac{1}{4}, 0)$

c. $\overline{z}_2 = (x_2, -y_2) = (4, 0)$

d. $\dfrac{z_1}{z_2} = z_1 z_2^{-1} = (12, 0)(\tfrac{1}{4}, 0) = (3, 0)$

12. a. $z_1 - z_2 = z_1 + (-z_2) = (\tfrac{1}{3}, \tfrac{1}{2}) + (3, 2) = (3\tfrac{1}{3}, 2\tfrac{1}{2})$

b. $z_2^{-1} = (\dfrac{x_2}{x_2^2 + y_2^2}, \dfrac{-y_2}{x_2^2 + y_2^2}) = (\dfrac{-3}{9 + 4}, \dfrac{2}{9 + 4}) = (-\tfrac{3}{13}, \tfrac{2}{13})$

c. $\overline{z}_2 = (x_2, -y_2) = (-3, 2)$

d. $\dfrac{z_1}{z_2} = z_1 \cdot z_2^{-1} = (\tfrac{1}{3}, \tfrac{1}{2})(-\tfrac{3}{13}, \tfrac{2}{13}) = (-\tfrac{1}{13} - \tfrac{1}{13}, \tfrac{2}{39} + -\tfrac{3}{26}) =$

$(-\tfrac{2}{13}, \tfrac{4}{78} - \tfrac{9}{78}) = (-\tfrac{2}{13}, -\tfrac{5}{78})$

B 13. $\dfrac{(3\sqrt{3}, 5\sqrt{3})}{(2\sqrt{3}, \sqrt{3})} = \dfrac{(3\sqrt{3}, 5\sqrt{3})(2\sqrt{3}, -\sqrt{3})}{(2\sqrt{3}, \sqrt{3})(2\sqrt{3}, -\sqrt{3})} = \dfrac{(18 + 15, -9 + 30)}{(12 + 3, 0)} = \dfrac{(33, 21)}{(15, 0)} =$

$(\tfrac{33}{15}, \tfrac{21}{15}) = (\tfrac{11}{5}, \tfrac{7}{5})$

14. $\dfrac{(\sqrt{5}, -2\sqrt{5})(-\sqrt{5}, -\sqrt{5})}{(-\sqrt{5}, \sqrt{5})(-\sqrt{5}, -\sqrt{5})} = \dfrac{(-5 - 10, -5 + 10)}{(5 + 5, 0)} = (-\tfrac{15}{10}, \tfrac{5}{10}) = (-\tfrac{3}{2}, \tfrac{1}{2})$

15. $\dfrac{(\sqrt{3}, \sqrt{2})(2\sqrt{3}, \sqrt{2})}{(2\sqrt{3}, -\sqrt{2})(2\sqrt{3}, \sqrt{2})} = \dfrac{(6 - 2, \sqrt{6} + 2\sqrt{6})}{(12 + 2, 0)} = (\dfrac{4}{14}, \dfrac{3\sqrt{6}}{14}) = (\dfrac{2}{7}, \dfrac{3\sqrt{6}}{14})$

16. $\dfrac{(\sqrt{5}, -\sqrt{3})(2\sqrt{3}, -\sqrt{5})}{(2\sqrt{3}, \sqrt{5})(2\sqrt{3}, -\sqrt{5})} = \dfrac{(2\sqrt{15} - \sqrt{15}, -5 - 6)}{(12 + 5, 0)} = (\dfrac{\sqrt{15}}{17}, -\dfrac{11}{17})$

17. $\dfrac{(15\sqrt{6}, -8)(7\sqrt{2}, -\sqrt{3})}{(7\sqrt{2}, \sqrt{3})(7\sqrt{2}, -\sqrt{3})} = \dfrac{(210\sqrt{3} - 8\sqrt{3}, -45\sqrt{2} - 56\sqrt{2})}{(98 + 3, 0)} = (\dfrac{202\sqrt{3}}{101}, \dfrac{-101\sqrt{2}}{101})$

$= (2\sqrt{3}, -\sqrt{2})$

18. $\dfrac{(\sqrt{2}, -\sqrt{2})(1, -\sqrt{2})}{(1, \sqrt{2})(1, -\sqrt{2})} = \dfrac{(\sqrt{2} - 2, -2 - \sqrt{2})}{(1 + 2, 0)} = (\dfrac{\sqrt{2} - 2}{3}, \dfrac{-\sqrt{2} - 2}{3})$

C   19. $\dfrac{z_1}{z_2} = \dfrac{z_1 \bar{z}_2}{z_2 \bar{z}_2} = \dfrac{(r_1 \cos \theta_1, r_1 \sin \theta_1)(r_2 \cos \theta_2, -r_2 \sin \theta_2)}{(r_2 \cos \theta_2, r_2 \sin \theta_2)(r_2 \cos \theta_2, -r_2 \sin \theta_2)} =$

$\dfrac{(r_1 r_2 \cos \theta_1 \cos \theta_2 + r_1 r_2 \sin \theta_1 \sin \theta_2, -r_1 r_2 \cos \theta_1 \sin \theta_1 + r_1 r_2 \sin \theta_1 \cos \theta_2)}{(r_2^2 \cos^2 \theta_2 + r_2^2 \sin^2 \theta_2, -r_2^2 \sin \theta_2 \cos \theta_2 + r_2^2 \sin \theta_2 \cos \theta_2)}$

$= \dfrac{[r_1 r_2 \cos (\theta_1 - \theta_2), r_1 r_2 \sin (\theta_1 - \theta_2)]}{[r_2^2 (\cos^2 \theta_2 + \sin^2 \theta_2), 0]}$

$= \dfrac{(r_1 r_2 \cos (\theta_1 - \theta_2), r_1 r_2 \sin (\theta_1 - \theta_2))}{(r_2^2, 0)}$

$= (\dfrac{r_1}{r_2} \cos (\theta_1 - \theta_2), \dfrac{r_1}{r_2} \sin (\theta_1 - \theta_2))$

Exercises 7-3 · Pages 248-249

A   1. a. $z_1 + z_2 = (3 + 2i) + (5 + 3i) = 8 + 5i$;

b. $z_1 - z_2 = (3 + 2i) - (5 + 3i) = -2 - i$;

c. $z_1 z_2 = (3 + 2i)(5 + 3i) = (15 - 6) + (10 + 9)i = 9 + 19i$;

d. $\dfrac{z_1}{z_2} = \dfrac{z_1 \bar{z}_2}{z_2 \bar{z}_2} = \dfrac{(3 + 2i)(5 - 3i)}{(5 + 3i)(5 - 3i)} = \dfrac{21 + i}{34} = \dfrac{21}{34} + \dfrac{1}{34}i$

2. a. $(5 - i) + (5 + i) = 10 + 0i = 10$;

b. $(5 - i) - (5 + i) = 0 - 2i = -2i$;

c. $(5 - i)(5 + i) = (25 + 1) + (5 - 5)i = 26 + 0i = 26$;

d. $\dfrac{5 - i}{5 + i} = \dfrac{(5 - i)(5 - i)}{(5 + i)(5 - i)} = \dfrac{24 - 10i}{26} = \dfrac{24}{26} - \dfrac{10}{26}i = \dfrac{12}{13} - \dfrac{5}{13}i$

3. a. $(7 + i) + (10 - 4i) = 17 - 3i$;

b. $(7 + i) - (10 - 4i) = -3 + 5i$;

c. $(7 + i)(10 - 4i) = (70 + 4) + (-28 + 10)i = 74 - 18i;$

d. $\dfrac{7 + i}{10 - 4i} = \dfrac{(7 + i)(10 + 4i)}{(10 - 4i)(10 + 4i)} = \dfrac{66 + 38i}{116} = \dfrac{66}{116} + \dfrac{38}{116}i = \dfrac{33}{58} + \dfrac{19}{58}i$

4. a. $(7 + i) + (7 - i) = 14 + 0i = 14;$

b. $(7 + i) - (7 - i) = 0 + 2i = 2i;$

c. $(7 + i)(7 - i) = (49 + 1) + (-7 + 7)i = 50 + 0i = 50;$

d. $\dfrac{7 + i}{7 - i} = \dfrac{(7 + i)(7 + i)}{(7 - i)(7 + i)} = \dfrac{48 + 14i}{50} = \dfrac{48}{50} + \dfrac{14}{50}i = \dfrac{24}{25} + \dfrac{7}{25}i$

5. a. $(0 + 3i) + (6 + 0i) = 6 + 3i;$

b. $(0 + 3i) - (6 + 0i) = -6 + 3i;$

c. $(0 + 3i)(6 + 0i) = (0 - 0) + (0 + 18)i = 0 + 18i = 18i$

d. $\dfrac{(0 + 3i)}{(6 + 0i)} = \dfrac{(0 + 3i)(6 - 0i)}{(6 + 0i)(6 - 0i)} = \dfrac{0 + 18i}{36} = \dfrac{0}{36} + \dfrac{18}{36}i = 0 + \dfrac{1}{2}i = \dfrac{1}{2}i$

6. a. $(-4 + 0i) + (0 + 2i) = -4 + 2i;$

b. $(-4 + 0i) - (0 + 2i) = -4 - 2i;$

c. $(-4 + 0i)(0 + 2i) = (0 - 0) + (-8 + 0)i = 0 - 8i = -8i;$

d. $\dfrac{(-4 + 0i)}{(0 + 2i)} = \dfrac{(-4 + 0i)(0 - 2i)}{(0 + 2i)(0 - 2i)} = \dfrac{0 + 8i}{4} = \dfrac{0}{4} + \dfrac{8i}{4} = 0 + 2i = 2i$

7. a. $(0 + \sqrt{3}\,i) + (0 - i) = 0 + (\sqrt{3} - 1)i = (\sqrt{3} - 1)i;$

b. $(0 + \sqrt{3}\,i) - (0 - i) = 0 + (\sqrt{3} + 1)i = (\sqrt{3} + 1)i;$

c. $(0 + \sqrt{3}\,i)(0 - i) = (0 + \sqrt{3}) + (0 + 0)i = \sqrt{3} + 0i = \sqrt{3};$

d. $\dfrac{(0 + \sqrt{3}\,i)}{(0 - i)} = \dfrac{(0 + \sqrt{3}\,i)(0 + i)}{(0 - i)(0 + i)} = \dfrac{-\sqrt{3} + 0i}{1} = -\sqrt{3} + 0i = -\sqrt{3}$

8. a. $(0 + \sqrt{2}\,i) + (0 - 3\sqrt{2}\,i) = 0 - 2\sqrt{2}\,i = -2\sqrt{2}\,i;$

b. $(0 + \sqrt{2}\,i) - (0 - 3\sqrt{2}\,i) = 0 + 4\sqrt{2}\,i = 4\sqrt{2}\,i;$

c. $(0 + \sqrt{2}\,i)(0 - 3\sqrt{2}\,i) = (0 + 6) + (0 + 0)i = 6 + 0i = 6;$

d. $\dfrac{0 + \sqrt{2}\,i}{0 - 3\sqrt{2}\,i} = \dfrac{(0 + \sqrt{2}\,i)(0 + 3\sqrt{2}\,i)}{(0 - 3\sqrt{2}\,i)(0 + 3\sqrt{2}\,i)} = \dfrac{-6 + 0i}{18} = \dfrac{-6}{18} + \dfrac{0}{18}i = \dfrac{-1}{3} + 0i = \dfrac{-1}{3}$

9. c   10. d   11. c   12. d   13. c   14. b

15. $6(\frac{1}{2}) + 8(-\frac{1}{2})i = 3 - 4i;$ c   16. $6(\frac{1}{2}) + 4(0)i = 3 + 0i = 3;$ b

17. $(1 - i)^2 = (1 - i)(1 - i) = (1 - 1) + (-1 - 1)i = -2i$

18. $(1 + i)^2 = (1 + i)(1 + i) = (1 - 1) + (1 + 1)i = 2i$

19. $(5 - 2i)^2 = (5 - 2i)(5 - 2i) = (25 - 4) + (-10 - 10)i = 21 - 20i$

20. $(3 - 4i)^2 = (3 - 4i)(3 - 4i) = (9 - 16) + (-12 - 12)i = -7 - 24i$

21. $(3 + 3i)^2 = (3 + 3i)(3 + 3i) = (9 - 9) + (9 + 9)i = 0 + 18i = 18i$

22. $(2 + 6i)^2 = (2 + 6i)(2 + 6i) = (4 - 36) + (12 + 12)i = -32 + 24i$

23. $(3i)^2 = (3i)(3i) = 9i^2 = -9$ 24. $(-7i)^2 = (-7i)(-7i) = 49i^2 = -49$

25. $(a + bi)^2 = (a + bi)(a + bi) = (a^2 - b^2) + (ab + ab)i = (a^2 - b^2) + 2abi$

26. $(a - bi)^2 = (a - bi)(a - bi) = (a^2 - b^2) + (-ab - ab)i = (a^2 - b^2) - 2abi$

B   27. $(x + 3i)^2 = 18i$; $(x^2 - 9) + 6xi = 18i$, $x^2 - 9 = 0$ and $6x = 18$, $(x + 3)(x - 3) = 0$
    and $6x = 18$, $x = 3$ or $x - 3$ and $x = 3$, the desired value of x is 3

28. $(x - i)^2 = x - 1$; $(x^2 - 1) - 2xi = x - 1$, $x^2 - 1 = x - 1$ and $-2x = 0$, $x^2 - x = 0$
    and $-2x = 0$, $x = 1$ or 0 and $x = 0$, the desired value of x is 0

29. $(2x + 5i) + (10 - i) = (4)(x + i)$; $(2x + 10) + 4i = 4x + 4i$; $2x + 10 = 4x$ and $4 = 4$;
    $10 = 2x$ and $4 = 4$; $x = 5$, the desired value of x is 5.

30. $(x + i)(x - i) = 10$; $(x^2 + 1) + 0i = 10 + 0i$; $x^2 + 1 = 10$, $x^2 = 9$, $x = \pm 3$, the
    desired values of x are 3 or -3.

31. $2 \cos x + 2i \sin x = 1 + \sqrt{3}i$, $2 \cos x = 1$ and $2 \sin x = \sqrt{3}$, $\cos x = \frac{1}{2}$ and
    $\sin x = \frac{\sqrt{3}}{2}$, $x = \frac{\pi}{3}$ or $\frac{5\pi}{3}$ and $x = \frac{\pi}{3}$ or $\frac{2\pi}{3}$, the desired value of x is $\frac{\pi}{3}$

32. $\sqrt{2} \cos x + \sqrt{2}i \sin x = 1 + i$, $\sqrt{2} \cos x = 1$ and $\sqrt{2} \sin x = 1$, $\cos x = \frac{1}{\sqrt{2}}$ and
    $\sin x = \frac{1}{\sqrt{2}}$, $x = \frac{\pi}{4}$ or $\frac{7\pi}{4}$ and $x = \frac{\pi}{4}$ and $\frac{3\pi}{4}$, the desired value of x is $\frac{\pi}{4}$.

33. $4 \cos x + 4i \sin x = -2 + 2\sqrt{3}i$, $4 \cos x = -2$ and $4 \sin x = 2\sqrt{3}$; $\cos x = -\frac{1}{2}$
    and $\sin x = \frac{\sqrt{3}}{2}$, $x = \frac{2\pi}{3}$ or $\frac{4\pi}{3}$ and $x = \frac{\pi}{3}$ or $\frac{2\pi}{3}$, the desired value of x is $\frac{2\pi}{3}$.

34. $(\cos x + i \sin x)^2 = 2i$, $(\cos^2 x - \sin^2 x) + 2i \sin x \cos x = 2i$, $\cos^2 - \sin^2 x = 0$
    and $2 \sin x \cos x = 2$, $\cos 2x = 0$ and $\sin 2x = 2$; $x = \frac{\pi}{4}, \frac{3\pi}{4}, \frac{5\pi}{4}, \frac{7\pi}{4}$ or $\emptyset$;
    there are no values for x.

35. $(\cos x + i \sin x)^2 = (\cos x - i \sin x)^2$, $(\cos^2 x - \sin^2 x) + 2i \sin x \cos x =$

$(\cos^2 x - \sin^2 x) - 2i \sin x \cos x$, $\cos 2x + i \sin 2x = \cos x - i \sin 2x$,

$\cos 2x = \cos 2x$ and $\sin 2x = -\sin 2x$, $\sin 2x = 0$, $x = 0$, $\frac{\pi}{2}$, π, $\frac{3\pi}{2}$

Exercises 7-4 · Page 254

A

1.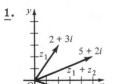

2.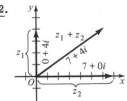

3.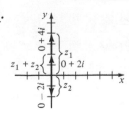

4.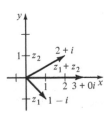

5.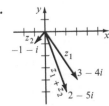

6.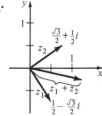

7.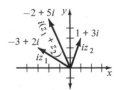

8.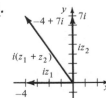

9.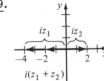

10.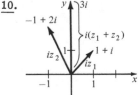

11.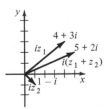

12.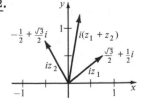

13. $i^7 = i^{4(1)+3} = i^3 = -i$

14. $i^{12} = i^{4(3)+0} = i^0 = 1$

15. $i^{18} = i^{4(4)+2} = i^2 = -1$

16. $i^{12345} = i^{4(3086)+1} = i$

17. $-i^{206} = -i^{4(51)+2} = -i^2 = 1$

18. $(-i)^{13} = (-i)^{4(3)+1} = -i$

19. $\frac{1}{i^3} = \frac{i^4}{i^3} = i$

20. $\frac{1}{i^5} = \frac{i^8}{i^5} = i^3 = -i$

21. $(1 - i)(2 + 7i) = (2 + 7) + (7 - 2)i =$

    $9 + 5i$

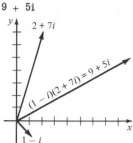

22. $(6 + 2i)(3 + 2i) = (18 - 4) + (12 + 6)i$

    $= 14 + 18i$

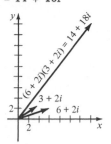

23. $(3 + 0i)(2 + 4i) = (6 - 0) + (12 + 0)i$

    $= 6 + 12i$

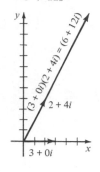

24. $(0 + 4i)(0 - 2i) = (0 + 8) + (0 + 0)i$

    $= 8$

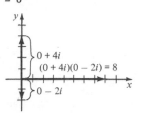

25. $(2 + 2i)(-1 + i) = (-2 - 2) + (2 - 2)i$

    $= -4$

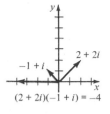

26. $(3 + i)(-2 + 6i) = (-6 - 6) + (18 - 2)i$

    $= -12 + 16i$

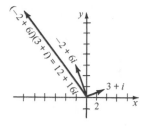

B   **27.** $\dfrac{1 + i}{2 + i} = \dfrac{(1 + i)(2 - i)}{(2 + i)(2 - i)} = \dfrac{3 + i}{5} =$

    $\dfrac{3}{5} + \dfrac{1}{5}i$

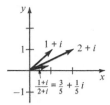

28. $\dfrac{2 + 2i}{1 + 2i} = \dfrac{(2 + 2i)(1 - 2i)}{(1 + 2i)(1 - 2i)} = \dfrac{6 - 2i}{5}$

    $\dfrac{6}{5} - \dfrac{2}{5}i$

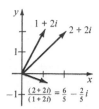

29. $\dfrac{1 + i}{1 - i} = \dfrac{(1 + i)(1 + i)}{(1 - i)(1 + i)} =$

$\dfrac{0 + 2i}{2} = i$

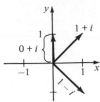

30. $\dfrac{6 - i}{3 + 4i} = \dfrac{(6 - i)(3 - 4i)}{(3 + 4i)(3 - 4i)} =$

$\dfrac{14 - 27i}{25} = \dfrac{14}{25} - \dfrac{27}{25}i$

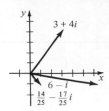

C   31. $(2 + i)(3 + 4i) = 2 + 11i$; $|2 + 11i| = |2 + i| \cdot |3 + 4i|$, $\sqrt{2^2 + 11^2} =$

$\sqrt{2^2 + 1^2} \cdot \sqrt{3^2 + 4^2}$, $\sqrt{125} = \sqrt{5} \cdot \sqrt{25}$, $\sqrt{125} = \sqrt{125}$; amp $(2 + 11i) = \tan^{-1} \dfrac{11}{2}$,

amp $(2 + i) = \tan^{-1} \dfrac{1}{2}$, and amp $(3 + 4i) = \tan^{-1} \dfrac{4}{3}$; $\tan^{-1} \dfrac{11}{2} = \tan^{-1} \dfrac{1}{2} + \tan^{-1} \dfrac{4}{3}$,

$\tan (\tan^{-1} \dfrac{11}{2}) = \tan [\tan^{-1} \dfrac{1}{2} + \tan^{-1} \dfrac{4}{3}]$; $\dfrac{11}{2} =$

$\dfrac{\tan (\tan^{-1} \frac{1}{2}) + \tan (\tan^{-1} \frac{4}{3})}{1 - \tan (\tan^{-1} \frac{1}{2}) \cdot \tan (\tan^{-1} \frac{4}{3})}$, $\dfrac{11}{2} = \dfrac{\frac{1}{2} + \frac{4}{3}}{1 - \frac{2}{3}}$, $\dfrac{11}{2} = \dfrac{3 + 8}{6 - 4}$, $\dfrac{11}{2} = \dfrac{11}{2}$

Exercises 7-5 · Pages 257-258

A   1. $1(\cos 30° + i \sin 30°) = 1(\dfrac{\sqrt{3}}{2}) + 1(\dfrac{1}{2})i = \dfrac{\sqrt{3}}{2} + \dfrac{1}{2}i$

2. $1(\cos 60° + i \sin 60°) = 1(\dfrac{1}{2}) + 1(\dfrac{\sqrt{3}}{2})i = \dfrac{1}{2} + \dfrac{\sqrt{3}}{2}i$

3. $1(\cos 90° + i \sin 90°) = 1(0) + 1(1)i = 0 + i$

4. $2(\cos 120° + i \sin 120°) = 2(-\dfrac{1}{2}) + 2(\dfrac{\sqrt{3}}{2})i = -1 + \sqrt{3}\,i$

5. $4(\cos 315° + i \sin 315°) = 4(\dfrac{1}{\sqrt{2}}) + 4(-\dfrac{1}{\sqrt{2}})i = 2\sqrt{2} - 2\sqrt{2}\,i$

6. $4[\cos (-60°) + i \sin (-60°)] = 4(\cos 60° - i \sin 60°) = 4(\dfrac{1}{2}) - 4(\dfrac{\sqrt{3}}{2})i = 2 - 2\sqrt{3}\,i$

7. $\rho = \sqrt{(\dfrac{1}{2})^2 + (\dfrac{\sqrt{3}}{2})^2} = 1$, $\cos \theta = \dfrac{1}{2}$ and $\sin \theta = \dfrac{\sqrt{3}}{2}$, $\theta = 60°$, $\dfrac{1}{2} + \dfrac{\sqrt{3}}{2}i =$

$\cos 60° + i \sin 60°$

8. $\rho = \sqrt{(-\dfrac{1}{2})^2 + (\dfrac{\sqrt{3}}{2})^2} = 1$, $\cos \theta = -\dfrac{1}{2}$, $\sin \theta = \dfrac{\sqrt{3}}{2}$, $\theta = 120°$, $-\dfrac{1}{2} + \dfrac{\sqrt{3}}{2}i =$

$\cos 120° + i \sin 120°$

9. $\rho = \sqrt{1^2 + (-\sqrt{3})^2} = 2$, $\sin \theta = -\dfrac{\sqrt{3}}{2}$, $\cos \theta = \dfrac{1}{2}$, $\theta = 300°$, $1 - \sqrt{3}\,i =$

$2(\cos 300° + i \sin 300°)$

10. $\rho = \sqrt{4^2 + (4\sqrt{3})^2} = 8$; $\cos \theta = \frac{4}{8} = \frac{1}{2}$, $\sin \theta = \frac{4\sqrt{3}}{8} = \frac{\sqrt{3}}{2}$, $\theta = 60$, $4 + 4\sqrt{3}\,i =$

$8(\cos 60° + i \sin.60°)$

11. $\rho = \sqrt{1^2 + 1^2} = \sqrt{2}$, $\cos \theta = \frac{1}{\sqrt{2}}$, $\sin \theta = \frac{1}{\sqrt{2}}$, $\theta = 45°$, $1 + i =$

$\sqrt{2}\,(\cos 45° + i \sin 45°)$

12. $\rho = \sqrt{0^2 + (-4)^2} = 4$, $\cos \theta = \frac{0}{4} = 0$, $\sin \theta = \frac{-4}{4} = -1$, $\theta = 270°$, $0 - 4i =$

$4(\cos 270° + i \sin 270°)$

13. $\rho = \sqrt{0^2 + 6^2} = 6$, $\cos \theta = \frac{0}{6} = 0$, $\sin \theta = \frac{6}{6} = 1$, $\theta = 90°$, $0 + 6i =$

$6(\cos 90° + i \sin 90°)$

14. $\rho = \sqrt{7^2 + 0^2} = 7$, $\cos \theta = \frac{7}{7} = 1$, $\sin \theta = \frac{0}{7} = 0$, $\theta = 0°$, $7 + 0i =$

$7(\cos 0° + i \sin 0°)$

15. $\rho = \sqrt{6^2 + (-3)^2} = 3\sqrt{5}$, $\cos \theta = \frac{6}{3\sqrt{5}} = \frac{2\sqrt{5}}{5}$, $\sin \theta = \frac{-3}{3\sqrt{5}} = -\frac{\sqrt{5}}{5}$, $\theta \doteq 333°30'$,

$6 - 3i = 3\sqrt{5}\,(\cos 333°30' + i \sin 333°30')$

16. $\rho = \sqrt{(-8)^2 + (-6)^2} = 10$, $\cos \theta = \frac{-8}{10} = -\frac{4}{5}$, $\sin \theta = -\frac{6}{10} = -\frac{3}{5}$, $\theta \doteq 216°50'$,

$-8 - 6i = 10\,(\cos 216°50' + i \sin 216°50')$

17. a. $z_1 z_2 = (2)(3)[\cos (60° + 30°) + i \sin (60° + 30°)] = 6(\cos 90° + i \sin 90°)$,

$6(0) + 6(1)i = 6i$;

b. $\frac{z_1}{z_2} = \frac{2}{3}[\cos (60° - 30°) + i \sin (60° - 30°)] = \frac{2}{3}(\cos 30° + i \sin 30°) =$

$\frac{2}{3}(\frac{\sqrt{3}}{2}) + i(\frac{2}{3})(\frac{1}{2}) = \frac{\sqrt{3}}{3} + \frac{1}{3}i$

18. a. $(6)(2)[\cos (90° - 60°) + i \sin (90° - 60°)] = 12(\cos 30° + i \sin 30°)$,

$12(\frac{\sqrt{3}}{2}) + 12(\frac{1}{2})i = 6\sqrt{3} + 6i$;

b. $\frac{z_1}{z_2} = \frac{6}{2}[\cos (90° + 60°) + i \sin (90° + 60°)] = 3(\cos 150° + i \sin 150°)$,

$3(-\frac{\sqrt{3}}{2}) + 3(\frac{1}{2})i = -\frac{3\sqrt{3}}{2} + \frac{3}{2}i$

19. a. $z_1 z_2 = (10)(5)[\cos (120° + 60°) + i \sin (120° + 60°)] = 50(\cos 180° + i \sin 180°)$,

$50(-1) + 50(0)i = -50$;

b. $\dfrac{z_1}{z_2} = \dfrac{10}{5}[\cos\ (120° - 60°) + i\ \sin\ (120° - 60°)] = 2(\cos\ 60° + i\ \sin\ 60°),$

$2(\tfrac{1}{2}) + 2(\dfrac{\sqrt{3}}{2})i = 1 + \sqrt{3}\,i$

20. a. $z_1 z_2 = (8)(6)[\cos\ (270° + 30°) + i\ \sin\ (270° + 30°)] = 48(\cos\ 300° + i\ \sin\ 300°),$

$48(\tfrac{1}{2}) + i(48)(-\dfrac{\sqrt{3}}{2}) = 24 - 24\sqrt{3}\,i;$

b. $\dfrac{z_1}{z_2} = \dfrac{8}{6}[\cos\ (270° - 30°) + i\ \sin\ (270° - 30°)] = \dfrac{4}{3}(\cos\ 240° + i\ \sin\ 240°),$

$\dfrac{4}{3}(-\tfrac{1}{2}) + \dfrac{4}{3}(-\dfrac{\sqrt{3}}{2})i = -\dfrac{2}{3} - \dfrac{2\sqrt{3}}{3}\,i$

21. a. $z_1 z_2 = (-2 - 2i)(-1 + i) = 4 = 4(1 + 0i) = 4(\cos\ 0° + i\ \sin\ 0°),$

b. $\dfrac{z_1}{z_2} = \dfrac{(-2 - 2i)}{(-1 + i)} = \dfrac{(-2 - 2i)(-1 - i)}{(-1 + i)(-1 - i)} = \dfrac{0 + 4i}{2} = 2i = 2(0 + 1i) =$

$2(\cos\ 90° + i\ \sin\ 90°)$

22. a. $z_1 z_2 = (-3 + 3i)(1 + i) = -6 + 0i = 6(-1 + 0i) = 6(\cos\ 180° + i\ \sin\ 180°);$

b. $\dfrac{z_1}{z_2} = \dfrac{(-3 + 3i)}{(1 + i)} = \dfrac{(-3 + 3i)(1 - i)}{(1 + i)(1 - i)} = \dfrac{0 + 6i}{2} = 0 + 3i = 3(0 + i) =$

$3(\cos\ 90° + i\ \sin\ 90°)$

23. a. $z_1 z_2 = -3i(5 - 5i) = -15 - 15i = 15(-1 - i) = 15\sqrt{2}\,(-\dfrac{1}{\sqrt{2}} - \dfrac{1}{\sqrt{2}}i) =$

$15\sqrt{2}\ (\cos\ 225° + i\ \sin\ 225°)$

b. $\dfrac{z_1}{z_2} = \dfrac{-3i}{5 - 5i} = \dfrac{-3i(5 + 5i)}{(5 - 5i)(5 + 5i)} = \dfrac{15 - 15i}{50} = \dfrac{3}{10}(1 + i) = \dfrac{3\sqrt{2}}{10}(\dfrac{1}{\sqrt{2}} - \dfrac{1}{\sqrt{2}}i) =$

$\dfrac{3\sqrt{2}}{10}(\cos\ 315° + i\ \sin\ 315°)$

24. a. $z_1 z_2 = 4i(1 - \sqrt{3}\,i) = 4\sqrt{3} + 4i = 4(\sqrt{3} + i) = 8(\dfrac{\sqrt{3}}{2} + \dfrac{1}{2}i) =$

$8(\cos\ 30° + i\ \sin\ 30°);$

b. $\dfrac{z_1}{z_2} = \dfrac{4i}{1 - \sqrt{3}\,i} = \dfrac{4i(1 + \sqrt{3}\,i)}{(1 - \sqrt{3}\,i)(1 + \sqrt{3}\,i)} = \dfrac{-4\sqrt{3} + 4i}{4} = -\sqrt{3} + 1i = 2(-\dfrac{\sqrt{3}}{2} + \dfrac{1}{2}i) =$

$2(\cos\ 150° + i\ \sin\ 150°)$

25. $z = \rho\,(\cos\ \theta + i\ \sin\ \theta),\ z^2 = z \cdot z = \rho\,(\cos\ \theta + i\ \sin\ \theta) \cdot \rho\,(\cos\ \theta + i\ \sin\ \theta) =$

$(\rho)(\rho)[\cos\ (\theta + \theta) + i\ \sin\ (\theta + \theta)] = \rho^2\,(\cos\ 2\theta + i\ \sin\ 2\theta)$

26. $z = \rho\,(\cos\ \theta + i\ \sin\ \theta),\ z^3 = z^2 \cdot z = \rho^2(\cos\ 2\theta + i\ \sin\ 2\theta) \cdot \rho\,(\cos\ \theta + i\ \sin\ \theta) =$

$\rho^2 \cdot \rho\,[\cos\ (2\theta + \theta) + i\ \sin\ (2\theta + \theta)] = \rho^3\,(\cos\ 3\theta + i\ \sin\ 3\theta)$

27. $\dfrac{1}{z} = \dfrac{1(\cos\ 0° + i\ \sin\ 0°)}{\rho(\cos\ \theta + i\ \sin\ \theta)} = \dfrac{1}{\rho}[\cos\ (0° - \theta) + i\ \sin\ (0° - \theta)] = \dfrac{1}{\rho}[\cos\ (-\theta) + i\ \sin\ (-\theta)]$

Exercises 7-6 · Pages 260-261

<u>A</u>   1. $[2(\cos 10° + i \sin 10°)]^6 = 2^6[\cos (6 · 10°) + i \sin (6 · 10°)] =$

$64(\cos 60° + i \sin 60°) = 64(\frac{1}{2}) + 64(\frac{\sqrt{3}}{2})i = 32 + 32\sqrt{3} i$

2. $[3(\cos 15° + i \sin 15°)]^4 = 3^4[\cos 4(15°) + i \sin 4(15°)] = 81(\cos 60° + i \sin 60°) =$

$81(\frac{1}{2}) + 81(\frac{\sqrt{3}}{2})i = \frac{81}{2} + \frac{81\sqrt{3}}{2} i$

3. $[\frac{1}{10}(\cos 20° + i \sin 20°)]^3 = (\frac{1}{10})^3[\cos 3(20°) + i \sin 3(20°)] =$

$\frac{1}{1000}(\cos 60° + i \sin 60°) = \frac{1}{1000}(\frac{1}{2}) + \frac{1}{1000}(\frac{\sqrt{3}}{2})i = \frac{1}{2000} + \frac{\sqrt{3}}{2000} i$

4. $[\frac{1}{2}(\cos 60° + i \sin 60°)]^{-5} = (\frac{1}{2})^{-5}[\cos (-5)(60°) + i \sin (-5)(60°)] =$

$32[\cos (-300°) + i \sin (-300°)] = 32(\frac{1}{2}) + 32(\frac{\sqrt{3}}{2})i = 16 + 16\sqrt{3} i$

5. $(1 - \sqrt{3}i)^3 = [2(\frac{1}{2} - \frac{\sqrt{3}}{2}i)]^3 = [2(\cos 300° + i \sin 300°)]^3 = 8(\cos 900° + i \sin 900°) =$

$8(\cos 180° + i \sin 180°) = 8(-1) + 8(0)i = -8$

6. $(3 + 3i)^4 = [3\sqrt{2}(\frac{1}{\sqrt{2}} + \frac{1}{\sqrt{2}} i)]^4 = [3\sqrt{2}(\cos 45° + i \sin 45°)]^4 =$

$324 (\cos 180° + i \sin 180°) = 324(-1) + 324(0)i = -324$

7. $(-\frac{1}{3} - \frac{1}{3}i)^{-4} = [\frac{\sqrt{2}}{3}(-\frac{1}{\sqrt{2}} - \frac{1}{\sqrt{2}} i)]^{-4} = [\frac{\sqrt{2}}{3}(\cos 225° + i \sin 225°)]^{-4} =$

$\frac{81}{4}[\cos (-900°) + i \sin (-900°)] = \frac{81}{4}(\cos 180° + i \sin 180°) = \frac{81}{4}(-1) + \frac{81}{4}(0)i = -\frac{81}{4}$

8. $(3 - \sqrt{3}i)^{-3} = [2\sqrt{3}(\frac{\sqrt{3}}{2} - \frac{1}{2}i)]^{-3} = [2\sqrt{3}(\cos 330° + i \sin 330°)]^{-3} =$

$\frac{1}{24\sqrt{3}}[\cos (-990°) + i \sin (-990°)] = \frac{1}{24\sqrt{3}}(\cos 90° + i \sin 90°) =$

$\frac{1}{24\sqrt{3}}(0) + \frac{1}{24\sqrt{3}}(1)i = +\frac{1}{24\sqrt{3}} i$

9. $[3(\cos 10° + i \sin 10°)]^3 [\frac{1}{3}(\cos 12° + i \sin 12°)]^5 = [27(\cos 30° + i \sin 30°)] ·$

$[\frac{1}{243}(\cos 60° + i \sin 60°)] = \frac{27}{243}(\cos 90° + i \sin 90°) = \frac{27}{243}(0) + \frac{27}{243}(1)i =$

$\frac{27}{243} i = \frac{1}{9}i$

10. $(\sqrt{3} + i)^4 (1 - i)^2 = [2(\cos 30° + i \sin 30°)]^4 · [\sqrt{2}(\cos 315° + i \sin 315°)]^2 =$

$[16(\cos 120° + i \sin 120°] · [2(\cos 630° + i \sin 630°)] = 32(\cos 750° + i \sin 750°) =$

$32(\cos 30° + i \sin 30°) = 32(\frac{\sqrt{3}}{2}) + 32(\frac{1}{2})i = 16\sqrt{3} + 16i$

11. $(1 + i)^4 (2 - 2i)^3 = [\sqrt{2}(\frac{1}{\sqrt{2}} + \frac{1}{\sqrt{2}}i)]^4 \cdot [2\sqrt{2}(\frac{1}{\sqrt{2}} - \frac{1}{\sqrt{2}}i)]^3 =$

$[\sqrt{2}(\cos 45° + i \sin 45°)]^4 \cdot [2\sqrt{2}(\cos 315° + i \sin 315°)]^3 =$

$[4(\cos 180° + i \sin 180°] \cdot [16\sqrt{2}(\cos 945° + i \sin 945°)] =$

$64\sqrt{2}(\cos 1125° + i \sin 1125°) = 64\sqrt{2}(\cos 45° + i \sin 45°) =$

$64\sqrt{2}(\frac{1}{\sqrt{2}}) + 64(\frac{1}{\sqrt{2}})i = 64 + 64i$

12. $\frac{(1 + i)^4}{(2 - 2i)^3} = \frac{[\sqrt{2}(\frac{1}{\sqrt{2}} + \frac{1}{\sqrt{2}}i)]^4}{[2\sqrt{2}(\frac{1}{\sqrt{2}} - \frac{1}{\sqrt{2}}i)]^3} = \frac{[\sqrt{2}(\cos 45° + i \sin 45°)]^4}{[2\sqrt{2}(\cos 315° + i \sin 315°)]^3} =$

$\frac{4(\cos 180° + i \sin 180°)}{16\sqrt{2}(\cos 945° + i \sin 945°)} = \frac{1}{4\sqrt{2}}[\cos (-765°) + i \sin (-765°)] =$

$\frac{1}{4\sqrt{2}}(\cos 45° - i \sin 45°) = \frac{1}{4\sqrt{2}}(\frac{1}{\sqrt{2}}) - (\frac{1}{4\sqrt{2}})(\frac{1}{\sqrt{2}})i = \frac{1}{8} - \frac{1}{8}i$

13. $\frac{(\sqrt{3} + i)^3}{(1 - i)^2} = \frac{[2(\frac{\sqrt{3}}{2} + \frac{1}{2}i)]^3}{[\sqrt{2}(\frac{1}{\sqrt{2}} - \frac{1}{\sqrt{2}}i)]^2} = \frac{[2(\cos 30° + i \sin 30°)]^3}{[\sqrt{2}(\cos 315° + i \sin 315°)]^2} =$

$\frac{8(\cos 90° + i \sin 90°)}{2(\cos 630° + i \sin 630°)} = \frac{4}{1}[\cos (-540°) + i \sin (-540°)] =$

$4(\cos 180° + i \sin 180°) = 4(-1) + 4(0)i = -4$

14. $\frac{(1 + \sqrt{3}i)^2}{(\sqrt{3} + i)^3} = \frac{[2(\frac{1}{2} + \frac{\sqrt{3}}{2}i)]^2}{[2(\frac{\sqrt{3}}{2} + \frac{1}{2}i)]^3} = \frac{[2(\cos 60° + i \sin 60°)]^2}{[2(\cos 30° + i \sin 30°)]^3} =$

$\frac{4(\cos 120° + i \sin 120°)}{8(\cos 90° + i \sin 90°)} = \frac{1}{2}(\cos 30° + i \sin 30°) = \frac{1}{2}(\frac{\sqrt{3}}{2}) + \frac{1}{2}(\frac{1}{2})i = \frac{\sqrt{3}}{4} + \frac{1}{4}i$

B   15. $(\cos \theta + i \sin \theta)^3 = \cos 3\theta + i \sin 3\theta$ by De Moivre's theorem; $(\cos \theta + i \sin \theta)^3$

= $(\cos \theta + i \sin \theta)^2 \cdot (\cos \theta + i \sin \theta)^3 = [(\cos^2 \theta - \sin^2 \theta) + i(2 \sin \theta \cos \theta)] \cdot$

$(\cos \theta + i \sin \theta) = (\cos^3 \theta - \sin^2 \theta \cos \theta - 2 \sin^2 \theta \cos \theta) +$

$[i(2 \sin \theta \cos^2 \theta) + i(\cos^2 \theta \sin \theta - \sin^3 \theta)] =$

$(\cos^3 \theta - \cos \theta + \cos^3 \theta - 2 \cos \theta + 2 \cos^3 \theta) +$

$i(2 \sin \theta - 2 \sin^3 \theta + \sin \theta - \sin^3 \theta - \sin^3 \theta) = (4 \cos^3 \theta - 3 \cos \theta) +$

$i(3 \sin \theta - 4 \sin^3 \theta)$; by definition of equality of complex numbers,

a. $\cos 3\theta = 4 \cos^3 \theta - 3 \cos \theta$ and b. $\sin 3\theta = 3 \sin \theta - 4 \sin^3 \theta$

16. $(\cos \theta + i \sin \theta)^4 = (\cos 4\theta + i \sin 4\theta)$ by De Moivre's theorem;

$(\cos \theta + i \sin \theta)^4 = (\cos \theta + i \sin \theta)^2(\cos \theta + i \sin \theta)^2 =$

$(\cos^2 \theta - \sin^2 \theta + 2i \sin \theta \cos \theta)(\cos^2 \theta - \sin^2 \theta + 2i \sin \theta \cos \theta) =$

$\cos^4 \theta - 2 \sin^2 \theta \cos^2 \theta + \sin^4 \theta - 4 \sin^2 \theta \cos^2 \theta +$

$i(2 \sin \theta \cos^3 \theta - 2 \sin^3 \theta \cos \theta + 2 \sin \theta \cos^3 \theta - 2 \sin^3 \theta \cos \theta) =$

$\cos^4 \theta - 6 \sin^2 \theta \cos^2 \theta + \sin^4 \theta + i(4 \sin \theta \cos^3 \theta - 4 \sin^3 \theta \cos \theta)$

By definition of equality of complex numbers $\cos 4\theta = \cos^4 \theta - 6 \sin^2 \theta \cos^2 \theta +$

$\sin^4 \theta$; $\sin 4\theta = 4 \sin \theta \cos^3 \theta - 4 \sin^3 \theta \cos \theta$

You may use trig identities $(\sin^2 \theta + \cos^2 \theta = 1)$ to further simplify,

$\cos 4\theta = \cos^4 \theta - 6(1 - \cos^2 \theta) \cos^2 \theta + (1 - \cos^2 \theta)^2$

$\cos 4\theta = 8 \cos^4 \theta - 8 \cos^2 \theta + 1$ and

$\sin 4\theta = 4 \sin \theta \cos^3 \theta - 4 \sin \theta (1 - \cos^2 \theta) \cos \theta$

$\sin 4\theta = 8 \sin \theta \cos^3 \theta - 4 \sin \theta \cos \theta$

<u>17.</u> $z^0 = [1(\cos \theta + i \sin \theta)]^0 = 1^0(\cos 0° + i \sin 0°) = 1(1) + 1(0)i = 1 + 0i = 1$

<u>18.</u> $z_1{}^3 = [1(\cos 0° + i \sin 0°)]^3 = 1(\cos 0° + i \sin 0°) = 1$;

$z_2{}^3 = [1(\cos 120° + i \sin 120°]^3 = 1(\cos 360° + i \sin 360°) =$

$1(\cos 0° + i \sin 0°) = 1$; $z_3{}^3 = [1(\cos 240° + i \sin 240°)]^3 =$

$1(\cos 720° + i \sin 720°) = 1(\cos 0° + i \sin 0°) = 1$; therefore $z_1{}^3 = z_2{}^3 = z_3{}^3 = 1.$

<u>C</u>   <u>19.</u> Let $z = \rho(\cos \theta + i \sin \theta)$ and prove that $z^n = \rho^n(\cos n\theta + i \sin n\theta)$.

For $n = 1$, $z^1 = \rho^1(\cos 1 \cdot \theta + i \sin 1 \cdot \theta) = \rho(\cos \theta + i \sin \theta)$.

Assume that for $n = k$, $z^n = \rho^n(\cos n \cdot \theta + i \sin n \cdot \theta)$

show that for $n = k + 1$, $z^{k+1} = \rho^{k+1}[\cos(k + 1) \cdot \theta + i \sin (k + 1) \cdot \theta]$

$z^{k+1} = (z^k)(z^1)$

$\qquad = [\rho^k(\cos k\theta + i \sin k\theta)][\rho(\cos \theta + i \sin \theta)]$

$\qquad = (\rho^k)(\rho)[\cos (k\theta + \theta) + i \sin (k\theta + \theta)]$

$\qquad = \rho^{k+1}[\cos (k + 1)\theta + i \sin (k + 1)\theta]$

<u>20.</u> $z^{-n} = \dfrac{1}{z^n} = (\dfrac{1}{z})^n = [\dfrac{1}{\rho}(\cos (-\theta) + i \sin (-\theta))]^n = (\dfrac{1}{\rho})^n[\cos n \cdot (-\theta) + i \sin n \cdot (-\theta)] =$

$\dfrac{1}{\rho^n}[\cos (-n\theta) + i \sin (-n\theta)] = \rho^{-n}[\cos (-n\theta) + i \sin (-n\theta)]$

Exercises 7-7 · Page 264

<u>A</u>   <u>1.</u> $1^{\frac{1}{3}} = (\cos\ 0° + i\ \sin\ 0°)^{\frac{1}{3}} = \cos \dfrac{0° + k\ 360°}{3} + i\ \sin \dfrac{0° + k\ 360°}{3};\ \cos\ 0° + i\ \sin\ 0°,$

$\cos\ 120° + i\ \sin\ 120°,\ \cos\ 240° + i\ \sin\ 240°,$ or $1,\ -\dfrac{1}{2} + \dfrac{\sqrt{3}}{2}i,\ -\dfrac{1}{2} - \dfrac{\sqrt{3}}{2}i$

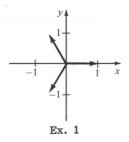

Ex. 1

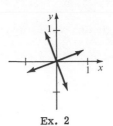

Ex. 2

<u>2.</u> $i^{\frac{1}{4}} = (\cos\ 90° + i\ \sin\ 90°)^{\frac{1}{4}} = \cos \dfrac{90° + k\ 360°}{4} + i\ \sin \dfrac{90° + k\ 360°}{4};$

$\cos\ 22.5° + i\ \sin\ 22.5°,\ \cos\ 112.5° + i\ \sin\ 112.5°,\ \cos\ 202.5° + i\ \sin\ 202.5°,$

$\cos\ 292.5° + i\ \sin\ 292.5°$

<u>3.</u> $[16(\cos\ 120° + i\ \sin\ 120°)]^{\frac{1}{4}} = 2(\cos \dfrac{120° + k\ 360°}{4} + i\ \sin \dfrac{120° + k\ 360°}{4});$

$2(\cos\ 30° + i\ \sin\ 30°),\ 2(\cos\ 120° + i\ \sin\ 120°),\ 2(\cos\ 210° + i\ \sin\ 210°),$

$2(\cos\ 300° + i\ \sin\ 300°);\ 2(\dfrac{\sqrt{3}}{2} + \dfrac{1}{2}i) = \sqrt{3} + i,\ 2(-\dfrac{1}{2} + \dfrac{\sqrt{3}}{2}i) = -1 + \sqrt{3}\,i,$

$2(-\dfrac{\sqrt{3}}{2} - \dfrac{1}{2}i) = -\sqrt{3} - i,\ 2(\dfrac{1}{2} - \dfrac{\sqrt{3}}{2}i) = 1 - \sqrt{3}\,i$

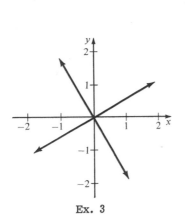

Ex. 3

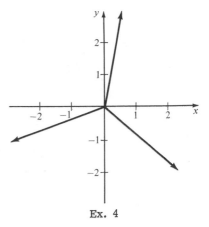

Ex. 4

<u>4.</u> $[27(\cos\ 240° + i\ \sin\ 240°)]^{\frac{1}{3}} = 3(\cos \dfrac{240° + k\ 360°}{3} + i\ \sin \dfrac{240° + k\ 360°}{3});$

$3(\cos\ 80° + i\ \sin\ 80°),\ 3(\cos\ 200° + i\ \sin\ 200°),\ 3(\cos\ 320° + i\ \sin\ 320°)$

5. $(16\sqrt{2} + 16\sqrt{2}\,i)^{\frac{1}{5}} = [32(\frac{1}{\sqrt{2}} + \frac{1}{\sqrt{2}}\,i)]^{\frac{1}{5}} = [32(\cos 45° + i \sin 45°)]^{\frac{1}{5}} =$

$2(\cos \dfrac{45° + k\,360°}{5} + i \sin \dfrac{45° + k\,360°}{5})$; $2(\cos 9° + i \sin 9°)$, $2(\cos 81° + i \sin 81°)$,

$2(\cos 153° + i \sin 153°)$, $2(\cos 225° + i \sin 225°)$, $2(\cos 297° + i \sin 297°)$,

$2(0.988 + 0.156i) \doteq 1.98 + 0.31i$, $2(0.156 + 0.988i) \doteq 0.31 + 1.98i$,

$2(-0.891 + 0.454i) \doteq -1.78 + 0.91i$, $2(-0.707 - 0.707i) \doteq -1.41 - 1.41i$,

$2(0.454 - 0.891i) \doteq 0.91 - 1.78i$

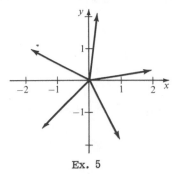

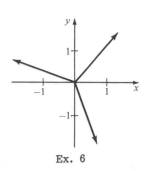

Ex. 5   Ex. 6

6. $(-4\sqrt{3} + 4i)^{\frac{1}{3}}$ 1 $[8(-\frac{\sqrt{3}}{2} + \frac{1}{2}i)]^{\frac{1}{3}} = [8(\cos 150° + i \sin 150°)]^{\frac{1}{3}} =$

$2(\cos \dfrac{150° + k\,360°}{3} + i \sin \dfrac{150° + k\,360°}{3})$; $2(\cos 50° + i \sin 50°)$,

$2(\cos 170° + i \sin 170°)$, $2(\cos 290° + i \sin 290°)$

7. $z^5 - 1 = 0$, $z^5 = 1$, $z = 1^{\frac{1}{5}} = (\cos 0° + i \sin 0°)^{\frac{1}{5}} = \cos \dfrac{0° + k\,360°}{5} +$

$i \sin \dfrac{0° + k\,360°}{5}$; $\{\cos 0° + i \sin 0°,\ \cos 72° + i \sin 72°,\ \cos 144° + i \sin 144°,$

$\cos 216° + i \sin 216°,\ \cos 288° + i \sin 288°\}$

8. $z^5 - i = 0$, $z^5 = i$, $z = i^{\frac{1}{5}} = (\cos 90° + i \sin 90°)^{\frac{1}{5}} = \cos \dfrac{90° + k\,360°}{5} +$

$i \sin \dfrac{90° + k\,360°}{5}$; $\{\cos 18° + i \sin 18°,\ \cos 90° + i \sin 90°,\ \cos 162° + i \sin 162°,$

$\cos 234° + i \sin 234°,\ \cos 306° + i \sin 306°\}$

9. $z^3 + 4 + 4\sqrt{3}\,i = 0$, $z^3 = -4 - 4\sqrt{3}\,i$, $z = (-4 - 4\sqrt{3}\,i)^{\frac{1}{3}} = [8(-\frac{1}{2} - \frac{\sqrt{3}}{2}i)]^{\frac{1}{3}} =$

$[8(\cos 240° + i \sin 240°)]^{\frac{1}{3}} = 2(\cos \dfrac{240° + k\,360°}{3} + i \sin \dfrac{240° + k\,360°}{3})$;

$\{2(\cos 80° + i \sin 80°),\ 2(\cos 200° + i \sin 200°),\ 2(\cos 320° + i \sin 320°)\}$

10. $z^4 + 8\sqrt{3} - 8i$, $z^4 = -8\sqrt{3} + 8i$, $z = (-8\sqrt{3} + 8i)^{\frac{1}{4}} = [16(-\frac{\sqrt{3}}{2} + \frac{1}{2}i)]^{\frac{1}{4}} =$

$[16(\cos 150° + i \sin 150°)]^{\frac{1}{4}} = 2(\cos \dfrac{150° + k\,360°}{4} + i \sin \dfrac{150° + k\,360°}{4})$;

{2(cos 37.5° + i sin 37.5°), 2(cos 127.5° + i sin 127.5°),

2(cos 217.5° + i sin 217.5°), 2(cos 307.5° + i sin 307.5°)}

<u>C</u> <u>11.</u> From problem 1, the three cube roots of 1 are cos 0° + i sin 0°,

cos 120° + i sin 120°, and cos 240° + i sin 240°. But cos 0° + i sin 0° = 1,

so let ω = cos 120° + i sin 120°, then ω^2 = (cos 120° + i sin 120°)^2 =

cos 240° + i sin 240°. Therefore the roots are in the form of 1, ω, and ω^2.

<u>12.</u> $1^{\frac{1}{4}}$ = (cos 0° + i sin 0°)$^{\frac{1}{4}}$ = cos $\dfrac{0° + k\,360°}{4}$ + i sin $\dfrac{0° + k\,360°}{4}$, the roots are

cos 0° + i sin 0°, cos 90° + i sin 90°, cos 180° + i sin 180°, and

cos 270° + i sin 270°. But cos 0° + i sin 0° = 1, so let ω =

(cos 90° + i sin 90°), ω^2 = (cos 90° + i sin 90°)^2 = cos 180° + i sin 180°,

ω^3 = (cos 90° + i sin 90°)^3 = cos 270° + i sin 270°; therefore the roots are in

the form of 1, ω, ω^2, and ω^3.

<u>13.</u> $\omega = -\dfrac{1}{2} + \dfrac{\sqrt{3}}{2}i$; $\omega^2 = -\dfrac{1}{2} - \dfrac{\sqrt{3}}{2}i$; $1 + \omega + \omega^2 = 1 - \dfrac{1}{2} + \dfrac{\sqrt{3}}{2}i - \dfrac{1}{2} - \dfrac{\sqrt{3}}{2}i = 0$;

$1 + \omega + \omega^3 = 0$; $\omega + \omega^2 + \omega\omega^2 = \omega(1 + \omega + \omega^2) = \omega \cdot 0 = 0$

<u>14.</u> $\omega = i$; $\omega^2 = -1$; $\omega^3 = -i$; $1 + \omega + \omega^2 + \omega^3 = 1 + i - 1 - i = 0$;

$1 + \omega + \omega^2 + \omega^3 = 0$; $\omega + \omega^2 + \omega^3 + \omega\omega^2 + \omega\omega^3 + \omega^2\omega^3 = i - 1 - i - i + 1 + i$

= 0, using ω = i.

<u>15.</u> $\omega^4 - 1 - i = 0$, $\omega^4 = 1 + i$, $\omega = (1 + i)^{\frac{1}{4}} = [\sqrt{2}(\dfrac{1}{\sqrt{2}} + \dfrac{1}{\sqrt{2}}i)]^{\frac{1}{4}}$ =

$[\sqrt{2}(\cos 45° + i \sin 45°)]^{\frac{1}{4}} = \sqrt[8]{2}\,(\cos \dfrac{45° + k\,360°}{4} + i \sin \dfrac{45° + k\,360°}{4})$,

$\omega = \{\sqrt[8]{2}\,[\cos(11.25 + k\,90) + i \sin(11.25 + k\,90)]\}$; $z = \omega^3$ =

$\{(\sqrt[8]{2}\,[\cos(11.25 + 90k) + i \sin(11.25 + 90k)])^3\}$, z =

$\{\sqrt[8]{8}\,(\cos 33.75° + i \sin 33.75°)°$, $\sqrt[8]{8}\,(\cos 303.75° + i \sin 303.75°)$;

$\sqrt[8]{8}\,(\cos 573.75° + i \sin 573.75°) = \sqrt[8]{8}\,(\cos (213.75°) + i \sin (213.75°))$;

$\sqrt[8]{8}\,(\cos 843.75° + i \sin 843.75°) = \sqrt[8]{8}\,(\cos (123.75°) + i \sin (123.75°))\}$

<u>16.</u> $\omega^3 + \dfrac{1}{2} - \dfrac{\sqrt{3}\,i}{2} = 0$, $\omega^3 = -\dfrac{1}{2} + \dfrac{\sqrt{3}}{2}i$, $\omega = (-\dfrac{1}{2} + \dfrac{\sqrt{3}}{2}i)^{\frac{1}{3}} = (\cos 120° + i \sin 120°)^{\frac{1}{3}}$ =

cos $\dfrac{120° + k\,360°}{3}$ + i sin $\dfrac{120° + k\,360°}{3} = \{\cos (40° + k\,120°) + i \sin (40° + k\,120°)\}$;

$z = \omega^4 = \{[\cos (40° + k\,120°) + i \sin (40° + k\,120°)]^4\}$, z =

{cos 160° + i sin 160°, cos 280° + i sin 280°, cos 40° + i sin 40°}

Chapter Test · Page 266

1. a. $z_1 + z_2 = (5, -2) + (1, 4) = (6, 2)$;

   b. $z_1 z_2 = (5, -2)(1, 4) = (5 + 8, 20 - 2) = (13, 18)$

2. a. $z_1 - z_2 = z_1 + (-z_2) = (5, -2) + (-1, -4) = (4, -6)$;

   b. $z_2^{-1} = (\dfrac{1}{1^2 + 4^2}, \dfrac{-4}{1^2 + 4^2}) = (\dfrac{1}{17}, -\dfrac{4}{17})$

   c. $\bar{z}_2 = (x_2, -y_2) = (1, -4)$;

   d. $\dfrac{z_1}{z_2} = z_1 z_2^{-1} = (5, -2)(\dfrac{1}{17}, -\dfrac{4}{17}) = (\dfrac{5}{17} - \dfrac{8}{17}, \dfrac{-20}{17} - \dfrac{2}{17}) = (-\dfrac{3}{17}, \dfrac{-22}{17})$

3. $(x + i)(x - 2i) = 6 + 2i$, $(x^2 + 2) - ix = 6 + 2i$, $x^2 + 2 = 6$ and $-x = 2$, $x^2 = 4$ and $x = -2$, $x = \pm 2$ and $x = -2$, the desired value of x is -2.

4.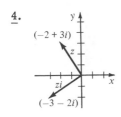

5. $z_1 z_2 = (1 + 4i)(2 + 3i) = (2 - 12) + (3 + 8)i = -10 + 11i$

   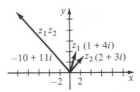

6. a. $4 - 4i = 4\sqrt{2}\left(\dfrac{1}{\sqrt{2}} - \dfrac{1}{\sqrt{2}}i\right) = 4\sqrt{2}(\cos 315° + i \sin 315°)$;

   b. $7i = 0 + 7i = 7(0 + i) = 7(\cos 90° + i \sin 90°)$

7. $z_1 = (1 + i) = \sqrt{2}\left(\dfrac{1}{\sqrt{2}} + \dfrac{1}{\sqrt{2}}i\right) = \sqrt{2}(\cos 45° + i \sin 45°)$, $z_2 = (\sqrt{3} - i) = 2(\dfrac{\sqrt{3}}{2} - \dfrac{1}{2}i) = 2(\cos 330° - i \sin 330°)$,

   a. $z_1 z_2 = [\sqrt{2}(\cos 45° + i \sin 45°)] \cdot [2(\cos 330° - i \sin 330°)] = 2\sqrt{2}(\cos 375° + i \sin 375°) = 2\sqrt{2}(\cos 15° + i \sin 15°)$;

   b. $\dfrac{z_1}{z_2} = \dfrac{\sqrt{2}(\cos 45° + i \sin 45°)}{2(\cos 330° + i \sin 330°)} = \dfrac{\sqrt{2}}{2}[\cos (-285°) + i \sin (-285°)] = \dfrac{\sqrt{2}}{2}(\cos 75° + i \sin 75°)$

8. a. $[\dfrac{1}{2}(\cos 130° + i \sin 130°)]^3 = \dfrac{1}{8}(\cos 390° + i \sin 390°) = \dfrac{1}{8}(\cos 30° + i \sin 30°) = \dfrac{1}{8}(\dfrac{\sqrt{3}}{2}) + \dfrac{1}{8}(\dfrac{1}{2})i = \dfrac{\sqrt{3}}{16} + \dfrac{1}{16}i$;

<u>b.</u> $\dfrac{(\sqrt{3} - i)^4}{i^3} = \dfrac{[2(\frac{\sqrt{3}}{2} - \frac{1}{2}i)]^4}{(0 + i)^3} = \dfrac{[2(\cos\ 330° + i\ \sin\ 330°)]^4}{(\cos\ 90° + i\ \sin\ 90°)^3} =$

$\dfrac{16(\cos\ 1320° + i\ \sin\ 1320°)}{(\cos\ 270° + i\ \sin\ 270°)} = 16\ (\cos\ 1050° + i\ \sin\ 1050°) =$

$16\ (\cos\ 330° + i\ \sin\ 330°) = 16(\frac{\sqrt{3}}{2}) + 16(-\frac{1}{2})i = 8\sqrt{3} - 8i$

<u>9.</u> $\omega^4 = \frac{1}{2} + \frac{\sqrt{3}}{2}i,\ \ \omega = (\frac{1}{2} + \frac{\sqrt{3}}{2}i)^{\frac{1}{4}} = (\cos\ 60° + i\ \sin\ 60°)^{\frac{1}{4}} = \cos\ \dfrac{60° + k\ 360°}{4} +$

$i\ \sin\ \dfrac{60° + k\ 360°}{4} = \cos\ (15° + k\ 90°) + i\ \sin\ (15° + k\ 90°),$

$\omega \in \{\cos\ 15° + i\ \sin\ 15°,\ \cos\ 105° + i\ \sin\ 105°,\ \cos\ 195° + i\ \sin\ 195°,$

$\cos\ 285° + i\ \sin\ 285°\}$

Exercises 8-1 · Pages 275-276

A 1. $A + B = \begin{bmatrix} 1 & -2 \\ 5 & 3 \end{bmatrix} + \begin{bmatrix} 0 & 5 \\ -4 & -6 \end{bmatrix} = \begin{bmatrix} 1 & 3 \\ 1 & -3 \end{bmatrix}$

2. $A - 2B = \begin{bmatrix} 1 & -2 \\ 5 & 3 \end{bmatrix} + \left(-2 \begin{bmatrix} 0 & 5 \\ -4 & -6 \end{bmatrix} \right) = \begin{bmatrix} 1 & -2 \\ 5 & 3 \end{bmatrix} + \begin{bmatrix} 0 & -10 \\ 8 & 12 \end{bmatrix} = \begin{bmatrix} 1 & -12 \\ 13 & 15 \end{bmatrix}$

3. $3A + B = \left(3 \begin{bmatrix} 1 & -2 \\ 5 & 3 \end{bmatrix} \right) + \begin{bmatrix} 0 & 5 \\ -4 & -6 \end{bmatrix} = \begin{bmatrix} 3 & -6 \\ 15 & 9 \end{bmatrix} + \begin{bmatrix} 0 & 5 \\ -4 & -6 \end{bmatrix} = \begin{bmatrix} 3 & -1 \\ 11 & 3 \end{bmatrix}$

4. $4A + 4B = \left(4 \begin{bmatrix} 1 & -2 \\ 5 & 3 \end{bmatrix} \right) + \left(4 \begin{bmatrix} 0 & 5 \\ -4 & -6 \end{bmatrix} \right) = \begin{bmatrix} 4 & -8 \\ 20 & 12 \end{bmatrix} + \begin{bmatrix} 0 & 20 \\ -16 & -24 \end{bmatrix} = \begin{bmatrix} 4 & 12 \\ 4 & -12 \end{bmatrix}$

5. $3(A + B) = 3A + 3B = \left(3 \begin{bmatrix} 1 & -2 \\ 5 & 3 \end{bmatrix} \right) + \left(3 \begin{bmatrix} 0 & 5 \\ -4 & -6 \end{bmatrix} \right) = \begin{bmatrix} 3 & -6 \\ 15 & 9 \end{bmatrix} + \begin{bmatrix} 0 & 15 \\ -12 & -18 \end{bmatrix} =$

$\begin{bmatrix} 3 & 9 \\ 3 & -9 \end{bmatrix}$

6. $-2(B - A) = -2 \left(\begin{bmatrix} 0 & 5 \\ -4 & -6 \end{bmatrix} - \begin{bmatrix} 1 & -2 \\ 5 & 3 \end{bmatrix} \right) = -2 \begin{bmatrix} -1 & 7 \\ -9 & -9 \end{bmatrix} = \begin{bmatrix} 2 & -14 \\ 18 & 18 \end{bmatrix}$

7. $\begin{bmatrix} 8 & 0 \\ 0 & 8 \end{bmatrix} = 8 \begin{bmatrix} 1 & 0 \\ 0 & 1 \end{bmatrix}$

8. $\begin{bmatrix} 2 & -2 \\ -2 & 2 \end{bmatrix} = 2 \begin{bmatrix} 1 & -1 \\ -1 & 1 \end{bmatrix}$

9. $\begin{bmatrix} 6 & 21 \\ -15 & 0 \end{bmatrix} = 3 \begin{bmatrix} 2 & 7 \\ -5 & 0 \end{bmatrix}$

10. $\begin{bmatrix} 3 & \frac{1}{2} \\ 12 & 6 \end{bmatrix} = \frac{1}{2} \begin{bmatrix} 6 & 1 \\ 24 & 12 \end{bmatrix}$

11. $\begin{bmatrix} -8 & \frac{1}{3} \\ 1 & \frac{3}{4} \end{bmatrix} = \frac{1}{12} \begin{bmatrix} -96 & 4 \\ 12 & 9 \end{bmatrix}$

12. $\begin{bmatrix} 168 & 84 \\ 252 & 378 \end{bmatrix} = 42 \begin{bmatrix} 4 & 2 \\ 6 & 9 \end{bmatrix}$

13. $X + \begin{bmatrix} 2 & 0 \\ 0 & 2 \end{bmatrix} = \begin{bmatrix} 5 & 0 \\ 0 & 5 \end{bmatrix}$, $X + 0 = \begin{bmatrix} 5 & 0 \\ 0 & 5 \end{bmatrix} + \begin{bmatrix} -2 & 0 \\ 0 & -2 \end{bmatrix}$, $X = \begin{bmatrix} 3 & 0 \\ 0 & 3 \end{bmatrix}$

14. $\begin{bmatrix} 2 & 1 \\ 1 & 3 \end{bmatrix} + X = \begin{bmatrix} 0 & 1 \\ 0 & 7 \end{bmatrix}$, $0 + X = \begin{bmatrix} -2 & -1 \\ -1 & -3 \end{bmatrix} + \begin{bmatrix} 0 & 1 \\ 0 & 7 \end{bmatrix}$, $X = \begin{bmatrix} -2 & 0 \\ -1 & 4 \end{bmatrix}$

15. $2X + \begin{bmatrix} 3 & 1 \\ -1 & 11 \end{bmatrix} = \begin{bmatrix} 0 & 0 \\ 0 & 0 \end{bmatrix}$, $2X + 0 = 0 + \begin{bmatrix} -3 & -1 \\ 1 & -11 \end{bmatrix}$, $2X = \begin{bmatrix} -3 & -1 \\ 1 & -11 \end{bmatrix}$,

$$X = \frac{1}{2}\begin{bmatrix} -3 & -1 \\ 1 & -11 \end{bmatrix}, \quad X = \begin{bmatrix} -\frac{3}{2} & -\frac{1}{2} \\ \frac{1}{2} & -\frac{11}{2} \end{bmatrix}$$

16. $3X - \begin{bmatrix} 1 & 7 \\ 6 & -2 \end{bmatrix} = \begin{bmatrix} 3 & 10 \\ 2 & -3 \end{bmatrix}$, $3X + 0 = \begin{bmatrix} 3 & 10 \\ 2 & -3 \end{bmatrix} + \begin{bmatrix} 1 & 7 \\ 6 & -2 \end{bmatrix}$, $3X = \begin{bmatrix} 4 & 17 \\ 8 & -5 \end{bmatrix}$,

$$X = \frac{1}{3}\begin{bmatrix} 4 & 17 \\ 8 & -5 \end{bmatrix}, \quad X = \begin{bmatrix} \frac{4}{3} & \frac{17}{3} \\ \frac{8}{3} & -\frac{5}{3} \end{bmatrix}$$

17. $2X + \frac{1}{2}\begin{bmatrix} -4 & 10 \\ 6 & 0 \end{bmatrix} = 4\begin{bmatrix} 1 & 0 \\ 0 & 1 \end{bmatrix}$, $2X + \begin{bmatrix} -2 & 5 \\ 3 & 0 \end{bmatrix} = \begin{bmatrix} 4 & 0 \\ 0 & 4 \end{bmatrix}$, $2X + 0 =$

$$\begin{bmatrix} 4 & 0 \\ 0 & 4 \end{bmatrix} + \begin{bmatrix} 2 & -5 \\ -3 & 0 \end{bmatrix}, \quad 2X = \begin{bmatrix} 6 & -5 \\ -3 & 4 \end{bmatrix}, \quad X = \frac{1}{2}\begin{bmatrix} 6 & -5 \\ -3 & 4 \end{bmatrix}, \quad X = \begin{bmatrix} 3 & -\frac{5}{2} \\ -\frac{3}{2} & 2 \end{bmatrix}$$

18. $4X = \begin{bmatrix} 8 & -16 \\ -16 & 8 \end{bmatrix}$, $X = \frac{1}{4}\begin{bmatrix} 8 & -16 \\ -16 & 8 \end{bmatrix}$, $X = \begin{bmatrix} 2 & -4 \\ -4 & 2 \end{bmatrix}$

19. $X + \begin{bmatrix} 1 & 0 \\ 0 & 1 \end{bmatrix} = \begin{bmatrix} 0 & 1 \\ 1 & 0 \end{bmatrix}$, $X + 0 = \begin{bmatrix} 0 & 1 \\ 1 & 0 \end{bmatrix} + \begin{bmatrix} -1 & 0 \\ 0 & -1 \end{bmatrix}$, $X = \begin{bmatrix} -1 & 1 \\ 1 & -1 \end{bmatrix}$

20. $6X = \begin{bmatrix} 20 & 10 \\ -10 & 20 \end{bmatrix} - 4X$, $6X + 4X = \begin{bmatrix} 20 & 10 \\ -10 & 20 \end{bmatrix}$, $10X = \begin{bmatrix} 20 & 10 \\ -10 & 20 \end{bmatrix}$,

$$X = \frac{1}{10}\begin{bmatrix} 20 & 10 \\ -10 & 20 \end{bmatrix}, \quad X = \begin{bmatrix} 2 & 1 \\ -1 & 2 \end{bmatrix}$$

B 21. $(A + B) + C = \left(\begin{bmatrix} a_1 & b_1 \\ c_1 & d_1 \end{bmatrix} + \begin{bmatrix} a_2 & b_2 \\ c_2 & d_2 \end{bmatrix} \right) + \begin{bmatrix} a_3 & b_3 \\ c_3 & d_3 \end{bmatrix} =$

$$\begin{bmatrix} a_1 + a_2 & b_1 + b_2 \\ c_1 + c_2 & d_1 + d_2 \end{bmatrix} + \begin{bmatrix} a_3 & b_3 \\ c_3 & d_3 \end{bmatrix} = \begin{bmatrix} (a_1 + a_2) + a_3 & (b_1 + b_2) + b_3 \\ (c_1 + c_2) + c_3 & (d_1 + d_2) + d_3 \end{bmatrix} =$$

$$\begin{bmatrix} a_1 + (a_2 + a_3) & b_1 + (b_2 + b_3) \\ c_1 + (c_2 + c_3) & d_1 + (d_2 + d_3) \end{bmatrix} = \begin{bmatrix} a_1 & b_1 \\ c_1 & d_1 \end{bmatrix} + \begin{bmatrix} a_2 + a_3 & b_2 + b_3 \\ c_2 + c_3 & d_2 + d_3 \end{bmatrix}$$

$$\begin{bmatrix} a_1 & b_1 \\ c_1 & d_1 \end{bmatrix} + \left(\begin{bmatrix} a_2 & b_2 \\ c_2 & d_2 \end{bmatrix} + \begin{bmatrix} a_3 & b_3 \\ c_3 & d_3 \end{bmatrix} \right) = A + (B + C)$$

22. $A + 0 = \begin{bmatrix} a_1 & b_1 \\ c_1 & d_1 \end{bmatrix} + \begin{bmatrix} 0 & 0 \\ 0 & 0 \end{bmatrix} = \begin{bmatrix} a_1 + 0 & b_1 + 0 \\ c_1 + 0 & d_1 + 0 \end{bmatrix} = \begin{bmatrix} a_1 & b_1 \\ c_1 & d_1 \end{bmatrix} = A$

23. $A + (-A) = \begin{bmatrix} a_1 & b_1 \\ c_1 & d_1 \end{bmatrix} + \begin{bmatrix} -a_1 & -b_1 \\ -c_1 & -d_1 \end{bmatrix} = \begin{bmatrix} a_1 - a_1 & b_1 - b_1 \\ c_1 - c_1 & d_1 - d_1 \end{bmatrix} = \begin{bmatrix} 0 & 0 \\ 0 & 0 \end{bmatrix} = 0$

24. $A + B = \begin{bmatrix} a_1 & b_1 \\ c_1 & d_1 \end{bmatrix} + \begin{bmatrix} a_2 & b_2 \\ c_2 & d_2 \end{bmatrix} = \begin{bmatrix} a_1 + a_2 & b_1 + b_2 \\ c_1 + c_2 & d_1 + d_2 \end{bmatrix} = \begin{bmatrix} a_2 + a_1 & b_2 + b_1 \\ c_2 + c_1 & d_2 + d_1 \end{bmatrix} =$

$\begin{bmatrix} a_2 & b_2 \\ c_2 & d_2 \end{bmatrix} + \begin{bmatrix} a_1 & b_1 \\ c_1 & d_1 \end{bmatrix} = B + A$

25. $cA = c \begin{bmatrix} a_1 & b_1 \\ c_1 & d_1 \end{bmatrix} = \begin{bmatrix} ca_1 & cb_1 \\ cc_1 & cd_1 \end{bmatrix}$; since multiplication in the set of real numbers is

closed, ca_1, cb_1, cc_1, and cd_1 are real numbers. Thus cA is a member of $S_{2 \times 2}$.

26. $c(dA) = c \left(d \begin{bmatrix} a_1 & b_1 \\ c_1 & d_1 \end{bmatrix} \right) = c \begin{bmatrix} da_1 & db_1 \\ dc_1 & dd_1 \end{bmatrix} = \begin{bmatrix} c(da_1) & c(db_1) \\ c(dc_1) & c(dd_1) \end{bmatrix} =$

$\begin{bmatrix} (cd)a_1 & (cd)b_1 \\ (cd)c_1 & (cd)d_1 \end{bmatrix} = (cd) \begin{bmatrix} a_1 & b_1 \\ c_1 & d_1 \end{bmatrix} = (cd)A$

27. $(c + d)A = (c + d) \begin{bmatrix} a_1 & b_1 \\ c_1 & d_1 \end{bmatrix} = \begin{bmatrix} (c + d)a_1 & (c + d)b_1 \\ (c + d)c_1 & (c + d)d_1 \end{bmatrix} =$

$\begin{bmatrix} ca_1 + da_1 & cb_1 + db_1 \\ cc_1 + dc_1 & cd_1 + dd_1 \end{bmatrix} = \begin{bmatrix} ca_1 & cb_1 \\ cc_1 & cd_1 \end{bmatrix} + \begin{bmatrix} da_1 & db_1 \\ dc_1 & dd_1 \end{bmatrix} =$

$c \begin{bmatrix} a_1 & b_1 \\ c_1 & d_1 \end{bmatrix} + d \begin{bmatrix} a_1 & b_1 \\ c_1 & d_1 \end{bmatrix} = cA + dA$

28. $c(A + B) = c \left(\begin{bmatrix} a_1 & b_1 \\ c_1 & d_1 \end{bmatrix} + \begin{bmatrix} a_2 & b_2 \\ c_2 & d_2 \end{bmatrix} \right) = c \begin{bmatrix} a_1 + a_2 & b_1 + b_2 \\ c_1 + c_2 & d_1 + d_2 \end{bmatrix} =$

$\begin{bmatrix} c(a_1 + a_2) & c(b_1 + b_2) \\ c(c_1 + c_2) & c(d_1 + d_2) \end{bmatrix} = \begin{bmatrix} ca_1 + ca_2 & cb_1 + cb_2 \\ cc_1 + cc_2 & cd_1 + cd_2 \end{bmatrix} =$

$\begin{bmatrix} ca_1 & cb_1 \\ cc_1 & cd_1 \end{bmatrix} + \begin{bmatrix} ca_2 & cb_2 \\ cc_2 & cd_2 \end{bmatrix} = c \begin{bmatrix} a_1 & b_1 \\ c_1 & d_1 \end{bmatrix} + c \begin{bmatrix} a_2 & b_2 \\ c_2 & d_2 \end{bmatrix} = cA + cB$

29. $1A = 1 \begin{bmatrix} a_1 & b_1 \\ c_1 & d_1 \end{bmatrix} = \begin{bmatrix} 1a_1 & 1b_1 \\ 1c_1 & 1d_1 \end{bmatrix} = \begin{bmatrix} a_1 & b_1 \\ c_1 & d_1 \end{bmatrix} = A$

30. $-1A = -1 \begin{bmatrix} a_1 & b_1 \\ c_1 & d_1 \end{bmatrix} = \begin{bmatrix} -1a_1 & -1b_1 \\ -1c_1 & -1d_1 \end{bmatrix} = \begin{bmatrix} -a_1 & -b_1 \\ -c_1 & -d_1 \end{bmatrix} = -A$

31. $0A = 0 \begin{bmatrix} a_1 & b_1 \\ c_1 & d_1 \end{bmatrix} = \begin{bmatrix} 0a_1 & 0b_1 \\ 0c_1 & 0d_1 \end{bmatrix} = \begin{bmatrix} 0 & 0 \\ 0 & 0 \end{bmatrix} = 0$

32. $c0 = c \begin{bmatrix} 0 & 0 \\ 0 & 0 \end{bmatrix} = \begin{bmatrix} c0 & c0 \\ c0 & c0 \end{bmatrix} = \begin{bmatrix} 0 & 0 \\ 0 & 0 \end{bmatrix} = 0$

Exercises 8-2 · Page 281

A 1. $\begin{bmatrix} 2 & -3 \\ -4 & 5 \end{bmatrix}\begin{bmatrix} 6 & -5 \\ 4 & -7 \end{bmatrix} = \begin{bmatrix} 12 - 12 & -10 + 21 \\ -24 + 20 & 20 - 35 \end{bmatrix} = \begin{bmatrix} 0 & 11 \\ -4 & -15 \end{bmatrix}$

2. $\begin{bmatrix} 1 & 2 \\ 3 & 4 \end{bmatrix}\begin{bmatrix} 1 & -1 \\ 0 & 0 \end{bmatrix} = \begin{bmatrix} 1 + 0 & -1 + 0 \\ 3 + 0 & -3 + 0 \end{bmatrix} = \begin{bmatrix} 1 & -1 \\ 3 & -3 \end{bmatrix}$

3. $\begin{bmatrix} 3 & 1 \\ 2 & -1 \end{bmatrix}\begin{bmatrix} 0 & 1 \\ 1 & 0 \end{bmatrix} = \begin{bmatrix} 0 + 1 & 3 + 0 \\ 0 - 1 & 2 + 0 \end{bmatrix} = \begin{bmatrix} 1 & 3 \\ -1 & 2 \end{bmatrix}$

4. $\begin{bmatrix} -1 & 0 \\ -2 & -1 \end{bmatrix}\begin{bmatrix} 1 & 0 \\ -2 & 1 \end{bmatrix} = \begin{bmatrix} -1 + 0 & 0 + 0 \\ -2 + 2 & 0 - 1 \end{bmatrix} = \begin{bmatrix} -1 & 0 \\ 0 & -1 \end{bmatrix}$

5. $\begin{bmatrix} 5 & 2 \\ 1 & 3 \end{bmatrix}\begin{bmatrix} 1 & 4 \\ 2 & -1 \end{bmatrix} = \begin{bmatrix} 5 + 4 & 20 - 2 \\ 1 + 6 & 4 - 3 \end{bmatrix} = \begin{bmatrix} 9 & 18 \\ 7 & 1 \end{bmatrix}$

6. $\begin{bmatrix} 3 & 1 \\ 0 & 3 \end{bmatrix}\begin{bmatrix} -1 & 2 \\ 2 & 4 \end{bmatrix} = \begin{bmatrix} -3 + 2 & 6 + 4 \\ 0 + 6 & 0 + 12 \end{bmatrix} = \begin{bmatrix} -1 & 10 \\ 6 & 12 \end{bmatrix}$

7. $\begin{bmatrix} 2 & 5 \\ 1 & 3 \end{bmatrix}\begin{bmatrix} -1 & 2 \\ 3 & -5 \end{bmatrix} = \begin{bmatrix} -2 + 15 & 4 - 25 \\ -1 + 9 & 2 - 15 \end{bmatrix} = \begin{bmatrix} 13 & -21 \\ 8 & -13 \end{bmatrix}$

8. $\begin{bmatrix} 3 & 4 \\ 2 & 9 \end{bmatrix}\begin{bmatrix} -6 & 4 \\ 5 & 0 \end{bmatrix} = \begin{bmatrix} -18 + 20 & 12 + 0 \\ -12 + 45 & 8 + 0 \end{bmatrix} = \begin{bmatrix} 2 & 12 \\ 33 & 8 \end{bmatrix}$

9. $(A - B)^2$, $A - B = \begin{bmatrix} 3 & -1 \\ 0 & 5 \end{bmatrix} - \begin{bmatrix} 6 & -2 \\ 8 & 1 \end{bmatrix} = \begin{bmatrix} -3 & 1 \\ -8 & 4 \end{bmatrix}$,

$(A - B)^2 = \begin{bmatrix} -3 & 1 \\ -8 & 4 \end{bmatrix}\begin{bmatrix} -3 & 1 \\ -8 & 4 \end{bmatrix} = \begin{bmatrix} 1 & 1 \\ -8 & 8 \end{bmatrix}$

10. $A^2 - B^2$, $A^2 = \begin{bmatrix} 3 & -1 \\ 0 & 5 \end{bmatrix}\begin{bmatrix} 3 & -1 \\ 0 & 5 \end{bmatrix} = \begin{bmatrix} 9 & -8 \\ 0 & 25 \end{bmatrix}$, $B^2 = \begin{bmatrix} 6 & -2 \\ 8 & 1 \end{bmatrix}\begin{bmatrix} 6 & -2 \\ 8 & 1 \end{bmatrix} = \begin{bmatrix} 20 & -14 \\ 56 & -15 \end{bmatrix}$,

$A^2 - B^2 = \begin{bmatrix} 9 & -8 \\ 0 & 25 \end{bmatrix} - \begin{bmatrix} 20 & -14 \\ 56 & -15 \end{bmatrix} = \begin{bmatrix} -11 & 6 \\ -56 & 40 \end{bmatrix}$

11. $(A + B)(A - B)$, $A + B = \begin{bmatrix} 3 & -1 \\ 0 & 5 \end{bmatrix} + \begin{bmatrix} 6 & -2 \\ 8 & 1 \end{bmatrix} = \begin{bmatrix} 9 & -3 \\ 8 & 6 \end{bmatrix}$, $A - B =$

$\begin{bmatrix} 3 & -1 \\ 0 & 5 \end{bmatrix} - \begin{bmatrix} 6 & -2 \\ 8 & 1 \end{bmatrix} = \begin{bmatrix} -3 & 1 \\ -8 & 4 \end{bmatrix}$, $(A + B)(A - B) = \begin{bmatrix} 9 & -3 \\ 8 & 6 \end{bmatrix}\begin{bmatrix} -3 & 1 \\ -8 & 4 \end{bmatrix} =$

$\begin{bmatrix} -3 & -3 \\ -72 & 32 \end{bmatrix}$

12. $(A - B)(A + B) = \begin{bmatrix} -3 & 1 \\ -8 & 4 \end{bmatrix}\begin{bmatrix} 9 & -3 \\ 8 & 6 \end{bmatrix} = \begin{bmatrix} -19 & 15 \\ -40 & 48 \end{bmatrix}$

13. $\begin{bmatrix} 5 & 1 \\ 0 & 4 \end{bmatrix}\begin{bmatrix} -2 & x_1 \\ 3 & x_2 \end{bmatrix} = \begin{bmatrix} -7 & 21 \\ 12 & 4 \end{bmatrix}$, $\begin{bmatrix} -7 & 5x_1 + x_2 \\ 12 & 4x_2 \end{bmatrix} = \begin{bmatrix} -7 & 21 \\ 12 & 4 \end{bmatrix}$, $5x_1 + x_2 = 21$

and $4x_2 = 4$, $x_2 = 1$, $5x_1 + 1 = 21$, $5x_1 = 20$, $x_1 = 4$; $x_1 = 4$ and $x_2 = 1$.

14. $\begin{bmatrix} -1 & 1 \\ 2 & 3 \end{bmatrix}\begin{bmatrix} x_1 & x_2 \\ 3 & 7 \end{bmatrix} = \begin{bmatrix} -1 & 5 \\ 17 & 25 \end{bmatrix}$, $\begin{bmatrix} -x_1 + 3 & -x_2 + 7 \\ 2x_1 + 9 & 2x_2 + 21 \end{bmatrix} = \begin{bmatrix} -1 & 5 \\ 17 & 25 \end{bmatrix}$,

$-x_1 + 3 = -1$ and $-x_2 + 7 = 5$; $x_1 = 4$ and $x_2 = 2$.

15. $\begin{bmatrix} x_1 & 3 \\ 1 & x_2 \end{bmatrix}^2 = \begin{bmatrix} 7 & -9 \\ -3 & 4 \end{bmatrix}$, $\begin{bmatrix} x_1^2 + 3 & 3x_1 + 3x_2 \\ x_1 + x_2 & 3 + x_2^2 \end{bmatrix} = \begin{bmatrix} 7 & -9 \\ -3 & 4 \end{bmatrix}$, $x_1^2 + 3 = 7$ and

$3 + x_2^2 = 4$, $x_1^2 = 4$ and $x_2^2 = 1$, $x_1 = \pm 2$ and $x_2 = \pm 1$, since $x_1 + x_2 = -3$ and

$3x_1 + 3x_2 = -9$, then $x_1 = -2$ and $x_2 = -1$.

16. $\begin{bmatrix} 3 & 5 \\ 4 & 7 \end{bmatrix}\begin{bmatrix} 7 & x_2 \\ x_1 & 3 \end{bmatrix} = \begin{bmatrix} 1 & 0 \\ 0 & 1 \end{bmatrix}$, $\begin{bmatrix} 21 + 5x_1 & 3x_2 + 15 \\ 28 + 7x_1 & 4x_2 + 21 \end{bmatrix} = \begin{bmatrix} 1 & 0 \\ 0 & 1 \end{bmatrix}$, $28 + 7x_1 = 0$

and $3x_2 + 15 = 0$, $7x_1 = -28$ and $3x_2 = -15$, $x_1 = -4$ and $x_2 = -5$

17. $\begin{bmatrix} -1 & 1 \\ 3 & -2 \end{bmatrix}\begin{bmatrix} x_1 & 1 \\ x_2 & 5 \end{bmatrix} = \begin{bmatrix} 4 & 4 \\ -6 & -7 \end{bmatrix}$, $\begin{bmatrix} -x_1 + x_2 & 4 \\ 3x_1 - 2x_2 & -7 \end{bmatrix} = \begin{bmatrix} 4 & 4 \\ -6 & -7 \end{bmatrix}$, $-x_1 + x_2 = 4$ and

$3x_1 - 2x_2 = -6$, $-2x_1 + 2x_2 = 8$ and $3x_1 - 2x_2 = -6$, $x_1 = 2$ and $x_2 = 6$.

18. $\begin{bmatrix} x_1 & x_2 \\ x_2 & x_1 \end{bmatrix}^2 = \begin{bmatrix} 2 & -2 \\ -2 & 2 \end{bmatrix}$, $\begin{bmatrix} x_1^2 + x_2^2 & 2x_1 x_2 \\ 2x_1 x_2 & x_2^2 + x_1^2 \end{bmatrix} = \begin{bmatrix} 2 & -2 \\ -2 & 2 \end{bmatrix}$, $x_1^2 + x_2^2 = 2$

and $2x_1 x_2 = -2$, $x_1 x_2 = -1$, $x_1 = \frac{-1}{x_2}$, $\frac{1}{x_2^2} + x_2^2 = 2$, $x_2^4 - 2x_2^2 + 1 = 0$,

$(x_2^2 - 1)^2 = 0$, $x_2^2 = 1$, $x_2 = \pm 1$, since $x_1 = \frac{-1}{x_2}$, then $x_1 = 1$ and $x_2 = -1$ or

$x_1 = -1$ and $x_2 = 1$.

B   19. $AB = \begin{bmatrix} a_1 & b_1 \\ c_1 & d_1 \end{bmatrix}\begin{bmatrix} a_2 & b_2 \\ c_2 & d_2 \end{bmatrix} = \begin{bmatrix} a_1 a_2 + b_1 c_2 & a_1 b_2 + b_1 d_2 \\ c_1 a_2 + d_1 c_2 & c_1 b_2 + d_1 d_2 \end{bmatrix}$. By closure addition

and multiplication in R, each entry is in R; $\therefore AB \in S_{2 \times 2}$.

<u>20.</u> $(AB)C = \left(\begin{bmatrix} a_1 & b_1 \\ c_1 & d_1 \end{bmatrix} \begin{bmatrix} a_2 & b_2 \\ c_2 & d_2 \end{bmatrix} \right) \begin{bmatrix} a_3 & b_3 \\ c_3 & d_3 \end{bmatrix} =$

$$\begin{bmatrix} a_1 a_2 + b_1 c_2 & a_1 b_2 + b_1 d_2 \\ c_1 a_2 + d_1 c_2 & c_1 b_2 + d_1 d_2 \end{bmatrix} \begin{bmatrix} a_3 & b_3 \\ c_3 & d_3 \end{bmatrix} =$$

$$\begin{bmatrix} (a_1 a_2 + b_1 c_2)a_3 + (a_1 b_2 + b_1 d_2)c_3 & (a_1 a_2 + b_1 c_2)b_3 + (a_1 b_2 + b_1 d_2)d_3 \\ (c_1 a_2 + d_1 c_2)a_3 + (c_1 b_2 + d_1 d_2)c_3 & (c_1 a_2 + d_1 c_2)b_3 + (c_1 b_2 + d_1 d_2)d_3 \end{bmatrix} =$$

$$\begin{bmatrix} a_1(a_2 a_3 + b_2 c_3) + b_1(c_2 a_3 + d_2 c_3) & a_1(a_2 b_3 + b_2 d_3) + b_1(c_2 b_3 + d_2 d_3) \\ c_1(a_2 a_3 + b_2 c_3) + d_1(c_2 a_3 + d_2 c_3) & c_1(a_2 b_3 + b_2 d_3) + d_1(c_2 b_3 + d_2 d_3) \end{bmatrix} =$$

$$\begin{bmatrix} a_1 & b_1 \\ c_1 & d_1 \end{bmatrix} \begin{bmatrix} a_2 a_3 + b_2 c_3 & a_2 b_3 + b_2 d_3 \\ c_2 a_3 + d_2 c_3 & c_2 b_3 + d_2 d_3 \end{bmatrix} = \begin{bmatrix} a_1 & b_1 \\ c_1 & d_1 \end{bmatrix} \left(\begin{bmatrix} a_2 & b_2 \\ c_2 & d_2 \end{bmatrix} \begin{bmatrix} a_3 & b_3 \\ c_3 & d_3 \end{bmatrix} \right) = A(BC)$$

<u>21.</u> $A(B + C) = \begin{bmatrix} a_1 & b_1 \\ c_1 & d_1 \end{bmatrix} \left(\begin{bmatrix} a_2 & b_2 \\ c_2 & d_2 \end{bmatrix} + \begin{bmatrix} a_3 & b_3 \\ c_3 & d_3 \end{bmatrix} \right) =$

$$\begin{bmatrix} a_1 & b_1 \\ c_1 & d_1 \end{bmatrix} \begin{bmatrix} a_2 + a_3 & b_2 + b_3 \\ c_2 + c_3 & d_2 + d_3 \end{bmatrix} =$$

$$\begin{bmatrix} a_1(a_2 + a_3) + b_1(c_2 + c_3) & a_1(b_2 + b_3) + b_1(d_2 + d_3) \\ c_1(a_2 + a_3) + d_1(c_2 + c_3) & c_1(b_2 + b_3) + d_1(d_2 + d_3) \end{bmatrix} =$$

$$\begin{bmatrix} (a_1 a_2 + b_1 c_2) + (a_1 a_3 + b_1 c_3) & (a_1 b_2 + b_1 d_2) + (a_1 b_3 + b_1 d_3) \\ (c_1 a_2 + d_1 c_2) + (c_1 a_3 + d_1 c_3) & (c_1 b_2 + d_1 d_2) + (c_1 b_3 + d_1 d_3) \end{bmatrix} =$$

$$\begin{bmatrix} a_1 a_2 + b_1 c_2 & a_1 b_2 + b_1 d_2 \\ c_1 a_2 + d_1 c_2 & c_1 b_2 + d_1 d_2 \end{bmatrix} + \begin{bmatrix} a_1 a_3 + b_1 c_3 & a_1 b_3 + b_1 d_3 \\ c_1 a_3 + d_1 c_3 & c_1 b_3 + d_1 d_3 \end{bmatrix} =$$

$$\begin{bmatrix} a_1 & b_1 \\ c_1 & d_1 \end{bmatrix} \begin{bmatrix} a_2 & b_2 \\ c_2 & d_2 \end{bmatrix} + \begin{bmatrix} a_1 & b_1 \\ c_1 & d_1 \end{bmatrix} \begin{bmatrix} a_3 & b_3 \\ c_3 & d_3 \end{bmatrix} = AB + AC$$

<u>22.</u> $(B + C)A = \left(\begin{bmatrix} a_2 & b_2 \\ c_2 & d_2 \end{bmatrix} + \begin{bmatrix} a_3 & b_3 \\ c_3 & d_3 \end{bmatrix} \right) \begin{bmatrix} a_1 & b_1 \\ c_1 & d_1 \end{bmatrix} =$

$$\begin{bmatrix} a_2 + a_3 & b_2 + b_3 \\ c_2 + c_3 & d_2 + d_3 \end{bmatrix} \begin{bmatrix} a_1 & b_1 \\ c_1 & d_1 \end{bmatrix} =$$

$$\begin{bmatrix} (a_2 + a_3)a_1 + (b_2 + b_3)c_1 & (a_2 + a_3)b_1 + (b_2 + b_3)d_1 \\ (c_2 + c_3)a_1 + (d_2 + d_3)c_1 & (c_2 + c_3)b_1 + (d_2 + d_3)d_1 \end{bmatrix} =$$

$$\begin{bmatrix} (a_2a_1 + b_2c_1) + (a_3a_1 + b_3c_1) & (a_2b_1 + b_2d_1) + (a_3b_1 + b_3d_1) \\ (c_2a_1 + d_2c_1) + (c_3a_1 + d_3c_1) & (c_2b_1 + d_2d_1) + (c_3b_1 + d_3d_1) \end{bmatrix} =$$

$$\begin{bmatrix} a_2a_1 + b_2c_1 & a_2b_1 + b_2d_1 \\ c_2a_1 + d_2c_1 & c_2b_1 + d_2d_1 \end{bmatrix} + \begin{bmatrix} a_3a_1 + b_3c_1 & a_3b_1 + b_3d_1 \\ c_3a_1 + d_3c_1 & c_3b_1 + d_3d_1 \end{bmatrix} =$$

$$\begin{bmatrix} a_2 & b_2 \\ c_2 & d_2 \end{bmatrix}\begin{bmatrix} a_1 & b_1 \\ c_1 & d_1 \end{bmatrix} + \begin{bmatrix} a_3 & b_3 \\ c_3 & d_3 \end{bmatrix}\begin{bmatrix} a_1 & b_1 \\ c_1 & d_1 \end{bmatrix} = BA + CA$$

<u>23.</u> $A(cB) = \begin{bmatrix} a_1 & b_1 \\ c_1 & d_1 \end{bmatrix}\left(c\begin{bmatrix} a_2 & b_2 \\ c_2 & d_2 \end{bmatrix}\right) = \begin{bmatrix} a_1 & b_1 \\ c_1 & d_1 \end{bmatrix}\begin{bmatrix} ca_2 & cb_2 \\ cc_2 & cd_2 \end{bmatrix} =$

$$\begin{bmatrix} a_1(ca_2) + b_1(cc_2) & a_1(cb_2) + b_1(cd_2) \\ c_1(ca_2) + d_1(cc_2) & c_1(cb_2) + d_1(cd_2) \end{bmatrix} =$$

$$\begin{bmatrix} (ca_1)a_2 + (cb_1)c_2 & (ca_1)b_2 + (cb_1)d_2 \\ (cc_1)a_2 + (cd_1)c_2 & (cc_1)b_2 + (cd_1)d_2 \end{bmatrix} =$$

$$\begin{bmatrix} ca_1 & cb_1 \\ cc_1 & cd_1 \end{bmatrix}\begin{bmatrix} a_2 & b_2 \\ c_2 & d_2 \end{bmatrix} = \left(c\begin{bmatrix} a_1 & b_1 \\ c_1 & d_1 \end{bmatrix}\right)\begin{bmatrix} a_2 & b_2 \\ c_2 & d_2 \end{bmatrix} = (cA)B$$

<u>24.</u> $AO = \begin{bmatrix} a_1 & b_1 \\ c_1 & d_1 \end{bmatrix}\begin{bmatrix} 0 & 0 \\ 0 & 0 \end{bmatrix} = \begin{bmatrix} a_1 0 + b_1 0 & a_1 0 + b_1 0 \\ c_1 0 + d_1 0 & c_1 0 + d_1 0 \end{bmatrix} = \begin{bmatrix} 0 & 0 \\ 0 & 0 \end{bmatrix} = 0,$

$$OA = \begin{bmatrix} 0 & 0 \\ 0 & 0 \end{bmatrix}\begin{bmatrix} a_1 & b_1 \\ c_1 & d_1 \end{bmatrix} = \begin{bmatrix} 0a_1 + 0c_1 & 0b_1 + 0d_1 \\ 0a_1 + 0c_1 & 0b_1 + 0d_1 \end{bmatrix} = \begin{bmatrix} 0 & 0 \\ 0 & 0 \end{bmatrix} = 0$$

<u>25.</u> $\begin{bmatrix} a & 0 \\ 0 & a \end{bmatrix}^2 = cI, \quad \begin{bmatrix} a & 0 \\ 0 & a \end{bmatrix}\begin{bmatrix} a & 0 \\ 0 & a \end{bmatrix} = c\begin{bmatrix} 1 & 0 \\ 0 & 1 \end{bmatrix}, \quad \begin{bmatrix} a^2 & 0 \\ 0 & a^2 \end{bmatrix} = c\begin{bmatrix} 1 & 0 \\ 0 & 1 \end{bmatrix},$

$$a^2\begin{bmatrix} 1 & 0 \\ 0 & 1 \end{bmatrix} = c\begin{bmatrix} 1 & 0 \\ 0 & 1 \end{bmatrix}, \quad c = a^2.$$

<u>26.</u> $(I + J)(I - J) = \left(\begin{bmatrix} 1 & 0 \\ 0 & 1 \end{bmatrix} + \begin{bmatrix} 0 & 1 \\ -1 & 0 \end{bmatrix}\right)\left(\begin{bmatrix} 1 & 0 \\ 0 & 1 \end{bmatrix} - \begin{bmatrix} 0 & 1 \\ -1 & 0 \end{bmatrix}\right) =$

$$\begin{bmatrix} 1 & 1 \\ -1 & 1 \end{bmatrix}\begin{bmatrix} 1 & -1 \\ 1 & 1 \end{bmatrix} = \begin{bmatrix} 1(1) + 1(1) & 1(-1) + 1(1) \\ -1(1) + 1(1) & -1(-1) + 1(1) \end{bmatrix} =$$

$$\begin{bmatrix} 2 & 0 \\ 0 & 2 \end{bmatrix} = 2\begin{bmatrix} 1 & 0 \\ 0 & 1 \end{bmatrix} = 2I$$

Exercises 8-3 · Pages 284-287

A   1. $\delta(A) = 5$, $A^{-1} = \frac{1}{5}\begin{bmatrix} 3 & -1 \\ 2 & 1 \end{bmatrix} = \begin{bmatrix} \frac{3}{5} & -\frac{1}{5} \\ \frac{2}{5} & \frac{1}{5} \end{bmatrix}$

2. $\delta(A) = -1$, $A^{-1} = -1\begin{bmatrix} 0 & -1 \\ -1 & 1 \end{bmatrix} = \begin{bmatrix} 0 & 1 \\ 1 & -1 \end{bmatrix}$

3. $\delta(A) = 0$, A is singular and does not have a multiplicative inverse.

4. $\delta(A) = -1$, $A^{-1} = -1\begin{bmatrix} 0 & -1 \\ -1 & 0 \end{bmatrix} = \begin{bmatrix} 0 & 1 \\ 1 & 0 \end{bmatrix}$

5. $\delta(A) = 1$, $A^{-1} = 1\begin{bmatrix} \frac{\sqrt{3}}{2} & \frac{1}{2} \\ -\frac{1}{2} & \frac{\sqrt{3}}{2} \end{bmatrix} = \begin{bmatrix} \frac{\sqrt{3}}{2} & \frac{1}{2} \\ -\frac{1}{2} & \frac{\sqrt{3}}{2} \end{bmatrix}$

6. $\delta(A) = 0$, A is singular and does not have a multiplicative inverse.

7. $\delta(A) = -\frac{5}{18}$, $A^{-1} = -\frac{18}{5}\begin{bmatrix} \frac{1}{3} & -\frac{2}{3} \\ -\frac{2}{3} & \frac{1}{2} \end{bmatrix} = \begin{bmatrix} -\frac{6}{5} & \frac{12}{5} \\ \frac{12}{5} & -\frac{9}{5} \end{bmatrix}$

8. $\begin{bmatrix} 1 & -1 \\ 2 & 1 \end{bmatrix}\begin{bmatrix} x \\ y \end{bmatrix} = \begin{bmatrix} 1 \\ 11 \end{bmatrix}$, $\begin{bmatrix} x \\ y \end{bmatrix} = \left(\frac{1}{3}\begin{bmatrix} 1 & 1 \\ -2 & 1 \end{bmatrix}\right)\begin{bmatrix} 1 \\ 11 \end{bmatrix} = \frac{1}{3}\left(\begin{bmatrix} 1 & 1 \\ -2 & 1 \end{bmatrix}\begin{bmatrix} 1 \\ 11 \end{bmatrix}\right) =$

$\frac{1}{3}\begin{bmatrix} 12 \\ 9 \end{bmatrix} = \begin{bmatrix} 4 \\ 3 \end{bmatrix}$, so x = 4 and y = 3.

9. $\begin{bmatrix} 5 & -2 \\ 6 & 3 \end{bmatrix}\begin{bmatrix} x \\ y \end{bmatrix} = \begin{bmatrix} 8 \\ -12 \end{bmatrix}$, $\begin{bmatrix} x \\ y \end{bmatrix} = \left(\frac{1}{27}\begin{bmatrix} 3 & 2 \\ -6 & 5 \end{bmatrix}\right)\begin{bmatrix} 8 \\ -12 \end{bmatrix} = \frac{1}{27}\left(\begin{bmatrix} 3 & 2 \\ -6 & 5 \end{bmatrix}\begin{bmatrix} 8 \\ -12 \end{bmatrix}\right) =$

$\frac{1}{27}\begin{bmatrix} 0 \\ -108 \end{bmatrix} = \begin{bmatrix} 0 \\ -4 \end{bmatrix}$, so x = 0 and y = -4.

10. $\begin{bmatrix} 9 & -5 \\ -2 & 3 \end{bmatrix}\begin{bmatrix} x \\ y \end{bmatrix} = \begin{bmatrix} 4 \\ 1 \end{bmatrix}$, $\begin{bmatrix} x \\ y \end{bmatrix} = \left(\frac{1}{17}\begin{bmatrix} 3 & 5 \\ 2 & 9 \end{bmatrix}\right)\begin{bmatrix} 4 \\ 1 \end{bmatrix} = \frac{1}{17}\left(\begin{bmatrix} 3 & 5 \\ 2 & 9 \end{bmatrix}\begin{bmatrix} 4 \\ 1 \end{bmatrix}\right) =$

$\frac{1}{17}\begin{bmatrix} 17 \\ 17 \end{bmatrix} = \begin{bmatrix} 1 \\ 1 \end{bmatrix}$, so x = 1 and y = 1.

11. $\begin{bmatrix} 2 & -5 \\ 7 & 9 \end{bmatrix}\begin{bmatrix} x \\ y \end{bmatrix} = \begin{bmatrix} 16 \\ 3 \end{bmatrix}$, $\begin{bmatrix} x \\ y \end{bmatrix} = \left(\dfrac{1}{53}\begin{bmatrix} 9 & 5 \\ -7 & 2 \end{bmatrix}\right)\begin{bmatrix} 16 \\ 3 \end{bmatrix} = \dfrac{1}{53}\left(\begin{bmatrix} 9 & 5 \\ -7 & 2 \end{bmatrix}\begin{bmatrix} 16 \\ 3 \end{bmatrix}\right) =$

$\dfrac{1}{53}\begin{bmatrix} 159 \\ -106 \end{bmatrix} = \begin{bmatrix} 3 \\ -2 \end{bmatrix}$, so x = 3 and y = -2.

12. $\begin{bmatrix} 8 & 3 \\ -3 & 2 \end{bmatrix}\begin{bmatrix} x \\ y \end{bmatrix} = \begin{bmatrix} 4 \\ 11 \end{bmatrix}$, $\begin{bmatrix} x \\ y \end{bmatrix} = \left(\dfrac{1}{25}\begin{bmatrix} 2 & -3 \\ 3 & 8 \end{bmatrix}\right)\begin{bmatrix} 4 \\ 11 \end{bmatrix} = \dfrac{1}{25}\left(\begin{bmatrix} 2 & -3 \\ 3 & 8 \end{bmatrix}\begin{bmatrix} 4 \\ 11 \end{bmatrix}\right) =$

$\dfrac{1}{25}\begin{bmatrix} -25 \\ 100 \end{bmatrix} = \begin{bmatrix} -1 \\ 4 \end{bmatrix}$, so x = -1 and y = 4.

13. $\begin{bmatrix} -5 & 2 \\ 3 & -1 \end{bmatrix}\begin{bmatrix} x \\ y \end{bmatrix} = \begin{bmatrix} 3 \\ 1 \end{bmatrix}$, $\begin{bmatrix} x \\ y \end{bmatrix} = \left(-1\begin{bmatrix} -1 & -2 \\ -3 & -5 \end{bmatrix}\right)\begin{bmatrix} 3 \\ 1 \end{bmatrix} = \begin{bmatrix} 1 & 2 \\ 3 & 5 \end{bmatrix}\begin{bmatrix} 3 \\ 1 \end{bmatrix} = \begin{bmatrix} 5 \\ 14 \end{bmatrix}$, so

x = 5 and y = 14.

14. $\delta(cA) = \delta\begin{bmatrix} ca_1 & cb_1 \\ ca_2 & cb_2 \end{bmatrix} = ca_1(cb_2) - ca_2(cb_1) = c^2(a_1b_2 - a_2b_1) = c^2\delta(A)$

15. $\delta(A^{-1}) = \delta\begin{bmatrix} \dfrac{b_2}{\delta(A)} & -\dfrac{b_1}{\delta(A)} \\ -\dfrac{a_2}{\delta(A)} & \dfrac{a_1}{\delta(A)} \end{bmatrix} = \dfrac{b_2a_1 - a_2b_1}{[\delta(A)]^2} = \dfrac{\delta(A)}{[\delta(A)]^2} = \dfrac{1}{\delta(A)}$

16. $A = \begin{bmatrix} a_1 & b_1 \\ c_1 & d_1 \end{bmatrix}$, $\delta(A) = a_1d_1 - b_1c_1$, $A^{-1} = \begin{bmatrix} \dfrac{d_1}{a_1d_1 - b_1c_1} & \dfrac{-b_1}{a_1d_1 - b_1c_1} \\ \dfrac{-c_1}{a_1d_1 - b_1c_1} & \dfrac{a_1}{a_1d_1 - b_1c_1} \end{bmatrix}$,

$\delta(A^{-1}) = \dfrac{1}{a_1d_1 - b_1c_1}$, $(A^{-1})^{-1} = a_1d_1 - b_1c_1\begin{bmatrix} \dfrac{a_1}{a_1d_1 - b_1c_1} & \dfrac{b_1}{a_1d_1 - b_1c_1} \\ \dfrac{c_1}{a_1d_1 - b_1c_1} & \dfrac{d_1}{a_1d_1 - b_1c_1} \end{bmatrix} =$

$\begin{bmatrix} a_1 & b_1 \\ c_1 & d_1 \end{bmatrix} = A$

17. a. $\delta(A) = 5$, $\delta(B) = 7$, $AB = \begin{bmatrix} 5 & 11 \\ -5 & -4 \end{bmatrix}$, $\delta(AB) = 35$, $\delta(AB) = \delta(A)\delta(B)$,

   35 = 7 × 5, 35 = 35.

b. $\delta(AB) = \delta\begin{bmatrix} a_1 & b_1 \\ c_1 & d_1 \end{bmatrix}\begin{bmatrix} a_2 & b_2 \\ c_2 & d_2 \end{bmatrix} = \delta\begin{bmatrix} a_1a_2 + b_1c_2 & a_1b_2 + b_1d_2 \\ c_1a_2 + d_1c_2 & c_1b_2 + d_1d_2 \end{bmatrix} =$

$$(a_1 a_2 + b_1 c_2)(c_1 b_2 + d_1 d_2) - (c_1 a_2 + d_1 c_2)(a_1 b_2 + b_1 d_2) =$$

$$a_1 a_2 d_1 d_2 + b_1 b_2 c_1 c_2 - b_1 c_1 a_2 d_2 - a_1 d_1 b_2 c_2 = (a_1 d_1 - b_1 c_1)(a_2 d_2 - b_2 c_2) = \delta(A)\delta(B)$$

<u>18</u>. <u>a</u>. $A^{-1} = \begin{bmatrix} \dfrac{4}{5} & \dfrac{1}{5} \\ \dfrac{3}{5} & \dfrac{2}{5} \end{bmatrix}$, $B^{-1} = \begin{bmatrix} \dfrac{5}{7} & -\dfrac{8}{7} \\ -\dfrac{1}{7} & \dfrac{3}{7} \end{bmatrix}$, $AB = \begin{bmatrix} 5 & 11 \\ -5 & -4 \end{bmatrix}$,

$(AB)^{-1} = \begin{bmatrix} -\dfrac{4}{35} & -\dfrac{11}{35} \\ \dfrac{5}{35} & \dfrac{5}{35} \end{bmatrix}$, $(AB)^{-1} = B^{-1}A^{-1}$, $\begin{bmatrix} -\dfrac{4}{35} & -\dfrac{11}{35} \\ \dfrac{5}{35} & \dfrac{5}{35} \end{bmatrix} = \begin{bmatrix} \dfrac{5}{7} & -\dfrac{8}{7} \\ -\dfrac{1}{7} & \dfrac{3}{7} \end{bmatrix}$

$\begin{bmatrix} \dfrac{4}{5} & \dfrac{1}{5} \\ \dfrac{3}{5} & \dfrac{2}{5} \end{bmatrix} = \begin{bmatrix} \dfrac{20}{35} - \dfrac{24}{35} & \dfrac{5}{35} - \dfrac{16}{35} \\ -\dfrac{4}{35} + \dfrac{9}{35} & \dfrac{-1}{35} + \dfrac{6}{35} \end{bmatrix} = \begin{bmatrix} \dfrac{-4}{35} & \dfrac{-11}{35} \\ \dfrac{5}{35} & \dfrac{5}{35} \end{bmatrix}$

<u>b</u>. $(AB)^{-1} = \left(\begin{bmatrix} a_1 & b_1 \\ c_1 & d_1 \end{bmatrix} \begin{bmatrix} a_2 & b_2 \\ c_2 & d_2 \end{bmatrix} \right)^{-1} = \begin{bmatrix} a_1 a_2 + b_1 c_2 & a_1 b_2 + b_1 d_2 \\ c_1 a_2 + d_1 c_2 & c_1 b_2 + d_1 d_2 \end{bmatrix}^{-1} =$

$\dfrac{1}{\delta(AB)} \begin{bmatrix} c_1 b_2 + d_1 d_2 & -a_1 b_2 - b_1 d_2 \\ -c_1 a_2 - d_1 c_2 & a_1 a_2 + b_1 c_2 \end{bmatrix} = \dfrac{1}{\delta(A)\delta(B)} \begin{bmatrix} d_2 & -b_2 \\ -c_2 & a_2 \end{bmatrix} \begin{bmatrix} d_1 & -b_1 \\ -c_1 & a_1 \end{bmatrix} =$

$B^{-1}A^{-1}$

<u>19</u>. <u>a</u>. $\delta(AB) = 35 \neq 0$, $BA = \begin{bmatrix} -18 & 29 \\ -13 & 19 \end{bmatrix}$, $\delta(BA) = -342 + 377 = 35 \neq 0$

<u>b</u>. We have $\delta(A) \neq 0$ and $\delta(B) \neq 0$. Then $\delta(A)\delta(B) \neq 0$, and $\therefore \delta(AB) \neq 0$ and $\delta(BA) \neq 0$, so that AB and BA are invertible.

<u>20</u>. $\delta \begin{bmatrix} \cos\theta & \sin\theta \\ -\sin\theta & \cos\theta \end{bmatrix} = \cos^2\theta - (-\sin\theta)(\sin\theta) = \cos^2\theta + \sin^2\theta = 1$

<u>21</u>. $\begin{bmatrix} \cos\theta & \sin\theta \\ -\sin\theta & \cos\theta \end{bmatrix} \begin{bmatrix} \cos\phi & \sin\phi \\ -\sin\phi & \cos\phi \end{bmatrix} =$

$\begin{bmatrix} \cos\theta\cos\phi - \sin\theta\sin\phi & \cos\theta\sin\phi + \sin\theta\cos\phi \\ -\sin\theta\cos\phi - \cos\theta\sin\phi & -\sin\theta\sin\phi + \cos\theta\cos\phi \end{bmatrix} =$

$\begin{bmatrix} \cos(\theta + \phi) & \sin(\theta + \phi) \\ -\sin(\theta + \phi) & \cos(\theta + \phi) \end{bmatrix}$

<u>C</u>   22. Since $\delta(A) = 1$, we have $aa - (-b)b = a^2 + b^2 = 1$. There is then some angle θ such that $\cos\theta = a$ and $\sin\theta = b$.

<u>23</u>. $\delta(A - xI) = \delta \begin{bmatrix} a & b \\ c & d \end{bmatrix} - x \begin{bmatrix} 1 & 0 \\ 0 & 1 \end{bmatrix} = \delta \begin{bmatrix} a-x & b \\ c & d-x \end{bmatrix} =$

$(a - x)(d - x) - bc = ad - ax - dx + x^2 - bc = x^2 - (a + d)x + (ad - bc) =$

$x^2 - (a + d)x + \delta(A)$

<u>24.</u> $\delta(A) = -10 + 6 = -4$; $x^2 - (a + d)x + \delta(A) = 0$, $x^2 - (5 - 2)x + (-4) = 0$,

$x^2 - 3x - 4 = 0$; $(x - 4)(x + 1) = 0$; $x = 4$ or $x = -1$.

<u>25.</u> $\delta(A) = 12 - 4 = 8$; $x^2 - (a + d)x + \delta(A) = 0$, $x^2 - (3 + 4)x + 8 = 0$,

$x^2 - 7x + 8 = 0$; $x = \dfrac{7 \pm \sqrt{49 - 32}}{2}$, $x = \dfrac{7 \pm \sqrt{17}}{2}$, $x = \dfrac{7 + \sqrt{17}}{2}$ or $\dfrac{7 - \sqrt{17}}{2}$

Exercises 8-4 · Page 290

<u>A</u>   <u>1.</u> $(2 + i) + (5 - 2i)$, $\begin{bmatrix} 2 & 1 \\ -1 & 2 \end{bmatrix} + \begin{bmatrix} 5 & -2 \\ 2 & 5 \end{bmatrix} = \begin{bmatrix} 7 & -1 \\ 1 & 7 \end{bmatrix}$; $7 - i$

<u>2.</u> $(1 + 3i) + (0 - 4i)$, $\begin{bmatrix} 1 & 3 \\ -3 & 1 \end{bmatrix} + \begin{bmatrix} 0 & -4 \\ 4 & 0 \end{bmatrix} = \begin{bmatrix} 1 & -1 \\ 1 & 1 \end{bmatrix}$; $1 - i$

<u>3.</u> $(1 - 2i) - (-4 + i)$, $\begin{bmatrix} 1 & -2 \\ 2 & 1 \end{bmatrix} - \begin{bmatrix} -4 & 1 \\ -1 & -4 \end{bmatrix} = \begin{bmatrix} 5 & -3 \\ 3 & 5 \end{bmatrix}$; $5 - 3i$

<u>4.</u> $(-5 + i) - (3 - 4i)$, $\begin{bmatrix} -5 & 1 \\ -1 & -5 \end{bmatrix} - \begin{bmatrix} 3 & -4 \\ 4 & 3 \end{bmatrix} = \begin{bmatrix} -8 & 5 \\ -5 & -8 \end{bmatrix}$; $-8 + 5i$

<u>5.</u> $(2 - i)(4 - 2i)$, $\begin{bmatrix} 2 & -1 \\ 1 & 2 \end{bmatrix}\begin{bmatrix} 4 & -2 \\ 2 & 4 \end{bmatrix} = \begin{bmatrix} 6 & -8 \\ 8 & 6 \end{bmatrix}$; $6 - 8i$

<u>6.</u> $(-1 + 5i)(3 - 2i)$, $\begin{bmatrix} -1 & 5 \\ -5 & -1 \end{bmatrix}\begin{bmatrix} 3 & -2 \\ 2 & 3 \end{bmatrix} = \begin{bmatrix} 7 & 17 \\ -17 & 7 \end{bmatrix}$; $7 + 17i$

<u>7.</u> $(1 + i)^2 = (1 + i)(1 + i)$, $\begin{bmatrix} 1 & 1 \\ -1 & 1 \end{bmatrix}\begin{bmatrix} 1 & 1 \\ -1 & 1 \end{bmatrix} = \begin{bmatrix} 0 & 2 \\ -2 & 0 \end{bmatrix}$; $0 + 2i = 2i$

<u>8.</u> $(1 - i)^3 = (1 - i)(1 - i)(1 - i)$, $\begin{bmatrix} 1 & -1 \\ 1 & 1 \end{bmatrix}\begin{bmatrix} 1 & -1 \\ 1 & 1 \end{bmatrix}\begin{bmatrix} 1 & -1 \\ 1 & 1 \end{bmatrix} = \begin{bmatrix} 0 & -2 \\ 2 & 0 \end{bmatrix}\begin{bmatrix} 1 & -1 \\ 1 & 1 \end{bmatrix} =$

$\begin{bmatrix} -2 & -2 \\ 2 & -2 \end{bmatrix}$; $-2 - 2i$

<u>9.</u> $3 \left(\cos \frac{\pi}{3} + i \sin \frac{\pi}{3}\right) \cdot 2 \left(\cos \frac{\pi}{6} + i \sin \frac{\pi}{6}\right)$,

$$3 \begin{bmatrix} \cos\frac{\pi}{3} & \sin\frac{\pi}{3} \\ -\sin\frac{\pi}{3} & \cos\frac{\pi}{3} \end{bmatrix} \cdot 2 \begin{bmatrix} \cos\frac{\pi}{6} & \sin\frac{\pi}{6} \\ -\sin\frac{\pi}{6} & \cos\frac{\pi}{6} \end{bmatrix} = 6 \begin{bmatrix} \cos(\frac{\pi}{3}+\frac{\pi}{6}) & \sin(\frac{\pi}{3}+\frac{\pi}{6}) \\ -\sin(\frac{\pi}{3}+\frac{\pi}{6}) & \cos(\frac{\pi}{3}+\frac{\pi}{6}) \end{bmatrix} =$$

$$6 \begin{bmatrix} \cos\frac{\pi}{2} & \sin\frac{\pi}{2} \\ -\sin\frac{\pi}{2} & \cos\frac{\pi}{2} \end{bmatrix} = 6 \begin{bmatrix} 0 & 1 \\ -1 & 0 \end{bmatrix}, \quad 6(\cos\frac{\pi}{2} + i\sin\frac{\pi}{2}) \text{ and } 0 + 6i.$$

10. $4(\cos\frac{\pi}{6} + i\sin\frac{\pi}{6}) \cdot 3(\cos\frac{2\pi}{3} + i\sin\frac{2\pi}{3})$,

$$4 \begin{bmatrix} \cos\frac{\pi}{6} & \sin\frac{\pi}{6} \\ -\sin\frac{\pi}{6} & \cos\frac{\pi}{6} \end{bmatrix} \cdot 3 \begin{bmatrix} \cos\frac{2\pi}{3} & \sin\frac{2\pi}{3} \\ -\sin\frac{2\pi}{3} & \cos\frac{2\pi}{3} \end{bmatrix} = 12 \begin{bmatrix} \cos\frac{5\pi}{6} & \sin\frac{5\pi}{6} \\ -\sin\frac{5\pi}{6} & \cos\frac{5\pi}{6} \end{bmatrix} =$$

$$12 \begin{bmatrix} -\frac{\sqrt{3}}{2} & \frac{1}{2} \\ -\frac{1}{2} & -\frac{\sqrt{3}}{2} \end{bmatrix}, \quad 12(\cos\frac{5\pi}{6} + i\sin\frac{5\pi}{6}) \text{ and } -6\sqrt{3} + 6i.$$

11. $-2(\cos 20° + i\sin 20°) \cdot (-2)(\cos 10° + i\sin 10°) =$

$2(-\cos 20° - i\sin 20°) \cdot 2(-\cos 10° - i\sin 10°) =$

$2(\cos 200° + i\sin 200°) \cdot 2(\cos 190° + i\sin 190°)$,

$$2 \begin{bmatrix} \cos 200° & \sin 200° \\ -\sin 200° & \cos 200° \end{bmatrix} \cdot 2 \begin{bmatrix} \cos 190° & \sin 190° \\ -\sin 190° & \cos 190° \end{bmatrix} =$$

$$4 \begin{bmatrix} \cos 390° & \sin 390° \\ -\sin 390° & \cos 390° \end{bmatrix} = 4 \begin{bmatrix} \cos 30° & \sin 30° \\ -\sin 30° & \cos 30° \end{bmatrix} = 4 \begin{bmatrix} \frac{\sqrt{3}}{2} & \frac{1}{2} \\ -\frac{1}{2} & \frac{\sqrt{3}}{2} \end{bmatrix},$$

$4(\cos 30° + i\sin 30°) \text{ and } 2\sqrt{3} + 2i$

12. $-4(\cos 40° + i\sin 40°) \cdot 3(\cos 80° + i\sin 80°) =$

$4(\cos 220° + i\sin 220°) \cdot 3(\cos 80° + i\sin 80°)$,

$$4 \begin{bmatrix} \cos 220° & \sin 220° \\ -\sin 220° & \cos 220° \end{bmatrix} \cdot 3 \begin{bmatrix} \cos 80° & \sin 80° \\ -\sin 80° & \cos 80° \end{bmatrix} = 12 \begin{bmatrix} \cos 300° & \sin 300° \\ -\sin 300° & \cos 300° \end{bmatrix} =$$

$$12 \begin{bmatrix} \frac{1}{2} & -\frac{\sqrt{3}}{2} \\ \frac{\sqrt{3}}{2} & \frac{1}{2} \end{bmatrix}, \quad 12(\cos 300° + i\sin 300°) \text{ and } 6 - 6\sqrt{3}\,i.$$

B   13. $|a + bi| = \sqrt{a^2 + b^2}$, $A = \begin{bmatrix} a & b \\ -b & a \end{bmatrix}$, $\delta(A) = a^2 - (-b^2) = a^2 + b^2$, $\sqrt{\delta(A)} =$

$\sqrt{a^2 + b^2}$, so $|a + bi| = \sqrt{\delta(A)}$.

14. $\begin{bmatrix} a & b \\ -b & a \end{bmatrix}$ corresponds to a + bi and $\begin{bmatrix} a & -b \\ b & a \end{bmatrix}$ corresponds to a - bi,

$\begin{bmatrix} a & b \\ -b & a \end{bmatrix}\begin{bmatrix} a & -b \\ b & a \end{bmatrix} = \begin{bmatrix} a^2 + b^2 & 0 \\ 0 & a^2 + b^2 \end{bmatrix}$ corresponds to (a + bi)(a - bi).

15. $\begin{bmatrix} \rho\cos\theta & \rho\sin\theta \\ -\rho\sin\theta & \rho\cos\theta \end{bmatrix}^3 = \left(\rho\begin{bmatrix} \cos\theta & \sin\theta \\ -\sin\theta & \cos\theta \end{bmatrix}\right)^3 =$

$\left(\rho\begin{bmatrix} \cos\theta & \sin\theta \\ -\sin\theta & \cos\theta \end{bmatrix}\right)^2 \cdot \rho\begin{bmatrix} \cos\theta & \sin\theta \\ -\sin\theta & \cos\theta \end{bmatrix} =$

$\rho^2\begin{bmatrix} \cos 2\theta & \sin 2\theta \\ -\sin 2\theta & \cos 2\theta \end{bmatrix} \cdot \rho\begin{bmatrix} \cos\theta & \sin\theta \\ -\sin\theta & \cos\theta \end{bmatrix} = \rho^3\begin{bmatrix} \cos 3\theta & \sin 3\theta \\ -\sin 3\theta & \cos 3\theta \end{bmatrix}$

16. $B^{-1} = \dfrac{1}{c^2 + d^2}\begin{bmatrix} c & -d \\ d & c \end{bmatrix}$, $AB^{-1} = \begin{bmatrix} a & b \\ -b & a \end{bmatrix} \cdot \left(\dfrac{1}{c^2 + d^2}\right)\begin{bmatrix} c & -d \\ d & c \end{bmatrix} =$

$\dfrac{1}{c^2 + d^2}\begin{bmatrix} ac + bd & -ad + bc \\ -bc + ad & bd + ac \end{bmatrix} = \dfrac{1}{c^2 + d^2}\begin{bmatrix} ac + bd & -ad + bc \\ -(-ad + bc) & ac + bd \end{bmatrix}$, thus

$\dfrac{(ac + bd) + (-ad + bc)i}{c^2 + d^2} = \dfrac{(a + bi)(c - di)}{(c + di)(c - di)} = \dfrac{a + bi}{c + di}$

17. (1 + 2i) corresponds to A = $\begin{bmatrix} 1 & 2 \\ -2 & 1 \end{bmatrix}$, (0 + i) corresponds to B = $\begin{bmatrix} 0 & 1 \\ -1 & 0 \end{bmatrix}$,

and $B^{-1} = \begin{bmatrix} 0 & -1 \\ 1 & 0 \end{bmatrix}$, $AB^{-1} = \begin{bmatrix} 1 & 2 \\ -2 & 1 \end{bmatrix}\begin{bmatrix} 0 & -1 \\ 1 & 0 \end{bmatrix} = \begin{bmatrix} 2 & -1 \\ 1 & 2 \end{bmatrix}$, so $\dfrac{1 + 2i}{i} = 2 - i$

18. (3 - 2i) corresponds to A = $\begin{bmatrix} 3 & -2 \\ 2 & 3 \end{bmatrix}$, (2 + i) corresponds to B = $\begin{bmatrix} 2 & 1 \\ -1 & 2 \end{bmatrix}$, and

$B^{-1} = \begin{bmatrix} \frac{2}{5} & -\frac{1}{5} \\ \frac{1}{5} & \frac{2}{5} \end{bmatrix}$, so $AB^{-1} = \begin{bmatrix} 3 & -2 \\ 2 & 3 \end{bmatrix}\begin{bmatrix} \frac{2}{5} & -\frac{1}{5} \\ \frac{1}{5} & \frac{2}{5} \end{bmatrix} = \begin{bmatrix} \frac{4}{5} & -\frac{7}{5} \\ \frac{7}{5} & \frac{4}{5} \end{bmatrix}$, so

$\dfrac{3 + 2i}{2 + i} = \dfrac{4}{5} - \dfrac{7}{5}i.$

19. 4(cos 60° + i sin 60°) corresponds to A = $\begin{bmatrix} 4\cos 60° & 4\sin 60° \\ -4\sin 60° & 4\cos 60° \end{bmatrix}$,

2(cos 15° + i sin 15°) corresponds to B = $\begin{bmatrix} 2\cos 15° & 2\sin 15° \\ -2\sin 15° & 2\cos 15° \end{bmatrix}$, and $B^{-1} = \dfrac{1}{4}\begin{bmatrix} 2\cos 15° & -2\sin 15° \\ 2\sin 15° & 2\cos 15° \end{bmatrix}$, $AB^{-1} =$

$\begin{bmatrix} 4\cos 60° & 4\sin 60° \\ -4\sin 60° & 4\cos 60° \end{bmatrix} \cdot \dfrac{1}{4}\begin{bmatrix} 2\cos 15° & -2\sin 15° \\ 2\sin 15° & 2\cos 15° \end{bmatrix} =$

$\dfrac{1}{4}\begin{bmatrix} 8\cos 45° & 8\sin 45° \\ -8\sin 45° & 8\cos 45° \end{bmatrix} = 2\begin{bmatrix} \cos 45° & \sin 45° \\ -\sin 45° & \cos 45° \end{bmatrix} = 2\begin{bmatrix} \dfrac{1}{\sqrt{2}} & \dfrac{1}{\sqrt{2}} \\ -\dfrac{1}{\sqrt{2}} & \dfrac{1}{\sqrt{2}} \end{bmatrix},$

so $\dfrac{4(\cos 60° + i\sin 60°)}{2(\cos 15° + i\sin 15°)} = \sqrt{2} + \sqrt{2}\,i.$

<u>20.</u> $8(\cos 90° + i\sin 90°)$ corresponds to $A = \begin{bmatrix} 8\cos 90° & 8\sin 90° \\ -8\sin 90° & 8\cos 90° \end{bmatrix},$

$2(\cos 30° + i\sin 30°)$ corresponds to $B = \begin{bmatrix} 2\cos 30° & 2\sin 30° \\ -2\sin 30° & 2\cos 30° \end{bmatrix},$

and $B^{-1} = \dfrac{1}{4}\begin{bmatrix} 2\cos 30° & -2\sin 30° \\ 2\sin 30° & 2\cos 30° \end{bmatrix}$, so $AB^{-1} =$

$\begin{bmatrix} 8\cos 90° & 8\sin 90° \\ -8\sin 90° & 8\cos 90° \end{bmatrix} \cdot \dfrac{1}{4}\begin{bmatrix} 2\cos 30° & -2\sin 30° \\ 2\sin 30° & 2\cos 30° \end{bmatrix} =$

$\dfrac{1}{4}\begin{bmatrix} 16\cos 60° & 16\sin 60° \\ -16\sin 60° & 16\cos 60° \end{bmatrix} = 4\begin{bmatrix} \cos 60° & \sin 60° \\ -\sin 60° & \cos 60° \end{bmatrix} = 4\begin{bmatrix} \dfrac{1}{2} & \dfrac{\sqrt{3}}{2} \\ -\dfrac{\sqrt{3}}{2} & \dfrac{1}{2} \end{bmatrix},$

thus $\dfrac{8(\cos 90° + i\sin 90°)}{2(\cos 30° + i\sin 30°)} = 2 + 2\sqrt{3}\,i.$

<u>Exercises 8-5 · Pages 296-298</u>

<u>A</u>   <u>1.</u> $\begin{bmatrix} 3 & 0 \\ 0 & 3 \end{bmatrix}\begin{bmatrix} x \\ y \end{bmatrix} = \begin{bmatrix} 3x \\ 3y \end{bmatrix}$; magnification.

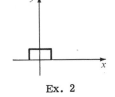

Ex. 1   Ex. 2

<u>2.</u> $\begin{bmatrix} -1 & 0 \\ 0 & 1 \end{bmatrix}\begin{bmatrix} x \\ y \end{bmatrix} = \begin{bmatrix} -x \\ y \end{bmatrix}$; reflection in the y-axis.

$\underline{3.}$ $\begin{bmatrix} -2 & 0 \\ 0 & -2 \end{bmatrix}\begin{bmatrix} x \\ y \end{bmatrix} = \begin{bmatrix} -2x \\ -2y \end{bmatrix}$; reflection in the origin and a magnification.

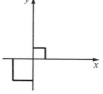

Ex. 3

Ex. 4

$\underline{4.}$ $\begin{bmatrix} 1 & 0 \\ 0 & -1 \end{bmatrix}\begin{bmatrix} x \\ y \end{bmatrix} = \begin{bmatrix} x \\ -y \end{bmatrix}$; reflection in the x-axis.

$\underline{5.}$ $\begin{bmatrix} 3 & 0 \\ 0 & -1 \end{bmatrix}\begin{bmatrix} x \\ y \end{bmatrix} = \begin{bmatrix} 3x \\ -y \end{bmatrix}$; reflection in the x-axis and a stretching.

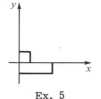

Ex. 5

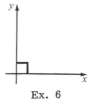

Ex. 6

$\underline{6.}$ $\begin{bmatrix} 0 & 1 \\ 1 & 0 \end{bmatrix}\begin{bmatrix} x \\ y \end{bmatrix} = \begin{bmatrix} y \\ x \end{bmatrix}$; reflection in the line x = y.

$\underline{7.}$ $\begin{bmatrix} 0 & -1 \\ 1 & 0 \end{bmatrix}\begin{bmatrix} x \\ y \end{bmatrix} = \begin{bmatrix} -y \\ x \end{bmatrix}$, reflection in the line y = x and in the y-axis.

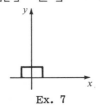

Ex. 7

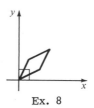

Ex. 8

$\underline{8.}$ $\begin{bmatrix} 2 & 1 \\ 1 & 2 \end{bmatrix}\begin{bmatrix} x \\ y \end{bmatrix} = \begin{bmatrix} 2x + y \\ x + 2y \end{bmatrix}$. Geometrically, this appears to be shearing along both

axes and a magnification. Actually, the transformation matrix $\begin{bmatrix} 2 & 1 \\ 1 & 2 \end{bmatrix}$ cannot be

written as the product of simple shearing matrices and a magnification matrix.

However, it can be written as the following product:

$$2 \cdot \begin{bmatrix} 1 & 0 \\ \frac{1}{2} & 1 \end{bmatrix}\begin{bmatrix} 1 & 0 \\ 0 & \frac{3}{4} \end{bmatrix}\begin{bmatrix} 1 & \frac{1}{2} \\ 0 & 1 \end{bmatrix} = \begin{bmatrix} 2 & 1 \\ 1 & 2 \end{bmatrix} ,$$

a shearing, a stretching, then another shearing.

Although not discussed in this section, shearing along both axes will also result from the following matrix:

$$\begin{bmatrix} 1 & a \\ b & 1 \end{bmatrix}, \text{ where } a \neq 0, \; b \neq 0.$$

9. $\begin{bmatrix} 4 & 0 \\ 0 & 4 \end{bmatrix}\begin{bmatrix} x \\ y \end{bmatrix} = \begin{bmatrix} 4x \\ 4y \end{bmatrix}$, magnification.

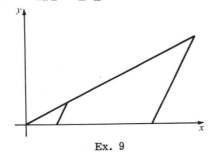

Ex. 9

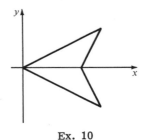

Ex. 10

10. $\begin{bmatrix} 1 & 0 \\ 0 & -1 \end{bmatrix}\begin{bmatrix} x \\ y \end{bmatrix} = \begin{bmatrix} x \\ -y \end{bmatrix}$, reflection in the x-axis.

11. $\begin{bmatrix} -1 & 0 \\ 0 & -1 \end{bmatrix}\begin{bmatrix} x \\ y \end{bmatrix} = \begin{bmatrix} -x \\ -y \end{bmatrix}$, reflection in the origin.

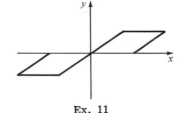

Ex. 11

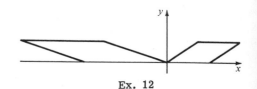

Ex. 12

12. $\begin{bmatrix} -2 & 0 \\ 0 & 1 \end{bmatrix}\begin{bmatrix} x \\ y \end{bmatrix} = \begin{bmatrix} -2x \\ y \end{bmatrix}$; reflection in the y-axis and a stretching.

13. $\begin{bmatrix} \frac{1}{2} & 0 \\ 0 & -\frac{1}{2} \end{bmatrix}\begin{bmatrix} x \\ y \end{bmatrix} = \begin{bmatrix} \frac{1}{2}x \\ -\frac{1}{2}y \end{bmatrix}$; shrinking and a reflection in the x-axis.

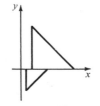

Ex. 13

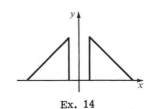

Ex. 14

14. $\begin{bmatrix} -1 & 0 \\ 0 & 1 \end{bmatrix} \begin{bmatrix} x \\ y \end{bmatrix} = \begin{bmatrix} -x \\ y \end{bmatrix}$, reflection in the y-axis.

15. $\begin{bmatrix} 0 & -1 \\ -1 & 0 \end{bmatrix} \begin{bmatrix} x \\ y \end{bmatrix} = \begin{bmatrix} -y \\ -x \end{bmatrix}$; reflection in the line y = -x.

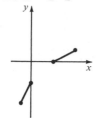

Ex. 15

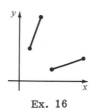

Ex. 16

16. $\begin{bmatrix} 0 & 1 \\ 1 & 0 \end{bmatrix} \begin{bmatrix} x \\ y \end{bmatrix} = \begin{bmatrix} y \\ x \end{bmatrix}$; reflection in the line y = x.

17. $\begin{bmatrix} 1 & 0 \\ 0 & -1 \end{bmatrix} \begin{bmatrix} x \\ y \end{bmatrix} = \begin{bmatrix} x \\ -y \end{bmatrix}$, $\begin{bmatrix} -1 & 0 \\ 0 & 1 \end{bmatrix} \begin{bmatrix} x \\ -y \end{bmatrix} = \begin{bmatrix} -x \\ -y \end{bmatrix}$, which is a reflection in the origin.

18. $\begin{bmatrix} 1 & 0 \\ 0 & -1 \end{bmatrix} \begin{bmatrix} x \\ y \end{bmatrix} = \begin{bmatrix} x \\ -y \end{bmatrix}$, $\begin{bmatrix} -1 & 0 \\ 0 & -1 \end{bmatrix} \begin{bmatrix} x \\ -y \end{bmatrix} = \begin{bmatrix} -x \\ y \end{bmatrix}$, which is a reflection in the y-axis.

19. $\begin{bmatrix} 0 & -1 \\ -1 & 0 \end{bmatrix} \begin{bmatrix} x \\ y \end{bmatrix} = \begin{bmatrix} -y \\ -x \end{bmatrix}$, $\begin{bmatrix} 0 & 1 \\ 1 & 0 \end{bmatrix} \begin{bmatrix} -y \\ -x \end{bmatrix} = \begin{bmatrix} -x \\ -y \end{bmatrix}$, which is a reflection in the origin.

B 20. $\begin{bmatrix} 3 & 0 \\ 0 & 3 \end{bmatrix} \begin{bmatrix} x \\ y \end{bmatrix} = \begin{bmatrix} 3x \\ 3y \end{bmatrix}$, magnification.

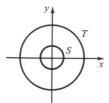

Ex. 20

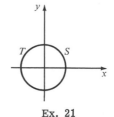

Ex. 21

21. $\begin{bmatrix} -1 & 0 \\ 0 & 1 \end{bmatrix} \begin{bmatrix} x \\ y \end{bmatrix} = \begin{bmatrix} -x \\ y \end{bmatrix}$, reflection in the y-axis.

22. $\begin{bmatrix} 0 & 1 \\ 1 & 0 \end{bmatrix} \begin{bmatrix} x \\ y \end{bmatrix} = \begin{bmatrix} y \\ x \end{bmatrix}$, reflection in the line y = x.

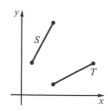

23. $\begin{bmatrix} 1 & 0 \\ 0 & -1 \end{bmatrix} \begin{bmatrix} x \\ y \end{bmatrix} = \begin{bmatrix} x \\ -y \end{bmatrix}$, reflection in the x-axis.

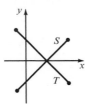

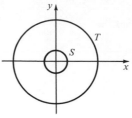

Ex. 23   Ex. 24

24. T is a magnification of S; therefore, the matrix is in the form $\begin{bmatrix} a & 0 \\ 0 & a \end{bmatrix}$

and must be $\begin{bmatrix} 4 & 0 \\ 0 & 4 \end{bmatrix}$.

25. $A\begin{bmatrix} x \\ y \end{bmatrix} = \begin{bmatrix} x \\ -y \end{bmatrix}$; therefore T is a reflection in the x-axis of S; the matrix is $\begin{bmatrix} 1 & 0 \\ 0 & -1 \end{bmatrix}$.

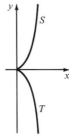

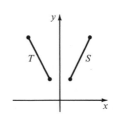

Ex. 25   Ex. 26

26. $A\begin{bmatrix} x \\ y \end{bmatrix} = \begin{bmatrix} -x \\ y \end{bmatrix}$, therefore T is the reflection in the y-axis of S; the matrix is the

form $\begin{bmatrix} -1 & 0 \\ 0 & 1 \end{bmatrix}$.

27. $A\begin{bmatrix} x \\ y \end{bmatrix} = \begin{bmatrix} -x \\ y \end{bmatrix}$, therefore T is a reflection in the y-axis of S; the matrix is in the

form $\begin{bmatrix} -1 & 0 \\ 0 & 1 \end{bmatrix}$.

28. $A\begin{bmatrix} x \\ y \end{bmatrix} = \begin{bmatrix} \frac{1}{3}x \\ \frac{1}{3}y \end{bmatrix}$, therefore T is a shrinking of S; the matrix is in the form $\begin{bmatrix} a & 0 \\ 0 & a \end{bmatrix}$

and must be $\begin{bmatrix} \frac{1}{3} & 0 \\ 0 & \frac{1}{3} \end{bmatrix}$.

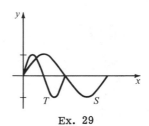

Ex. 28   Ex. 29

29. $A\begin{bmatrix} x \\ y \end{bmatrix} = \begin{bmatrix} \frac{1}{2}x \\ y \end{bmatrix}$; therefore T is a contraction of S; the matrix is in the form $\begin{bmatrix} a & 0 \\ 0 & 1 \end{bmatrix}$

and must be $\begin{bmatrix} \frac{1}{2} & 0 \\ 0 & 1 \end{bmatrix}$.

30. a. $\begin{bmatrix} 1 & 1 \\ 1 & 1 \end{bmatrix}\begin{bmatrix} x \\ y \end{bmatrix} = \begin{bmatrix} x+y \\ x+y \end{bmatrix}$, each vector is transformed into the line $y = x$;

b. $\begin{bmatrix} 1 & 0 \\ 0 & 0 \end{bmatrix}\begin{bmatrix} x \\ y \end{bmatrix} = \begin{bmatrix} x \\ 0 \end{bmatrix}$, each vector is transformed into the x-axis;

c. $\begin{bmatrix} 0 & 0 \\ 1 & 1 \end{bmatrix}\begin{bmatrix} x \\ y \end{bmatrix} = \begin{bmatrix} 0 \\ x+y \end{bmatrix}$, each vector is transformed into the y-axis.

31. $\begin{bmatrix} a_1 & a_2 \\ a_3 & a_4 \end{bmatrix}\begin{bmatrix} x \\ y \end{bmatrix} = \begin{bmatrix} a_1x + a_2y \\ a_3x + a_4y \end{bmatrix} = \begin{bmatrix} x' \\ y' \end{bmatrix}$, $A^{-1} = \frac{1}{a_1a_4 - a_2a_3}\begin{bmatrix} a_4 & -a_2 \\ -a_3 & a_1 \end{bmatrix}$,

$\frac{1}{a_1a_4 - a_2a_3}\begin{bmatrix} a_4 & -a_2 \\ -a_3 & a_1 \end{bmatrix}\begin{bmatrix} x' \\ y' \end{bmatrix} = \frac{1}{a_1a_4 - a_2a_3}\begin{bmatrix} a_4 & -a_2 \\ -a_3 & a_1 \end{bmatrix}\begin{bmatrix} a_1x + a_2y \\ a_3x + a_4y \end{bmatrix} =$

$\frac{1}{a_1a_4 - a_2a_3}\begin{bmatrix} a_4(a_1x + a_2y) - a_2(a_3x + a_4y) \\ -a_3(a_1x + a_2y) + a_1(a_3x + a_4y) \end{bmatrix} = \frac{1}{a_1a_4 - a_2a_3}\begin{bmatrix} (a_1a_4 - a_2a_3)x \\ (a_1a_4 - a_2a_3)y \end{bmatrix} = \begin{bmatrix} x \\ y \end{bmatrix}$

Exercises 8-6 · Pages 301-302

A   1. $\begin{bmatrix} \cos 30° & -\sin 30° \\ \sin 30° & \cos 30° \end{bmatrix} = \begin{bmatrix} \frac{\sqrt{3}}{2} & -\frac{1}{2} \\ \frac{1}{2} & \frac{\sqrt{3}}{2} \end{bmatrix}$, $\begin{bmatrix} \frac{\sqrt{3}}{2} & -\frac{1}{2} \\ \frac{1}{2} & \frac{\sqrt{3}}{2} \end{bmatrix}\begin{bmatrix} 0 \\ 0 \end{bmatrix} = \begin{bmatrix} 0 \\ 0 \end{bmatrix}$;

$$\begin{bmatrix} \frac{\sqrt{3}}{2} & -\frac{1}{2} \\ \frac{1}{2} & \frac{\sqrt{3}}{2} \end{bmatrix} \begin{bmatrix} \sqrt{3} \\ 0 \end{bmatrix} = \begin{bmatrix} \frac{3}{2} \\ \frac{\sqrt{3}}{2} \end{bmatrix} ; \quad \begin{bmatrix} \frac{\sqrt{3}}{2} & -\frac{1}{2} \\ \frac{1}{2} & \frac{\sqrt{3}}{2} \end{bmatrix} \begin{bmatrix} 2\sqrt{3} \\ 4 \end{bmatrix} = \begin{bmatrix} 1 \\ 3\sqrt{3} \end{bmatrix}$$

Ex. 1   Ex. 2

<u>2.</u> $\begin{bmatrix} \cos 45° & -\sin 45° \\ \sin 45° & \cos 45° \end{bmatrix} = \begin{bmatrix} \frac{1}{\sqrt{2}} & -\frac{1}{\sqrt{2}} \\ \frac{1}{\sqrt{2}} & \frac{1}{\sqrt{2}} \end{bmatrix} ; \quad \begin{bmatrix} \frac{1}{\sqrt{2}} & -\frac{1}{\sqrt{2}} \\ \frac{1}{\sqrt{2}} & \frac{1}{\sqrt{2}} \end{bmatrix} \begin{bmatrix} \sqrt{2} \\ 0 \end{bmatrix} = \begin{bmatrix} 1 \\ 1 \end{bmatrix} ;$

$$\begin{bmatrix} \frac{1}{\sqrt{2}} & -\frac{1}{\sqrt{2}} \\ \frac{1}{\sqrt{2}} & \frac{1}{\sqrt{2}} \end{bmatrix} \begin{bmatrix} 1 \\ 3 \end{bmatrix} = \begin{bmatrix} -\sqrt{2} \\ 2\sqrt{3} \end{bmatrix} ; \quad \begin{bmatrix} \frac{1}{\sqrt{2}} & -\frac{1}{\sqrt{2}} \\ \frac{1}{\sqrt{2}} & \frac{1}{\sqrt{2}} \end{bmatrix} \begin{bmatrix} 2 \\ 4 \end{bmatrix} = \begin{bmatrix} -\sqrt{2} \\ 3\sqrt{2} \end{bmatrix}$$

<u>3.</u> $\begin{bmatrix} \cos 60° & -\sin 60° \\ \sin 60° & \cos 60° \end{bmatrix} = \begin{bmatrix} \frac{1}{2} & -\frac{\sqrt{3}}{2} \\ \frac{\sqrt{3}}{2} & \frac{1}{2} \end{bmatrix} ; \quad \begin{bmatrix} \frac{1}{2} & -\frac{\sqrt{3}}{2} \\ \frac{\sqrt{3}}{2} & \frac{1}{2} \end{bmatrix} \begin{bmatrix} 0 \\ 0 \end{bmatrix} = \begin{bmatrix} 0 \\ 0 \end{bmatrix} ;$

$$\begin{bmatrix} \frac{1}{2} & -\frac{\sqrt{3}}{2} \\ \frac{\sqrt{3}}{2} & \frac{1}{2} \end{bmatrix} \begin{bmatrix} 2 \\ 0 \end{bmatrix} = \begin{bmatrix} 1 \\ \sqrt{3} \end{bmatrix} ; \quad \begin{bmatrix} \frac{1}{2} & -\frac{\sqrt{3}}{2} \\ \frac{\sqrt{3}}{2} & \frac{1}{2} \end{bmatrix} \begin{bmatrix} 2 \\ 2 \end{bmatrix} = \begin{bmatrix} 1 - \sqrt{3} \\ 1 + \sqrt{3} \end{bmatrix} ;$$

$$\begin{bmatrix} \frac{1}{2} & -\frac{\sqrt{3}}{2} \\ \frac{\sqrt{3}}{2} & \frac{1}{2} \end{bmatrix} \begin{bmatrix} 0 \\ 2 \end{bmatrix} = \begin{bmatrix} -\sqrt{3} \\ 1 \end{bmatrix}$$

<u>4.</u> $\begin{bmatrix} \cos(-30°) & -\sin(-30°) \\ \sin(-30°) & \cos(-30°) \end{bmatrix} = \begin{bmatrix} \cos 30° & \sin 30° \\ -\sin 30° & \cos 30° \end{bmatrix} = \begin{bmatrix} \frac{\sqrt{3}}{2} & \frac{1}{2} \\ -\frac{1}{2} & \frac{\sqrt{3}}{2} \end{bmatrix} ;$

$$\begin{bmatrix} \frac{\sqrt{3}}{2} & \frac{1}{2} \\ -\frac{1}{2} & \frac{\sqrt{3}}{2} \end{bmatrix} \begin{bmatrix} 0 \\ 0 \end{bmatrix} = \begin{bmatrix} 0 \\ 0 \end{bmatrix} ; \quad \begin{bmatrix} \frac{\sqrt{3}}{2} & \frac{1}{2} \\ -\frac{1}{2} & \frac{\sqrt{3}}{2} \end{bmatrix} \begin{bmatrix} 2 \\ 0 \end{bmatrix} = \begin{bmatrix} \sqrt{3} \\ -1 \end{bmatrix} ;$$

$$\begin{bmatrix} \frac{\sqrt{3}}{2} & \frac{1}{2} \\ -\frac{1}{2} & \frac{\sqrt{3}}{2} \end{bmatrix} \begin{bmatrix} 2 \\ 2 \end{bmatrix} = \begin{bmatrix} \sqrt{3} + 1 \\ \sqrt{3} - 1 \end{bmatrix} ; \quad \begin{bmatrix} \frac{\sqrt{3}}{2} & \frac{1}{2} \\ -\frac{1}{2} & \frac{\sqrt{3}}{2} \end{bmatrix} \begin{bmatrix} 0 \\ 2 \end{bmatrix} = \begin{bmatrix} 1 \\ \sqrt{3} \end{bmatrix}$$

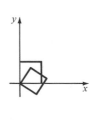

5. $\begin{bmatrix} \cos(-135°) & -\sin(-135°) \\ \sin(-135°) & \cos(-135°) \end{bmatrix} = \begin{bmatrix} \cos 135° & \sin 135° \\ -\sin 135° & \cos 135° \end{bmatrix} = \begin{bmatrix} -\dfrac{1}{\sqrt{2}} & \dfrac{1}{\sqrt{2}} \\ -\dfrac{1}{\sqrt{2}} & -\dfrac{1}{\sqrt{2}} \end{bmatrix}$

$\begin{bmatrix} -\dfrac{1}{\sqrt{2}} & \dfrac{1}{\sqrt{2}} \\ -\dfrac{1}{\sqrt{2}} & -\dfrac{1}{\sqrt{2}} \end{bmatrix}\begin{bmatrix} \sqrt{2} \\ 0 \end{bmatrix} = \begin{bmatrix} -1 \\ -1 \end{bmatrix};$ $\begin{bmatrix} -\dfrac{1}{\sqrt{2}} & \dfrac{1}{\sqrt{2}} \\ -\dfrac{1}{\sqrt{2}} & -\dfrac{1}{\sqrt{2}} \end{bmatrix}\begin{bmatrix} \sqrt{2} \\ \sqrt{2} \end{bmatrix} = \begin{bmatrix} 0 \\ -2 \end{bmatrix};$

$\begin{bmatrix} -\dfrac{1}{\sqrt{2}} & \dfrac{1}{\sqrt{2}} \\ -\dfrac{1}{\sqrt{2}} & -\dfrac{1}{\sqrt{2}} \end{bmatrix}\begin{bmatrix} 0 \\ \sqrt{2} \end{bmatrix} = \begin{bmatrix} 1 \\ -1 \end{bmatrix}$

6. $\begin{bmatrix} \cos 120° & -\sin 120° \\ \sin 120° & \cos 120° \end{bmatrix} = \begin{bmatrix} -\dfrac{1}{2} & -\dfrac{\sqrt{3}}{2} \\ \dfrac{\sqrt{3}}{2} & -\dfrac{1}{2} \end{bmatrix};$ $\begin{bmatrix} -\dfrac{1}{2} & -\dfrac{\sqrt{3}}{2} \\ \dfrac{\sqrt{3}}{2} & -\dfrac{1}{2} \end{bmatrix}\begin{bmatrix} -2 \\ \sqrt{3} \end{bmatrix} = \begin{bmatrix} -\dfrac{1}{2} \\ -\dfrac{3}{2}\sqrt{3} \end{bmatrix};$

$\begin{bmatrix} -\dfrac{1}{2} & -\dfrac{\sqrt{3}}{2} \\ \dfrac{\sqrt{3}}{2} & -\dfrac{1}{2} \end{bmatrix}\begin{bmatrix} -2 \\ 2\sqrt{3} \end{bmatrix} = \begin{bmatrix} -2 \\ -2\sqrt{3} \end{bmatrix};$ $\begin{bmatrix} -\dfrac{1}{2} & -\dfrac{\sqrt{3}}{2} \\ \dfrac{\sqrt{3}}{2} & -\dfrac{1}{2} \end{bmatrix}\begin{bmatrix} 3 \\ \sqrt{3} \end{bmatrix} =$

$\begin{bmatrix} -3 \\ \sqrt{3} \end{bmatrix};$ $\begin{bmatrix} -\dfrac{1}{2} & -\dfrac{\sqrt{3}}{2} \\ \dfrac{\sqrt{3}}{2} & -\dfrac{1}{2} \end{bmatrix}\begin{bmatrix} 3 \\ 2\sqrt{3} \end{bmatrix} = \begin{bmatrix} -\dfrac{9}{2} \\ \dfrac{\sqrt{3}}{2} \end{bmatrix}$

7. $\begin{bmatrix} \dfrac{\sqrt{3}}{2} & \dfrac{1}{2} \\ -\dfrac{1}{2} & \dfrac{\sqrt{3}}{2} \end{bmatrix}\begin{bmatrix} 0 \\ 0 \end{bmatrix} = \begin{bmatrix} 0 \\ 0 \end{bmatrix};$ $\begin{bmatrix} \dfrac{\sqrt{3}}{2} & \dfrac{1}{2} \\ -\dfrac{1}{2} & \dfrac{\sqrt{3}}{2} \end{bmatrix}\begin{bmatrix} \sqrt{3} \\ 0 \end{bmatrix} = \begin{bmatrix} \dfrac{3}{2} \\ -\dfrac{\sqrt{3}}{2} \end{bmatrix};$

$\begin{bmatrix} \dfrac{\sqrt{3}}{2} & \dfrac{1}{2} \\ -\dfrac{1}{2} & \dfrac{\sqrt{3}}{2} \end{bmatrix}\begin{bmatrix} 2\sqrt{3} \\ 4 \end{bmatrix} = \begin{bmatrix} 5 \\ \sqrt{3} \end{bmatrix}$

8. $\begin{bmatrix} \dfrac{1}{\sqrt{2}} & \dfrac{1}{\sqrt{2}} \\ -\dfrac{1}{\sqrt{2}} & \dfrac{1}{\sqrt{2}} \end{bmatrix}\begin{bmatrix} \sqrt{2} \\ 0 \end{bmatrix} = \begin{bmatrix} 1 \\ -1 \end{bmatrix},$ $\begin{bmatrix} \dfrac{1}{\sqrt{2}} & \dfrac{1}{\sqrt{2}} \\ -\dfrac{1}{\sqrt{2}} & \dfrac{1}{\sqrt{2}} \end{bmatrix}\begin{bmatrix} 1 \\ 3 \end{bmatrix} = \begin{bmatrix} 2\sqrt{2} \\ \sqrt{2} \end{bmatrix},$

$\begin{bmatrix} \dfrac{1}{\sqrt{2}} & \dfrac{1}{\sqrt{2}} \\ -\dfrac{1}{\sqrt{2}} & \dfrac{1}{\sqrt{2}} \end{bmatrix}\begin{bmatrix} 2 \\ 4 \end{bmatrix} = \begin{bmatrix} 3\sqrt{2} \\ \sqrt{2} \end{bmatrix}$

9. $\begin{bmatrix} \frac{1}{2} & \frac{\sqrt{3}}{2} \\ -\frac{\sqrt{3}}{2} & \frac{1}{2} \end{bmatrix}\begin{bmatrix} 0 \\ 0 \end{bmatrix} = \begin{bmatrix} 0 \\ 0 \end{bmatrix}$; $\begin{bmatrix} \frac{1}{2} & \frac{\sqrt{3}}{2} \\ -\frac{\sqrt{3}}{2} & \frac{1}{2} \end{bmatrix}\begin{bmatrix} 2 \\ 0 \end{bmatrix} = \begin{bmatrix} 1 \\ -\sqrt{3} \end{bmatrix}$;

$\begin{bmatrix} \frac{1}{2} & \frac{\sqrt{3}}{2} \\ -\frac{\sqrt{3}}{2} & \frac{1}{2} \end{bmatrix}\begin{bmatrix} 2 \\ 2 \end{bmatrix} = \begin{bmatrix} 1 + \sqrt{3} \\ 1 - \sqrt{3} \end{bmatrix}$; $\begin{bmatrix} \frac{1}{2} & \frac{\sqrt{3}}{2} \\ -\frac{\sqrt{3}}{2} & \frac{1}{2} \end{bmatrix}\begin{bmatrix} 0 \\ 2 \end{bmatrix} = \begin{bmatrix} \sqrt{3} \\ 1 \end{bmatrix}$

10. $\begin{bmatrix} \frac{\sqrt{3}}{2} & -\frac{1}{2} \\ \frac{1}{2} & \frac{\sqrt{3}}{2} \end{bmatrix}\begin{bmatrix} 0 \\ 0 \end{bmatrix} = \begin{bmatrix} 0 \\ 0 \end{bmatrix}$; $\begin{bmatrix} \frac{\sqrt{3}}{2} & -\frac{1}{2} \\ \frac{1}{2} & \frac{\sqrt{3}}{2} \end{bmatrix}\begin{bmatrix} 2 \\ 0 \end{bmatrix} = \begin{bmatrix} \sqrt{3} \\ 1 \end{bmatrix}$;

$\begin{bmatrix} \frac{\sqrt{3}}{2} & -\frac{1}{2} \\ \frac{1}{2} & \frac{\sqrt{3}}{2} \end{bmatrix}\begin{bmatrix} 2 \\ 2 \end{bmatrix} = \begin{bmatrix} \sqrt{3} - 1 \\ \sqrt{3} + 1 \end{bmatrix}$; $\begin{bmatrix} \frac{\sqrt{3}}{2} & -\frac{1}{2} \\ \frac{1}{2} & \frac{\sqrt{3}}{2} \end{bmatrix}\begin{bmatrix} 0 \\ 2 \end{bmatrix} = \begin{bmatrix} -1 \\ \sqrt{3} \end{bmatrix}$

11. $\begin{bmatrix} -\frac{1}{\sqrt{2}} & -\frac{1}{\sqrt{2}} \\ \frac{1}{\sqrt{2}} & -\frac{1}{\sqrt{2}} \end{bmatrix}\begin{bmatrix} \sqrt{2} \\ 0 \end{bmatrix} = \begin{bmatrix} -1 \\ 1 \end{bmatrix}$; $\begin{bmatrix} -\frac{1}{\sqrt{2}} & -\frac{1}{\sqrt{2}} \\ \frac{1}{\sqrt{2}} & -\frac{1}{\sqrt{2}} \end{bmatrix}\begin{bmatrix} \sqrt{2} \\ \sqrt{2} \end{bmatrix} = \begin{bmatrix} -2 \\ 0 \end{bmatrix}$;

$\begin{bmatrix} -\frac{1}{\sqrt{2}} & -\frac{1}{\sqrt{2}} \\ \frac{1}{\sqrt{2}} & -\frac{1}{\sqrt{2}} \end{bmatrix}\begin{bmatrix} 0 \\ \sqrt{2} \end{bmatrix} = \begin{bmatrix} -1 \\ -1 \end{bmatrix}$

12. $\begin{bmatrix} -\frac{1}{2} & \frac{\sqrt{3}}{2} \\ -\frac{\sqrt{3}}{2} & -\frac{1}{2} \end{bmatrix}\begin{bmatrix} -2 \\ \sqrt{3} \end{bmatrix} = \begin{bmatrix} \frac{5}{2} \\ \frac{\sqrt{3}}{2} \end{bmatrix}$; $\begin{bmatrix} -\frac{1}{2} & \frac{\sqrt{3}}{2} \\ -\frac{\sqrt{3}}{2} & -\frac{1}{2} \end{bmatrix}\begin{bmatrix} -2 \\ 2\sqrt{3} \end{bmatrix} = \begin{bmatrix} 4 \\ 0 \end{bmatrix}$;

$\begin{bmatrix} -\frac{1}{2} & \frac{\sqrt{3}}{2} \\ -\frac{\sqrt{3}}{2} & -\frac{1}{2} \end{bmatrix}\begin{bmatrix} 3 \\ \sqrt{3} \end{bmatrix} = \begin{bmatrix} 0 \\ -2\sqrt{3} \end{bmatrix}$; $\begin{bmatrix} -\frac{1}{2} & \frac{\sqrt{3}}{2} \\ -\frac{\sqrt{3}}{2} & -\frac{1}{2} \end{bmatrix}\begin{bmatrix} 3 \\ 2\sqrt{3} \end{bmatrix} = \begin{bmatrix} \frac{3}{2} \\ -\frac{5}{2}\sqrt{3} \end{bmatrix}$

B   13. $2 \cos \theta - \sin \theta = -1$, $2 = 2 \sin \theta + \cos \theta$, $\sin \theta = 1$, $\cos \theta = 0$,

$A = \begin{bmatrix} \cos \theta & -\sin \theta \\ \sin \theta & \cos \theta \end{bmatrix} = \begin{bmatrix} 0 & -1 \\ 1 & 0 \end{bmatrix}$

14. $4 = -4 \cos \theta + 7 \sin \theta$, $7 = -4 \sin \theta - 7 \cos \theta$, $\cos \theta = -1$, $\sin \theta = 0$,

$$A = \begin{bmatrix} \cos\theta & -\sin\theta \\ \sin\theta & \cos\theta \end{bmatrix} = \begin{bmatrix} -1 & 0 \\ 0 & -1 \end{bmatrix}$$

15. $-4\sqrt{2} = 3\cos\theta - 5\sin\theta$, $-3\sqrt{2} = 3\sin\theta + 5\cos\theta$,

$$\cos\theta = \frac{-27\sqrt{2}}{34}, \ \sin\theta = \frac{11\sqrt{2}}{34}, \ A = \begin{bmatrix} \cos\theta & -\sin\theta \\ \sin\theta & \cos\theta \end{bmatrix} = \begin{bmatrix} \frac{-27\sqrt{2}}{34} & \frac{-11\sqrt{2}}{34} \\ \frac{11\sqrt{2}}{4} & \frac{-27\sqrt{2}}{34} \end{bmatrix}$$

16. $1 = 5\cos\theta + 5\sin\theta$, $7 = 5\sin\theta - 5\cos\theta$, $\sin\theta = \frac{4}{5}$, $\cos\theta = -\frac{3}{5}$,

$$A = \begin{bmatrix} \cos\theta & -\sin\theta \\ \sin\theta & \cos\theta \end{bmatrix} = \begin{bmatrix} -\frac{3}{5} & -\frac{4}{5} \\ \frac{4}{5} & -\frac{3}{5} \end{bmatrix}$$

C   17. $x' = x\cos\theta - y\sin\theta$, $y' = x\sin\theta + y\cos\theta$, $\|(x', y')\| =$

$$\sqrt{(x\cos\theta - y\sin\theta)^2 + (x\sin\theta + y\cos\theta)^2} =$$

$$\sqrt{x^2(\cos^2\theta + \sin^2\theta) + y^2(\sin^2\theta + \cos^2\theta)} = \sqrt{x^2 + y^2} = \|(x, y)\|.$$

18. $x' = x\cos\theta + y\sin\theta$, $y' = -x\sin\theta + y\cos\theta$.   Then

$x'\cos\theta - y'\sin\theta = (x\cos^2\theta + y\sin\theta\cos\theta) -$

$(-x\sin^2\theta + y\cos\theta\sin\theta) = x\cos^2\theta + x\sin^2\theta = x$, and

$x'\sin\theta + y'\cos\theta = (x\cos\theta\sin\theta + y\sin^2\theta) +$

$(-x\sin\theta\cos\theta + y\cos^2\theta) = y\sin^2\theta + y\cos^2\theta = y.$

19. Let (x_1, y_1) and (x_2, y_2) be the two points.   Then $d =$

$\sqrt{(x_2 - x_1)^2 + (y_2 - y_1)^2}$.   Let (x_1', y_1') and (x_2', y_2') be the respective images

of these points.   Then we have $d' = \sqrt{(x_2' - x_1')^2 + (y_2' - y_1')^2} =$

$$\sqrt{[(x_2\cos\theta - y_2\sin\theta) - (x_1\cos\theta - y_1\sin\theta)]^2 +}$$

$$\overline{[(x_2\sin\theta + y_2\cos\theta) - (x_1\sin\theta + y_1\cos\theta)]^2} =$$

$$\sqrt{[(x_2^2\cos^2\theta - 2x_2y_2\cos\theta\sin\theta + y_2^2\sin^2\theta) -}$$

$$\overline{2(x_2\cos\theta - y_2\sin\theta)(x_1\cos\theta - y_1\sin\theta) +}$$

$$\overline{(x_1^2\cos^2\theta - 2x_1y_1\cos\theta\sin\theta + y_1^2\sin^2\theta)] +}$$

$$\overline{[(x_2^2\sin^2\theta + 2x_2y_2\sin\theta\cos\theta + y_2^2\cos^2\theta) -}$$

$$\overline{2(x_2\sin\theta + y_2\cos\theta)(x_1\sin\theta + y_1\cos\theta) +}$$

$$\overline{(x_1^2\sin^2\theta + 2x_1y_1\sin\theta\cos\theta + y_1^2\cos^2\theta)] =}$$

$$\sqrt{x_2^2 \cos^2 \theta + y_2^2 \sin^2 \theta - 2(x_1 x_2 \cos^2 \theta - x_2 y_1 \cos \theta \sin \theta - }$$

$$\overline{x_1 y_2 \sin \theta \cos \theta + y_1 y_2 \sin^2 \theta) + x_1^2 \cos^2 \theta + y_1^2 \sin^2 \theta +}$$

$$\overline{x_2^2 \sin^2 \theta + y_2^2 \cos^2 \theta - 2(x_1 x_2 \sin^2 \theta + x_2 y_1 \sin \theta \cos \theta +}$$

$$\overline{y_2 x_1 \cos \theta \sin \theta + y_1 y_2 \cos^2 \theta) + x_1^2 \sin^2 \theta + y_1^2 \cos^2 \theta} =$$

$$\sqrt{x_2^2 (\cos^2 \theta + \sin^2 \theta) + y_2^2 (\cos^2 \theta + \sin^2 \theta) - }$$

$$\overline{2x_1 x_2 (\cos^2 \theta + \sin^2 \theta) - 2y_1 y_2 (\sin^2 \theta + \cos^2 \theta) +}$$

$$\overline{x_1^2 (\cos^2 \theta + \sin^2 \theta) + y_1^2 (\sin^2 \theta + \cos^2 \theta)} =$$

$$\sqrt{x_2^2 + y_2^2 - 2x_1 x_2 - 2y_1 y_2 + x_1^2 + y_1^2} = \sqrt{(x_2 - x_1)^2 + (y_2 - y_1)^2} = d.$$

20. x' = x cos π - y sin π = -x, y' = x sin π + y cos π = -y, so

$$\begin{bmatrix} x' \\ y' \end{bmatrix} = \begin{bmatrix} -x \\ -y \end{bmatrix} = - \begin{bmatrix} x \\ y \end{bmatrix}.$$

21. A = $\begin{bmatrix} \cos \theta & -\sin \theta \\ \sin \theta & \cos \theta \end{bmatrix}$; we have then $x^2 - (2 \cos \theta)x + \delta(A) = x^2 - (2 \cos \theta)x + 1$

= 0, where the roots are the eigenvalues of A, so that x =

$\dfrac{2 \cos \theta \pm \sqrt{4 \cos^2 \theta - 4}}{2}$ = cos $\theta \pm \sqrt{\cos^2 \theta - 1}$ = cos $\theta \pm \sqrt{-\sin^2 \theta}$, which is

real if and only if $\sin^2 \theta = 0$, sin $\theta = 0$, $\theta = n \cdot 180°$.   If $\theta = n \cdot 180°$,

x = cos $\theta = \pm 1$.

Chapter Test · Page 303

1. A - 3B = $\begin{bmatrix} 2 & 5 \\ 0 & 1 \end{bmatrix} - 3 \begin{bmatrix} 0 & -1 \\ 2 & 3 \end{bmatrix} = \begin{bmatrix} 2 & 5 \\ 0 & 1 \end{bmatrix} + \begin{bmatrix} 0 & 3 \\ -6 & -9 \end{bmatrix} = \begin{bmatrix} 2 & 8 \\ -6 & -8 \end{bmatrix}$

2. AB = $\begin{bmatrix} 2 & 5 \\ 0 & 1 \end{bmatrix} \begin{bmatrix} 0 & -1 \\ 2 & 3 \end{bmatrix} = \begin{bmatrix} 10 & 13 \\ 2 & 3 \end{bmatrix}$

3. $A^2 - 2A = \begin{bmatrix} 2 & 5 \\ 0 & 1 \end{bmatrix} \begin{bmatrix} 2 & 5 \\ 0 & 1 \end{bmatrix} - 2 \begin{bmatrix} 2 & 5 \\ 0 & 1 \end{bmatrix} = \begin{bmatrix} 4 & 15 \\ 0 & 1 \end{bmatrix} + \begin{bmatrix} -4 & -10 \\ 0 & -2 \end{bmatrix} = \begin{bmatrix} 0 & 5 \\ 0 & -1 \end{bmatrix}$

4. $A^{-1} = \dfrac{1}{2} \begin{bmatrix} 1 & -5 \\ 0 & 2 \end{bmatrix}$, $BA^{-1} = \begin{bmatrix} 0 & -1 \\ 2 & 3 \end{bmatrix} \cdot \dfrac{1}{2} \begin{bmatrix} 1 & -5 \\ 0 & 2 \end{bmatrix} = \dfrac{1}{2} \begin{bmatrix} 0 & -2 \\ 2 & -4 \end{bmatrix} = \begin{bmatrix} 0 & -1 \\ 1 & -2 \end{bmatrix}$

5. $\begin{bmatrix} 6 & 1 \\ 12 & 4 \end{bmatrix} \begin{bmatrix} -1 & x_1 \\ 7 & x_2 \end{bmatrix} = \begin{bmatrix} 1 & 7 \\ 16 & -8 \end{bmatrix}$, $\begin{bmatrix} 1 & 6x_1 + x_2 \\ 16 & 12x_1 + 4x_2 \end{bmatrix} = \begin{bmatrix} 1 & 7 \\ 16 & -8 \end{bmatrix}$,

$6x_1 + x_2 = 7$, $12x_1 + 4x_2 = -8$, $x_1 = 3$, $x_2 = -11$

$\underline{6.}$ $(4 + i)(3 - 6i)$ corresponds to $\begin{bmatrix} 4 & 1 \\ -1 & 4 \end{bmatrix}\begin{bmatrix} 3 & -6 \\ 6 & 3 \end{bmatrix} = \begin{bmatrix} 18 & -21 \\ 21 & 18 \end{bmatrix}$, so $18 - 21i$

$\underline{7.}$ $\begin{bmatrix} -2 & 0 \\ 0 & -2 \end{bmatrix}\begin{bmatrix} x \\ y \end{bmatrix} = \begin{bmatrix} -2x \\ -2y \end{bmatrix}$, the result is a magnification and a reflection in the origin.

$\underline{8.}$ $\begin{bmatrix} 0 & 1 \\ 1 & 0 \end{bmatrix}\begin{bmatrix} 0 \\ 0 \end{bmatrix} = \begin{bmatrix} 0 \\ 0 \end{bmatrix}$, $\begin{bmatrix} 0 & 1 \\ 1 & 0 \end{bmatrix}\begin{bmatrix} 1 \\ 4 \end{bmatrix} = \begin{bmatrix} 4 \\ 1 \end{bmatrix}$,

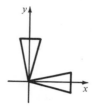

$\begin{bmatrix} 0 & 1 \\ 1 & 0 \end{bmatrix}\begin{bmatrix} -1 \\ 4 \end{bmatrix} = \begin{bmatrix} 4 \\ -1 \end{bmatrix}$

$\underline{9.}$ $\begin{bmatrix} 1 & 0 \\ 0 & -1 \end{bmatrix}\begin{bmatrix} x \\ y \end{bmatrix} = \begin{bmatrix} x \\ -y \end{bmatrix}$, $\begin{bmatrix} -1 & 0 \\ 0 & -1 \end{bmatrix}\begin{bmatrix} x \\ -y \end{bmatrix} = \begin{bmatrix} -x \\ y \end{bmatrix}$; the result is a reflection in the y-axis.

$\underline{10.}$ $\begin{bmatrix} \cos 60° & -\sin 60° \\ \sin 60° & \cos 60° \end{bmatrix} = \begin{bmatrix} \frac{1}{2} & -\frac{\sqrt{3}}{2} \\ \frac{\sqrt{3}}{2} & \frac{1}{2} \end{bmatrix}$, $\begin{bmatrix} \frac{1}{2} & -\frac{\sqrt{3}}{2} \\ \frac{\sqrt{3}}{2} & \frac{1}{2} \end{bmatrix}\begin{bmatrix} 4 \\ 0 \end{bmatrix} = \begin{bmatrix} 2 \\ 2\sqrt{3} \end{bmatrix}$,

$\begin{bmatrix} \frac{1}{2} & -\frac{\sqrt{3}}{2} \\ \frac{\sqrt{3}}{2} & \frac{1}{2} \end{bmatrix}\begin{bmatrix} 1 \\ \sqrt{3} \end{bmatrix} = \begin{bmatrix} -1 \\ \sqrt{3} \end{bmatrix}$

Exercises 9-1 · Page 311

<u>A</u>   **<u>1</u>.** 3, 6, 9, 12

<u>2</u>. 1, -2, -5, -8

<u>3</u>. 3, $\frac{5}{3}$, $\frac{7}{5}$, $\frac{9}{7}$

<u>4</u>. 2, $\frac{5}{2}$, $\frac{10}{3}$, $\frac{17}{4}$

<u>5</u>. $\frac{1}{2}$, $\frac{1}{4}$, $\frac{1}{8}$, $\frac{1}{16}$

<u>6</u>. $\frac{\sqrt{3}}{2}$, $\frac{3}{4}$, $\frac{3\sqrt{3}}{8}$, $\frac{9}{16}$

<u>7</u>. i, -1, -i, 1

<u>8</u>. -1, 0, 1, 0

<u>9</u>. $\lim\limits_{n\to\infty} 1 = 1$

<u>10</u>. $\lim\limits_{n\to\infty} \frac{-6}{n} = (\lim\limits_{n\to\infty} -6)(\lim\limits_{n\to\infty} \frac{1}{n}) = -6(0) = 0$

<u>11</u>. $\lim\limits_{n\to\infty} \frac{3^n}{4^n} = \lim\limits_{n\to\infty} (\frac{3}{4})^n = 0$

<u>12</u>. $\lim\limits_{n\to\infty} \frac{7}{n} = (\lim\limits_{n\to\infty} 7)(\lim\limits_{n\to\infty} \frac{1}{n}) = 7(0) = 0$

<u>13</u>. $\lim\limits_{n\to\infty} (-\frac{3}{5})^n = 0$

<u>14</u>. $\lim\limits_{n\to\infty} \frac{(-1)^n}{3^n} = \lim\limits_{n\to\infty} (-\frac{1}{3})^n = 0$

<u>15</u>. $\lim\limits_{n\to\infty} \frac{2n-1}{n} = \lim\limits_{n\to\infty} \frac{2 - \frac{1}{n}}{1} = \frac{\lim\limits_{n\to\infty} 2 - \lim\limits_{n\to\infty} \frac{1}{n}}{\lim\limits_{n\to\infty} 1} = \frac{2-0}{1} = 2$

<u>16</u>. $\lim\limits_{n\to\infty} (3 - \frac{1}{n}) = \lim\limits_{n\to\infty} 3 - \lim\limits_{n\to\infty} \frac{1}{n} = 3 - 0 = 3$

<u>17</u>. $\lim\limits_{n\to\infty} (1 + (\frac{3}{4})^n) = \lim\limits_{n\to\infty} 1 + \lim\limits_{n\to\infty} (\frac{3}{4})^n = 1 + 0 = 1$

<u>18</u>. $\lim\limits_{n\to\infty} \frac{n}{3n-2} = \lim\limits_{n\to\infty} \frac{1}{3 - \frac{2}{n}} = \frac{\lim\limits_{n\to\infty} 1}{\lim\limits_{n\to\infty} 3 - \lim\limits_{n\to\infty} \frac{2}{n}} = \frac{1}{3-0} = \frac{1}{3}$

<u>19</u>. $\lim\limits_{n\to\infty} \frac{(n-1)^2}{n^2} = \lim\limits_{n\to\infty} \frac{n^2 - 2n + 1}{n^2} = \lim\limits_{n\to\infty} \frac{1 - \frac{2}{n} + \frac{1}{n^2}}{1} = \frac{\lim\limits_{n\to\infty} (1 - \frac{2}{n} + \frac{1}{n^2})}{\lim\limits_{n\to\infty} 1} = $

$\frac{1 - 0 + 0}{1} = 1$

<u>20</u>. limit does not exist

Exercises 9-2 · Pages 314-315

<u>A</u>   **<u>1</u>.** $A_1 = 2$, $A_2 = 2 + \frac{3}{2} = \frac{7}{2}$, $A_3 = 2 + \frac{3}{2} + \frac{4}{3} = \frac{29}{6}$, $A_4 = 2 + \frac{3}{2} + \frac{4}{3} + \frac{5}{4} = \frac{73}{12}$

<u>2</u>. $A_1 = \frac{1}{2}$, $A_2 = \frac{1}{2} + \frac{1}{4} = \frac{3}{4}$, $A_3 = \frac{1}{2} + \frac{1}{4} + \frac{1}{8} = \frac{7}{8}$, $A_4 = \frac{1}{2} + \frac{1}{4} + \frac{1}{8} + \frac{1}{16} = \frac{15}{16}$

<u>3</u>. $A_1 = -1$, $A_2 = -1 + \frac{1}{2} = -\frac{1}{2}$, $A_3 = -1 + \frac{1}{2} - \frac{1}{3} = -\frac{5}{6}$, $A_4 = -1 + \frac{1}{2} - \frac{1}{3} + \frac{1}{4} = -\frac{7}{12}$

<u>4</u>. $A_1 = 1$, $A_2 = 1 + 5 = 6$, $A_3 = 1 + 5 + 19 = 25$, $A_4 = 1 + 5 + 19 + 65 = 90$

<u>5</u>. $A_1 = 2$, $A_2 = 2 + 1 = 3$, $A_3 = 2 + 1 + \frac{2}{3} = 3\frac{2}{3}$, $A_4 = 2 + 1 + \frac{2}{3} + \frac{1}{2} = 4\frac{1}{6}$

<u>6</u>. $A_1 = -1$, $A_2 = -1 + 1 = 0$, $A_3 = -1 + 1 - 1 = -1$, $A_4 = -1 + 1 - 1 + 1 = 0$

<u>7</u>. $A_1 = 1$, $A_2 = 1 + 2 = 3$, $A_3 = 1 + 2 + 4 = 7$, $A_4 = 1 + 2 + 4 + 8 = 15$

<u>8</u>. $A_1 = 1$, $A_2 = 1 - 1 = 0$, $A_3 = 1 - 1 - 3 = -3$, $A_4 = 1 - 1 - 3 - 5 = -8$

<u>9</u>. $A_1 = 1$, $A_2 = 1 + 4 = 5$, $A_3 = 1 + 4 + 27 = 32$, $A_4 = 1 + 4 + 27 + 256 = 288$

<u>10</u>. $A_1 = \frac{1}{5}$, $A_2 = \frac{1}{5} + \frac{1}{3} = \frac{8}{15}$, $A_3 = \frac{1}{5} + \frac{1}{3} + \frac{3}{7} = \frac{101}{105}$, $A_4 = \frac{1}{5} + \frac{1}{3} + \frac{3}{7} + \frac{1}{2} = \frac{307}{210}$

<u>11</u>. $A_1 = 1$, $A_2 = 1 - 3 = -2$, $A_3 = 1 - 3 + 7 = 5$, $A_4 = 1 - 3 + 7 - 15 = -10$

<u>12</u>. $A_1 = -1$, $A_2 = -1 + 1 = 0$, $A_3 = -1 + 1 - 1 = -1$, $A_4 = -1 + 1 - 1 + 1 = 0$

<u>13</u>. $A_5 = 1 + 4 + 9 + 16 + 25 = 55$ <u>14</u>. $A_5 = 0 + \frac{1}{2} + \frac{2}{3} + \frac{3}{4} + \frac{4}{5} = \frac{163}{60}$

<u>15</u>. $A_5 = \frac{2}{3} + \frac{4}{9} + \frac{8}{27} + \frac{16}{81} + \frac{32}{243} = \frac{422}{243}$ <u>16</u>. $A_5 = -\frac{1}{2} + \frac{1}{4} - \frac{1}{8} + \frac{1}{16} - \frac{1}{32} = -\frac{11}{32}$

<u>17</u>. $A_5 = \frac{2}{3} + \frac{3}{8} + \frac{4}{15} + \frac{5}{24} + \frac{6}{35} = \frac{709}{420}$ <u>18</u>. $A_5 = -1 - \frac{2}{3} - \frac{1}{2} - \frac{2}{5} - \frac{1}{3} = -\frac{29}{10}$

<u>B</u> <u>19</u>. $\sum_{i=1}^{n} as_i = as_1 + as_2 + \cdots + as_n = a(s_1 + s_2 + \cdots + s_n) = a \sum_{i=1}^{n} s_i$

<u>20</u>. $\sum_{i=1}^{n} s_i + \sum_{i=1}^{n} r_i = (s_1 + s_2 + \cdots + s_n) + (r_1 + r_2 + \cdots + r_n) =$

$(s_1 + r_1) + (s_2 + r_2) + \cdots + (s_n + r_n) = \sum_{i=1}^{n} (s_i + r_i)$

<u>21</u>. Answers will vary.   Let $s_i = 1$ for all i, and $r_i = 1$ for all i.   Then

$$\left(\sum_{i=1}^{n} s_i \right) \left(\sum_{i=1}^{n} r_i \right) = \left(\sum_{i=1}^{n} 1 \right) \left(\sum_{i=1}^{n} 1 \right) = (n)(n) = n^2 \text{ and } \sum_{i=1}^{n} s_i r_i = \sum_{i=1}^{n} 1 = n.$$

Exercises 9-3 · Pages 318-319

<u>A</u> <u>1</u>. $a_5 = 12\left(\frac{1}{3}\right)^4 = 12\left(\frac{1}{81}\right) = \frac{12}{81} = \frac{4}{27}$ <u>2</u>. $a_7 = 24\left(\frac{1}{2}\right)^6 = 24\left(\frac{1}{64}\right) = \frac{24}{64} = \frac{3}{8}$

<u>3</u>. $S_4 = \dfrac{1[1 - \left(\frac{1}{3}\right)^4]}{1 - \frac{1}{3}} = \dfrac{1 - \frac{1}{81}}{1 - \frac{1}{3}} = \dfrac{81 - 1}{81 - 27} = \dfrac{80}{54} = \dfrac{40}{27}$

<u>4</u>. $S_5 = \dfrac{5[1 - 4^5]}{1 - 4} = \dfrac{5(1 - 1024)}{-3} = \dfrac{-5115}{-3} = +1705$

<u>5</u>. $S_6 = \dfrac{27[1 - \left(\frac{1}{3}\right)^6]}{1 - \frac{1}{3}} = \dfrac{27(1 - \frac{1}{729})}{\frac{2}{3}} = \dfrac{27 - \frac{1}{27}}{\frac{2}{3}} = \dfrac{729 - 1}{18} = \dfrac{728}{18} = \dfrac{364}{9}$

<u>6</u>. $S = \dfrac{64}{1 - \frac{1}{2}} = \dfrac{64}{\frac{1}{2}} = 128$ <u>7</u>. $S = \dfrac{12}{1 - \frac{1}{2}} = \dfrac{12}{\frac{1}{2}} = 24$

8. $S = \dfrac{-4}{1 + \frac{1}{3}} = \dfrac{-4}{\frac{4}{3}} = -3$

9. $S = \dfrac{18}{1 - \frac{2}{3}} = \dfrac{18}{\frac{1}{3}} = 54$

10. $S = \dfrac{\frac{2}{3}}{1 - \frac{3}{4}} = \dfrac{\frac{2}{3}}{\frac{1}{4}} = \dfrac{8}{3}$

11. $S = \dfrac{\frac{1}{3}}{1 + \frac{3}{6}} = \dfrac{\frac{1}{3}}{\frac{9}{6}} = \dfrac{2}{9}$

12. $0.666\ldots = \dfrac{6}{10} + \dfrac{6}{100} + \dfrac{6}{1000} + \cdots, \quad 0.666\ldots = \sum_{i=1}^{\infty} \dfrac{6}{10}\left(\dfrac{1}{10}\right)^{i-1}$

$S = \dfrac{\frac{6}{10}}{1 - \frac{1}{10}} = \dfrac{\frac{6}{10}}{\frac{9}{10}} = \dfrac{2}{3}$

13. $0.273273\ldots = \dfrac{273}{1000} + \dfrac{273}{1000000} + \cdots, \quad 0.273273\ldots = \sum_{i=1}^{\infty} \left(\dfrac{273}{1000}\right)\left(\dfrac{1}{1000}\right)^{i-1},$

$S = \dfrac{\frac{273}{1000}}{1 - \frac{1}{1000}} = \dfrac{\frac{273}{1000}}{\frac{999}{1000}} = \dfrac{273}{999} = \dfrac{91}{333}$

14. $0.363363\ldots = \dfrac{363}{1000} + \dfrac{363}{1000000} + \cdots, \quad 0.363363\ldots = \sum_{i=1}^{\infty} \left(\dfrac{363}{1000}\right)\left(\dfrac{1}{1000}\right)^{i-1},$

$1.363363\ldots = S = 1 + \dfrac{\frac{363}{1000}}{1 - \frac{1}{1000}} = 1 + \dfrac{\frac{363}{1000}}{\frac{999}{1000}} = 1 + \dfrac{363}{999} = 1 + \dfrac{121}{333} = \dfrac{454}{333}$

15. $0.324324\ldots = \dfrac{324}{1000} + \dfrac{324}{100000} + \cdots, \quad 0.324324\ldots = \sum_{i=1}^{\infty} \left(\dfrac{324}{1000}\right)\left(\dfrac{1}{1000}\right)^{i-1},$

$2.324324\ldots = S = 2 + \dfrac{\frac{324}{1000}}{1 - \frac{1}{1000}} = 2 + \dfrac{324}{999} = 2 + \dfrac{12}{37} = \dfrac{86}{37}$

16. $0.00888\ldots = \sum_{i=1}^{\infty} \left(\dfrac{8}{1000}\right)\left(\dfrac{1}{10}\right)^{i-1} = \dfrac{8}{1000} + \dfrac{8}{10000} + \cdots, \quad S = .12 + \dfrac{\frac{8}{1000}}{1 - \frac{8}{10}} =$

$.12 + \dfrac{8}{900} = \dfrac{3}{25} + \dfrac{2}{225} = \dfrac{29}{225}$

17. $0.00333 = \dfrac{3}{1000} + \dfrac{3}{10000} + \cdots, \quad S = .82 + \dfrac{\frac{3}{1000}}{1 - \frac{1}{10}} = .82 + \dfrac{3}{900} = \dfrac{82}{100} + \dfrac{1}{300} = \dfrac{247}{300}$

B **18.** $\sum_{i=1}^{\infty} a_1 r_1^{i-1} + \sum_{i=1}^{\infty} a_2 r_2^{i-1} = \dfrac{a_1}{1 - r_1} + \dfrac{a_2}{1 - r_2} = \dfrac{a_1(1 - r_2) + a_2(1 - r_1)}{(1 - r_1)(1 - r_2)} =$

$\dfrac{a_1 + a_2 - (a_1 r_2 + a_2 r_1)}{(1 - r_1)(1 - r_2)}$

19. $\dfrac{\sum_{j=1}^{\infty} a_1 r_1^{i-1}}{\sum_{j=1}^{\infty} a_2 r_2^{i-1}} = \dfrac{\frac{a_1}{1 - r_1}}{\frac{a_2}{1 - r_2}} = \dfrac{a_1(1 - r_2)}{a_2(1 - r_1)}$

20. $\left(\sum_{i=1}^{\infty} a_1 r_1^{i-1}\right)\left(\sum_{i=1}^{\infty} a_2 r_2^{i-1}\right) = \left(\dfrac{a_1}{1 - r_1}\right)\left(\dfrac{a_2}{1 - r_2}\right) = \dfrac{a_1 a_2}{(1 - r_1)(1 - r_2)}$

<u>C</u>   21. Let a_1a_2 denote $10a_1 + a_2$, and a_2a_1 denote $10a_2 + a_1$.   Then

$$0 \cdot a_1a_2a_1a_2 \cdots = \frac{\frac{a_1a_2}{100}}{1 - \frac{1}{100}} = \frac{a_1a_2}{99}; \text{ also } 0.a_1a_2a_1a_2a_1 \cdots =$$

$$\frac{a_1}{10} + \frac{\frac{a_2a_1}{1000}}{1 - \frac{1}{100}} = \frac{a_1}{10} + \frac{a_2a_1}{990} = \frac{99a_1 + a_2a_1}{990} = \frac{99a_1 + 10a_2 + a_1}{990} =$$

$$\frac{100a_1 + 10a_2}{990} = \frac{10a_1 + a_2}{99} = \frac{a_1a_2}{99}$$

<u>22.</u> Let $a_1a_2a_3$ denote $100a_1 + 10a_2 + a_3$, and similarly for $a_2a_3a_1$ and $a_3a_1a_2$.

$$\text{Then } \frac{\frac{a_1a_2a_3}{1000}}{1 - \frac{1}{1000}} = \frac{a_1a_2a_3}{999}; \text{ also } \frac{a_1}{10} + \frac{\frac{a_2a_3a_1}{10,000}}{1 - \frac{1}{1000}} = \frac{a_1}{10} + \frac{a_2a_3a_1}{9990} =$$

$$\frac{999a_1 + 100a_2 + 10a_3 + a_1}{9990} = \frac{1000a_1 + 100a_2 + 10a_3}{9990} = \frac{100a_1 + 10a_2 + a_3}{999} =$$

$$\frac{a_1a_2a_3}{999}, \text{ and } \frac{a_1a_2}{100} + \frac{\frac{a_3a_1a_2}{100,000}}{1 - \frac{1}{1000}} = \frac{a_1a_2}{100} + \frac{a_3a_1a_2}{99,900} = \frac{999(a_1a_2) + a_3a_1a_2}{99,900} =$$

$$\frac{9990a_1 + 999a_2 + 100a_3 + 10a_1 + a_2}{99,900} = \frac{10,000a_1 + 1000a_2 + 100a_3}{99,900} =$$

$$\frac{100a_1 + 10a_2 + a_3}{999} = \frac{a_1a_2a_3}{999}$$

Exercises 9-4 · Page 323

<u>A</u>   1. $\dfrac{a_{n+1}}{a_n} = \dfrac{x^{n+1}}{x^n} = x, \quad |x| < 1$

<u>2.</u> $\dfrac{a_{n+1}}{a_n} = \dfrac{x^{-n-1}}{x^{-n}} = x^{-1}, \quad \left|\dfrac{1}{x}\right| < 1, \quad |x| > 1$

<u>3.</u> $\lim\limits_{n \to \infty} \dfrac{\frac{x^{n+1}}{(n+1)(n+2)}}{\frac{x^n}{n(n+1)}} = \lim\limits_{n \to \infty} \dfrac{nx}{n+2} = x, \quad |x| < 1$

<u>4.</u> $\lim\limits_{n \to \infty} \dfrac{\frac{(n+1)x^n}{2^n}}{\frac{nx^{n-1}}{2^{n-1}}} = \lim\limits_{n \to \infty} \dfrac{(n+1)x}{2n} = \dfrac{x}{2}, \quad \left|\dfrac{x}{2}\right| < 1, \quad |x| < 2$

<u>5.</u> $\lim\limits_{n \to \infty} \dfrac{(n+1)x^{n+1}}{nx^n} = \lim\limits_{n \to \infty} \dfrac{(n+1)x}{n} = x, \quad |x| < 1$

6. $\lim\limits_{n\to\infty} \dfrac{\dfrac{x^n}{(n+1)4^n}}{\dfrac{x^{n-1}}{n(4^{n-1})}} = \lim\limits_{n\to\infty} \dfrac{nx}{(n+1)4} = \dfrac{x}{4}, \; |\dfrac{x}{4}| < 1, \; |x| < 4$

7. $\lim\limits_{n\to\infty} \dfrac{\dfrac{(2n+1)}{2^n}}{\dfrac{(2n-1)x^{n-1}}{2^{n-1}}} = \lim\limits_{n\to\infty} \dfrac{(2n+1)x}{2(2n-1)} = \dfrac{x}{2}, \; |\dfrac{x}{2}| < 1, \; |x| < 2$

8. $\lim\limits_{n\to\infty} \dfrac{\dfrac{x^{n+1}}{(2[n+1])!}}{\dfrac{x^n}{(2n)!}} = \lim\limits_{n\to\infty} \dfrac{x}{(2n+2)(2n+1)} = 0, \; x \in \mathcal{R}.$

9. $\lim\limits_{n\to\infty} \dfrac{\dfrac{x^{2(n+1)}}{(2[n+1])!}}{\dfrac{x^{2n}}{(2n)!}} = \lim\limits_{n\to\infty} \dfrac{x^2}{(2n+2)(2n+1)} = 0, \; x \in \mathcal{R}.$

B 10. a. $f(x) + g(x) = (\frac{1}{2} + \frac{1}{3}) + (\frac{1}{4} + \frac{1}{6})x + (\frac{1}{6} + \frac{1}{9})x^2 + (\frac{1}{8} + \frac{1}{12})x^3 + \cdots =$

$\frac{5}{6} + \frac{5}{12}x + \frac{5}{18}x^2 + \frac{5}{24}x^3 + \cdots;$

b. $f(x) \cdot g(x) = (\frac{1}{2} + \frac{1}{4}x + \frac{1}{6}x^2 + \frac{1}{8}x^3 + \cdots) \cdot (\frac{1}{3} + \frac{1}{6}x + \frac{1}{9}x^2 + \frac{1}{12}x^3 + \cdots) =$

$\frac{1}{2}(\frac{1}{3}) + [\frac{1}{4}(\frac{1}{3}) + \frac{1}{2}(\frac{1}{6})]x + [\frac{1}{6}(\frac{1}{3}) + \frac{1}{4}(\frac{1}{6}) + \frac{1}{2}(\frac{1}{9})]x^2 +$

$[\frac{1}{8}(\frac{1}{3}) + \frac{1}{6}(\frac{1}{6}) + \frac{1}{4}(\frac{1}{9}) + \frac{1}{2}(\frac{1}{12})]x^3 + \cdots = \frac{1}{6} + [\frac{1}{12} + \frac{1}{12}]x +$

$[\frac{1}{18} + \frac{1}{24} + \frac{1}{18}]x^2 + [\frac{1}{24} + \frac{1}{36} + \frac{1}{36} + \frac{1}{24}]x^3 + \cdots = \frac{1}{6} + \frac{1}{6}x + \frac{11}{72}x^2 +$

$\frac{5}{36}x^3 + \cdots$

11. a. $f(x) + g(x) = (\frac{1}{2} + \frac{1}{1}) + (\frac{1}{6} + \frac{1}{4})x + (\frac{1}{12} + \frac{1}{9})x^2 + (\frac{1}{20} + \frac{1}{16})x^3 + \cdots =$

$\frac{3}{2} + \frac{5}{12}x + \frac{7}{36}x^2 + \frac{9}{80}x^3 + \cdots;$

b. $f(x)g(x) = \frac{1}{2}(\frac{1}{1}) + [\frac{1}{6}(\frac{1}{1}) + \frac{1}{2}(\frac{1}{4})]x + [\frac{1}{12}(\frac{1}{1}) + \frac{1}{6}(\frac{1}{4}) + \frac{1}{2}(\frac{1}{9})]x^2 +$

$[\frac{1}{20}(\frac{1}{1}) + \frac{1}{12}(\frac{1}{4}) + \frac{1}{6}(\frac{1}{9}) + \frac{1}{2}(\frac{1}{16})]x^3 + \cdots = \frac{1}{2} + \frac{7}{24}x + \frac{13}{72}x^2 + \frac{521}{4320}x^3 + \cdots$

12. a. $f(x) + g(x) = 2f(x) = 2[\frac{1}{1} + \frac{-1}{2!}x + \frac{1}{4!}x^2 + \frac{-1}{6!}x^3 + \cdots] =$

$2[1 - \frac{1}{2}x + \frac{1}{24}x^2 - \frac{1}{720}x^3 + \cdots] = 2 - x + \frac{1}{12}x^2 - \frac{1}{360}x^3 + \cdots;$

b. $f(x)g(x) = f^2(x) = [1 - \frac{x}{2} + \frac{x^2}{24} - \frac{x^3}{720} + \cdots]^2 =$

$1 + x[-\frac{1}{2} - \frac{1}{2}] + x^2[2(\frac{1}{24}) + (\frac{1}{2})^2] + x^3[2(-\frac{1}{720}) - 2(\frac{1}{48})] + \cdots =$

$1 - x + \frac{1}{3}x^2 - \frac{2}{45}x^3 + \cdots$

Exercises 9-5 · Pages 326-327

A　　1. $\frac{n!}{(n-2)!} = \frac{n(n-1)(n-2)!}{(n-2)!} = n(n-1)$

　　　2. $\frac{(n+2)!}{n!} = \frac{(n+2)(n+1)n!}{n!} = (n+2)(n+1)$

　　　3. $\frac{(n+2)(n+3)!}{(n+4)!} = \frac{(n+2)(n+3)!}{(n+4)(n+3)!} = \frac{n+2}{n+4}$

　　　4. $\frac{(2n+2)!}{(2n)!} = \frac{(2n+2)(2n+1)(2n)!}{(2n)!} = (2n+2)(2n+1)$

　　　5. $\frac{(2n+2)!}{(2n-1)!\,2n} = \frac{(2n+2)(2n+1)(2n)(2n-1)!}{(2n-1)!\,2n} = (2n+2)(2n+1)$

　　　6. $\frac{(n+1)!}{(n-2)!(n^2-n)} = \frac{(n+1)(n)(n-1)(n-2)!}{(n-2)!(n)(n-1)} = n+1$

　　　7. $(1+x)^8 = 1 + 8(x) + \frac{8(7)x^2}{2!} + \frac{8(7)(6)}{3!}x^3 + \cdots = 1 + 8x + 28x^2 + 56x^3 + \cdots$

　　　8. $(1-x)^{12} = 1 - (12)(x) + \frac{12(11)x^2}{2!} - \frac{(12)(11)(10)}{3!}x^3 + \cdots =$

　　　　　$1 - 12x + 66x^2 - 220x^3 + \cdots$

　　　9. $(1 - \frac{1}{x})^{10} = 1 - (10)(\frac{1}{x}) + \frac{10(9)}{2!}(\frac{1}{x})^2 - \frac{10(9)(8)}{3!}(\frac{1}{x})^3 + \cdots =$

　　　　　$1 - \frac{10}{x} + \frac{45}{x^2} - \frac{120}{x^3} + \cdots$

　　10. $(1 + \frac{1}{2})^{\frac{1}{4}} = 1 + \frac{1}{4}(\frac{1}{2}) + \frac{(\frac{1}{4})(-\frac{3}{4})(\frac{1}{2})^2}{2!} + \frac{(\frac{1}{4})(-\frac{3}{4})(-\frac{7}{4})(\frac{1}{2})^3}{3!} + \cdots =$

　　　　　$1 + \frac{1}{8} - \frac{3}{128} + \frac{21}{3072} + \cdots$

　　11. $\sqrt{1.03} = (1 + 0.03)^{\frac{1}{2}} \doteq 1 + (\frac{1}{2})(0.03) + \frac{(\frac{1}{2})(-\frac{1}{2})}{2!}(0.03)^2 \doteq 1.01$

　　12. $(1.01)^{\frac{1}{3}} = (1 + 0.01)^{\frac{1}{3}} \doteq 1 + \frac{1}{3}(0.01) + \frac{(\frac{1}{3})(-\frac{2}{3})}{2!}(0.01)^2 \doteq 1.00$

　　13. $(0.99)^{20} = (1 - 0.01)^{20} \doteq 1 - (20)(.01) + \frac{(20)(19)}{2!}(.01)^2 \doteq 0.82$

　　14. $(1.02)^{20} = (1 + 0.02)^{20} \doteq 1 + 20(0.02) + \frac{(20)(19)}{2!}(0.02)^2 + \frac{(20)(19)(18)}{3!}(0.02)^3 \doteq 1.49$

<u>B</u> <u>15.</u> $\lim\limits_{n\to\infty} \dfrac{\dfrac{(x+2)^n}{(n+1)(n+2)}}{\dfrac{(x+2)^{n-1}}{(n)(n+1)}} = \lim\limits_{n\to\infty} \dfrac{n(x+2)}{n+2} = x + 2, \ |x+2| < 1, \ -3 < x < -1$

<u>16.</u> $\lim\limits_{n\to\infty} \dfrac{\sqrt{n+1}\,(x-3)^{n+1}}{\sqrt{n}\,(x-3)^n} = \lim\limits_{n\to\infty} \sqrt{\dfrac{n+1}{n}}\,(x-3) = x - 3, \ |x-3| < 1, \ 2 < x < 4$

<u>17.</u> $\lim\limits_{n\to\infty} \dfrac{\dfrac{(x-2)^{n+1}}{(3n+4)}}{\dfrac{(x-2)^n}{(3n+1)}} = \lim\limits_{n\to\infty} \dfrac{(3n+1)(x-2)}{3n+4} = x - 2, \ |x-2| < 1, \ 1 < x < 3$

<u>18.</u> $\lim\limits_{n\to\infty} \dfrac{\dfrac{(x+1)^{n+1}}{(n+1)!}}{\dfrac{(x+1)^n}{n!}} = \lim\limits_{n\to\infty} \dfrac{(x+1)}{n+1} = 0, \ x \in \mathcal{R}$

<u>19.</u> $e = \lim\limits_{n\to\infty} (1 + \frac{1}{n})^n \doteq 1 + 1 + \frac{1}{2!} + \frac{1}{3!} + \frac{1}{4!} + \frac{1}{5!} + \frac{1}{6!} + \frac{1}{7!} \doteq$

$1 + 1 + 0.5 + 0.16667 + 0.04167 + 0.00833 + 0.00139 + 0.00020 \doteq 2.7183$

<u>20.</u> Let $x \in \mathcal{R}$. $e^x = \sum\limits_{i=1}^{\infty} \dfrac{x^{i-1}}{(i-1)!}$, $\lim\limits_{n\to\infty} \dfrac{\dfrac{x^n}{n!}}{\dfrac{x^{n-1}}{(n-1)!}} = \lim\limits_{n\to\infty} \dfrac{x}{n} = 0, \ 0 < 1$

<u>21.</u> $e^{-x} = \lim\limits_{n\to\infty} (1 + \frac{1}{n})^{-nx} = 1 + (-x) + \dfrac{(-x)^2}{2!} + \dfrac{(-x)^3}{3!} + \cdots + \dfrac{(-x)^r}{r!} + \cdots =$

$1 - x + \dfrac{x^2}{2!} - \dfrac{x^3}{3!} + \cdots + \dfrac{(-1)^r x^r}{r!} + \cdots$

<u>22.</u> $e^{-x} = \sum\limits_{i=1}^{\infty} \dfrac{(-1)^{i-1} x^{i-1}}{(i-1)!}$, $\lim\limits_{n\to\infty} \dfrac{\dfrac{(-1)^n x^n}{n!}}{\dfrac{(-1)^{n-1} x^{n-1}}{(n-1)!}} = \lim\limits_{n\to\infty} \dfrac{-x}{n} = 0, \ 0 < 1.$ Series

converges for $x \in \mathcal{R}$.

<u>23.</u> The general form for the rth term is $\dfrac{n(n-1)\cdots[n-(r-1)]}{r!} x^r$. If $r > n + 1$,

then $n - (r-1) < 0$ and the rth term is

$\dfrac{n(n-1)\cdots 2 \cdot 1 \cdot 0 \cdot (-1) \cdots [n-(r-1)]}{r!} x^r = \dfrac{0}{r!} x^r = 0$

<u>24.</u> $e^{i\pi} \approx \sum\limits_{j=1}^{4} \dfrac{(i\pi)^{j-1}}{(j-1)!} = \dfrac{(i\pi)^0}{0!} + \dfrac{(i\pi)^1}{1!} + \dfrac{(i\pi)^2}{2!} + \dfrac{(i\pi)^3}{3!} = 1 + i\pi - \dfrac{\pi^2}{2} - \dfrac{i\pi^3}{6};$

using $\pi \doteq 3.14159$, $e^{i\pi} \doteq 1 + i(3.14159) - \dfrac{9.86959}{2} - i\dfrac{30.99620}{6} =$

$1 + i(3.14159) - 4.93479 - i(5.16603) \doteq -3.9348 - i(2.0244)$

Exercises 9-6 · Pages 331-332

<u>A</u> <u>1.</u> $\sin 0.1 \doteq 0.1 - \dfrac{(0.1)^3}{3!} \doteq 0.10$ <u>2.</u> $\cos 0.1 \doteq 1 - \dfrac{(0.1)^2}{2!} = 1.00$

<u>3.</u> $\cos 1 \doteq 1 - \dfrac{(1)^2}{2!} + \dfrac{(1)^4}{4!} \doteq 1 - \dfrac{1}{2} + \dfrac{1}{24} \doteq 0.54$

$\underline{4.}$ $\sin 1 \doteq 1 - \dfrac{(1)^3}{3!} + \dfrac{(1)}{5!} \doteq 1 - \dfrac{1}{6} + \dfrac{1}{120} \doteq 0.84$

$\underline{5.}$ $\sin \dfrac{1}{2} \doteq \dfrac{1}{2} - \dfrac{\left(\frac{1}{2}\right)^3}{3!} \doteq 0.48$ $\qquad$ $\underline{6.}$ $\cos \dfrac{1}{2} \doteq 1 - \dfrac{\left(\frac{1}{2}\right)^2}{2!} \doteq 0.88$

$\underline{7.}$ $\cos \dfrac{\pi}{6} = 1 - \dfrac{\left(\frac{\pi}{6}\right)^2}{2!} + \dfrac{\left(\frac{\pi}{6}\right)^4}{4!} \doteq 0.87$ $\qquad$ $\underline{8.}$ $\sin \dfrac{\pi}{6} = \dfrac{\pi}{6} - \dfrac{\left(\frac{\pi}{6}\right)^3}{3!} \doteq 0.50$

$\underline{B}$ $\underline{9.}$ $\sin \dfrac{x}{2} = \dfrac{x}{2} - \dfrac{x^3}{2^3 \cdot 3!} + \cdots + (-1)^{n+1} \dfrac{x^{2n-1}}{2^{2n-1}(2n-1)!}$

$\underline{10.}$ $\cos \dfrac{x}{2} = 1 - \dfrac{x^2}{2^2 2!} + \cdots + (-1)^{n+1} \dfrac{x^{2n-2}}{2^{2n-2}(2n-2)!}$

$\underline{11.}$ $\sin 2x = 2x - \dfrac{2^3 x^3}{3!} + \cdots + (-1)^{n+1} \dfrac{2^{2n-1} x^{2n-1}}{(2n-1)!}$

$\underline{12.}$ $\cos 2x = 1 - \dfrac{2^2 x^2}{2!} + \cdots + (-1)^{n+1} \dfrac{2^{2n-2} x^{2n-2}}{(2n-2)!}$

$\underline{13.}$ $\cos x = \displaystyle\sum_{i=1}^{\infty} (-1)^{i+1} \dfrac{x^{2i-2}}{(2i-2)!}$, $\lim \dfrac{\dfrac{(-1)^{n+2} x^{2n}}{(2n)!}}{\dfrac{(-1)^{n+1} x^{2n-2}}{(2n-2)!}} = \displaystyle\lim_{n\to\infty} \dfrac{-x^2}{2n(2n-1)} = 0 < 1$

$\underline{14.}$ $\cos x = 1 + 0x - \dfrac{x^2}{2!} + 0x + \dfrac{x^4}{4!} + 0x - \dfrac{x^6}{6!} + 0x + \dfrac{x^8}{8!} + \cdots$, by II,

$\cos^2 x = 1(1) + [0(1) + 1(0)]x + [-\dfrac{1}{2!}(1) + 0(0) + 1(-\dfrac{1}{2!})]x^2 +$

$[0(1) - \dfrac{1}{2!}(0) + 0(-\dfrac{1}{2!}) + 1(0)]x^3 + [\dfrac{1}{4!}(1) + 0(0) - \dfrac{1}{2!}(-\dfrac{1}{2!}) + 0(0) + 1(\dfrac{1}{4!})]x^4 +$

$[0(1) + \dfrac{1}{4!}(0) + 0(-\dfrac{1}{2!}) + 0(-\dfrac{1}{2!}) + 0(\dfrac{1}{4!}) + 1(0)]x^5 +$

$[-\dfrac{1}{6!}(1) + 0(0) + \dfrac{1}{4!}(-\dfrac{1}{2!}) + 0(0) - \dfrac{1}{2!}(\dfrac{1}{4!}) + 0(0) + 1(-\dfrac{1}{6!})]x^6 + \cdots =$

$1 - [\dfrac{1}{2!} + \dfrac{1}{2!}]x^2 + [\dfrac{1}{4!} + \dfrac{1}{2!2!} + \dfrac{1}{4!}]x^4 - [\dfrac{1}{6!} + \dfrac{1}{4!2!} + \dfrac{1}{2!4!} + \dfrac{1}{6!}]x^6 + \cdots;$

hence counting terms with nonzero coefficients only, $\cos^2 x =$

$\dfrac{1}{0!0!} - (\dfrac{1}{2!0!} + \dfrac{1}{0!2!})x^2 + (\dfrac{1}{4!0!} + \dfrac{1}{2!2!} + \dfrac{1}{0!4!})x^4 + \cdots +$

$(-1)^{n+1}(\dfrac{1}{(2n-2)!0!} + \dfrac{1}{(2n-4)!2!} + \cdots + \dfrac{1}{2!(2n-4)!} + \dfrac{1}{0!(2n-2)!})x^{2n-2} + \cdots$

Exercises 9-7 · Page 335

$\underline{A}$ $\underline{1.}$ $\cosh(-x) = \dfrac{e^{-x} + e^{-(-x)}}{2} = \dfrac{e^{-x} + e^x}{2} = \dfrac{e^x + e^{-x}}{2} = \cosh x$

$\underline{2.}$ $\sinh(-x) = \dfrac{e^{-x} - e^{-(-x)}}{2} = \dfrac{e^{-x} - e^x}{2} = -(\dfrac{e^x - e^{-x}}{2}) = -\sinh x$

<u>3.</u> $\tanh^2 x + \text{sech}^2 x = \dfrac{\sinh^2 x}{\cosh^2 x} + \dfrac{1}{\cosh^2 x} = \dfrac{\sinh^2 x + 1}{\cosh^2 x} = \dfrac{\cosh^2 x}{\cosh^2 x} = 1$

<u>4.</u> $\coth^2 x - \text{csch}^2 x = \dfrac{\cosh^2 x}{\sinh^2 x} - \dfrac{1}{\sinh^2 x} = \dfrac{\cosh^2 x - 1}{\sinh^2 x} = \dfrac{\sinh^2 x}{\sinh^2 x} = 1$

<u>5.</u> $\cosh x + \sinh x = \dfrac{e^x + e^{-x}}{2} + \dfrac{e^x - e^{-x}}{2} = \dfrac{2e^x}{2} = e^x$

<u>6.</u> $\cosh x - \sinh x = \dfrac{e^x + e^{-x}}{2} - \dfrac{e^x - e^{-x}}{2} = \dfrac{2e^{-x}}{2} = e^{-x}$

<u>7.</u> $\cosh x = \dfrac{e^x + e^{-x}}{2} =$

$$\dfrac{\left(1 + x + \dfrac{x^2}{2!} + \dfrac{x^3}{3!} + \cdots + \dfrac{x^r}{r!} + \cdots\right) + \left(1 - x + \dfrac{x^2}{2!} - \dfrac{x^3}{3!} + \cdots + \dfrac{(-1)^r x^r}{r!} + \cdots\right)}{2} =$$

$$\dfrac{2 + \dfrac{2x^2}{2!} + \dfrac{2x^4}{4!} + \cdots + \dfrac{2x^{2n-2}}{(2n-2)!} + \cdots}{2} = 1 + \dfrac{x^2}{2!} + \dfrac{x^4}{4!} + \cdots + \dfrac{x^{2n-2}}{(2n-2)!} + \cdots$$

<u>8.</u> $\sinh x = \dfrac{e^x - e^{-x}}{2} =$

$$\dfrac{\left(1 + x + \dfrac{x^2}{2!} + \dfrac{x^3}{3!} + \cdots + \dfrac{x^r}{r!} + \cdots\right) - \left(1 - x + \dfrac{x^2}{2!} - \dfrac{x^3}{3!} + \cdots + \dfrac{(-1)^r x^r}{r!} + \cdots\right)}{2} =$$

$$\dfrac{2x + \dfrac{2x^3}{3!} + \dfrac{2x^5}{5!} + \cdots + \dfrac{2x^{2n-1}}{(2n-1)!} + \cdots}{2} = x + \dfrac{x^3}{3!} + \dfrac{x^5}{5!} + \cdots + \dfrac{x^{2n-1}}{(2n-1)!} + \cdots$$

<u>9.</u> $\cosh 1 \doteq 1 + \dfrac{1^2}{2!} + \dfrac{1^4}{4!} + \dfrac{1^6}{6!} \doteq 1 + 0.5 + 0.042 + 0.001 = 1.543$

<u>10.</u> $\sinh 1 \doteq 1 + \dfrac{1^3}{3!} + \dfrac{1^5}{5!} \doteq 1 + 0.167 + 0.008 = 1.175$

<u>11.</u> $\sinh\left(-\dfrac{1}{2}\right) \doteq -\dfrac{1}{2} + \dfrac{\left(-\dfrac{1}{2}\right)^3}{3!} \doteq -0.5 - 0.021 = -0.521$

<u>12.</u> $\cosh\left(-\dfrac{1}{2}\right) \doteq 1 + \dfrac{\left(-\dfrac{1}{2}\right)^2}{2!} + \dfrac{\left(-\dfrac{1}{2}\right)^4}{4!} \doteq 1 + 0.125 + 0.003 = 1.128$

Chapter Test · Page 337

<u>1.</u> $\lim\limits_{n\to\infty} \dfrac{n}{3n-2} = \lim\limits_{n\to\infty} \dfrac{1}{3 - \dfrac{2}{n}} = \dfrac{1}{3}$

<u>2.</u> $\lim\limits_{n\to\infty} (-1)^n \dfrac{1}{n} = 0$ since given $d > 0$, let $m > \dfrac{1}{d}$ then for all $n > m$

$\left| 0 - (-1)^n \dfrac{1}{n} \right| = \dfrac{1}{n} < \dfrac{1}{m} < d.$

$\underline{3.}$ $\underline{a.}$ $A_1 = \frac{1}{3}$, $A_2 = \frac{1}{3} + \frac{1}{2}$. $\quad A_3 = \frac{1}{3} + \frac{1}{2} + \frac{3}{5} = \frac{43}{30}$

$\underline{b.}$ $\displaystyle\sum_{i=1}^{6} \frac{i}{i+2} - \sum_{i=1}^{5} \frac{i}{i+2} = \frac{6}{6+2} = \frac{3}{4}$

$\underline{4.}$ $L_1 = 9\left(\frac{2}{3}\right)^5 = 9 \cdot \frac{32}{243} = \frac{32}{27}$

$\underline{5.}$ $\underline{a.}$ $A_1 = 6$, $A_2 = 6 + \frac{3}{2} = \frac{15}{2}$, $A_3 = 6 + \frac{3}{2} + \frac{3}{8} = \frac{63}{8}$;

$\underline{b.}$ $S = \dfrac{6}{1 - \frac{1}{4}} = \dfrac{6}{\frac{3}{4}} = \frac{24}{3} = 8$

$\underline{6.}$ $\displaystyle\lim_{n\to\infty} \frac{\dfrac{(n+1)(x^{n+1})}{3^{n+1}}}{\dfrac{nx^n}{3^n}} = \lim_{n\to\infty} \frac{(n+1)x}{3n} = \frac{1}{3}|x|$, $\frac{1}{3}|x| < 1$, $|x| < 3$

$\underline{7.}$ $(1 + .06)^{\frac{1}{3}} = 1 + \frac{1}{3}(.06) + \dfrac{\left(\frac{1}{3}\right)\left(-\frac{2}{3}\right)}{2!}(.06)^2 \doteq 1 + .02 - .0004 \doteq 1.02$

$\underline{8.}$ $\dfrac{n!\,(n+3)!}{(n+2)!\,(n-1)!} = \dfrac{n(n-1)!\,(n+3)(n+2)!}{(n+2)!\,(n-1)!} = n(n+3)$

$\underline{9.}$ $e^{\frac{\pi}{2}i} = \cos\frac{\pi}{2} + i\sin\frac{\pi}{2} = 0 + i$ $\underline{10.}$

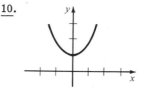

Cumulative Review: Chapters 7-9 · Pages 338-339

Chapter 7

$\underline{1.}$ $z_1 + z_2 = (-4,\ 2) + (6,\ -5) = (2,\ -3)$

$\underline{2.}$ $z_1 - z_2 = (-4,\ 2) - (6,\ -5) = (-4,\ 2) + (-6,\ 5) = (-10,\ 7)$

$\underline{3.}$ $z_1 z_2 = (-4,\ 2)(6,\ -5) = (-24 + 10,\ 20 + 12) = (-14,\ 32)$

$\underline{4.}$ $\dfrac{z_1}{z_2} = \dfrac{(-4,\ 2)}{(6,\ -5)} = \dfrac{(-4,\ 2)(6,\ 5)}{(6,\ -5)(6,\ 5)} = \dfrac{(-34,\ -8)}{(61,\ 0)} = \left(-\frac{34}{61},\ -\frac{8}{61}\right)$

$\underline{5.}$ $\dfrac{z_2}{z_1} = \dfrac{(6,\ -5)}{(-4,\ 2)} = \dfrac{(6,\ -5)(-4,\ -2)}{(-4,\ 2)(-4,\ -2)} = \dfrac{(-34,\ 8)}{(20,\ 0)} = \left(-\frac{34}{20},\ \frac{8}{20}\right) = \left(-\frac{17}{10},\ \frac{2}{5}\right)$

$\underline{6.}$ $(3 - i)(4 + i) = 12 - 4i + 3i - i^2 = 13 - i$

$\underline{7.}$ $\dfrac{(2 - i)}{(1 + 3i)} = \dfrac{(2 - i)(1 - 3i)}{(1 + 3i)(1 - 3i)} = \dfrac{-1 - 7i}{10} = -\frac{1}{10} - \frac{7}{10}i$

$\underline{8}$. $2(\cos 5° + i \sin 5°)^6 = (2)^6(\cos 30° + i \sin 30°) = (2)^6(\frac{\sqrt{3}}{2}) + (2)^6(\frac{1}{2})i =$

$32\sqrt{3} + 32i$

$\underline{9}$. $-\frac{1}{2} - \frac{\sqrt{3}}{2} = \cos 240° + i \sin 240°$

$\underline{10}$. $z_1 z_2 = 5(\cos 30° + i \sin 30°) \cdot 2[\cos(-30°) + i \sin (-30°)] = 10(\cos 0° + i \sin 0°)$

$\underline{11}$. $\frac{z_1}{z_2} = \frac{5(\cos 30° + i \sin 30°)}{2[\cos (-30°) + i \sin (-30°)]} = \frac{5}{2}(\cos 60° + i \sin 60°)$

$\underline{12}$. $z^3 = -8 = 8(\cos 180° + i \sin 180°)$, $z = [8(\cos 180° + i \sin 180°)]^{\frac{1}{3}} =$

$2(\cos \frac{180° + k\,360°}{3} + i \sin \frac{180° + k\,360°}{3}) = 2[\cos (60° + 120°k) +$

$i \sin (60° + 120°k)$; $2(\cos 60° + i \sin 60°)$, $2(\cos 180° + i \sin 180°)$, and

$2(\cos 300° + i \sin 300°)$

Chapter 8

$\underline{13}$. $A + B = \begin{bmatrix} -3 & 7 \\ 1 & 2 \end{bmatrix} + \begin{bmatrix} 4 & 7 \\ 1 & 2 \end{bmatrix} = \begin{bmatrix} 1 & 14 \\ 2 & 4 \end{bmatrix}$

$\underline{14}$. $A - 2B = \begin{bmatrix} -3 & 7 \\ 1 & 2 \end{bmatrix} - 2\begin{bmatrix} 4 & 7 \\ 1 & 2 \end{bmatrix} = \begin{bmatrix} -3 & 7 \\ 1 & 2 \end{bmatrix} + \begin{bmatrix} -8 & -14 \\ -2 & -4 \end{bmatrix} = \begin{bmatrix} -11 & -7 \\ -1 & -2 \end{bmatrix}$

$\underline{15}$. $AB = \begin{bmatrix} -3 & 7 \\ 1 & 2 \end{bmatrix}\begin{bmatrix} 4 & 7 \\ 1 & 2 \end{bmatrix} = \begin{bmatrix} -12 + 7 & -12 + 14 \\ 4 + 2 & 7 + 4 \end{bmatrix} = \begin{bmatrix} -5 & -7 \\ 6 & 11 \end{bmatrix}$

$\underline{16}$. $B(A + I) = \begin{bmatrix} 4 & 7 \\ 1 & 2 \end{bmatrix}\left(\begin{bmatrix} -3 & 7 \\ 1 & 2 \end{bmatrix} + \begin{bmatrix} 1 & 0 \\ 0 & 1 \end{bmatrix}\right) = \begin{bmatrix} 4 & 7 \\ 1 & 2 \end{bmatrix}\begin{bmatrix} -2 & 7 \\ 1 & 3 \end{bmatrix} =$

$\begin{bmatrix} -8 + 7 & +28 + 21 \\ -2 + 2 & 7 + 6 \end{bmatrix} = \begin{bmatrix} -1 & 49 \\ 0 & 13 \end{bmatrix}$

$\underline{17}$. $\begin{bmatrix} -4 & -2 \\ 1 & 1 \end{bmatrix}^{-1} = -\frac{1}{2}\begin{bmatrix} 1 & 2 \\ -1 & -4 \end{bmatrix} = \begin{bmatrix} -\frac{1}{2} & -1 \\ \frac{1}{2} & 2 \end{bmatrix}$

$\underline{18}$. $\begin{bmatrix} 3 & -1 \\ 1 & 2 \end{bmatrix}\begin{bmatrix} x \\ y \end{bmatrix} = \begin{bmatrix} -5 \\ 3 \end{bmatrix}$, $\begin{bmatrix} x \\ y \end{bmatrix} = \frac{1}{7}\begin{bmatrix} 2 & 1 \\ -1 & 3 \end{bmatrix}\begin{bmatrix} -5 \\ 3 \end{bmatrix} = \frac{1}{7}\begin{bmatrix} -7 \\ 14 \end{bmatrix} = \begin{bmatrix} -1 \\ 2 \end{bmatrix}$, $x = -1$ and $y = 2$

$\underline{19}$. $\begin{bmatrix} 4 & -1 \\ 1 & 4 \end{bmatrix}\begin{bmatrix} 1 & 3 \\ -3 & 1 \end{bmatrix} = \begin{bmatrix} 7 & 11 \\ -11 & 7 \end{bmatrix} = 7 + 11i$

$\underline{20.}$ $2\begin{bmatrix} \cos\frac{\pi}{3} & \sin\frac{\pi}{3} \\ -\sin\frac{\pi}{3} & \cos\frac{\pi}{3} \end{bmatrix} \cdot (-3)\begin{bmatrix} \cos\frac{2\pi}{3} & \sin\frac{2\pi}{3} \\ -\sin\frac{2\pi}{3} & \cos\frac{2\pi}{3} \end{bmatrix} = -6\begin{bmatrix} \cos\pi & \sin\pi \\ -\sin\pi & \cos\pi \end{bmatrix} =$

$-6(\cos\pi + i\sin\pi)$

$\underline{21.}$ $\begin{bmatrix} -2 & 0 \\ 0 & 2 \end{bmatrix}\begin{bmatrix} x \\ y \end{bmatrix} = \begin{bmatrix} -2x \\ 2y \end{bmatrix}$, the result is a magnification and a reflection in the y-axis.

$\begin{bmatrix} -2 & 0 \\ 0 & 2 \end{bmatrix}\begin{bmatrix} 0 \\ 0 \end{bmatrix} = \begin{bmatrix} 0 \\ 0 \end{bmatrix}$, $\begin{bmatrix} -2 & 0 \\ 0 & 2 \end{bmatrix}\begin{bmatrix} 1 \\ 0 \end{bmatrix} = \begin{bmatrix} -2 \\ 0 \end{bmatrix}$,

$\begin{bmatrix} -2 & 0 \\ 0 & 2 \end{bmatrix}\begin{bmatrix} 1 \\ 1 \end{bmatrix} = \begin{bmatrix} -2 \\ 2 \end{bmatrix}$, $\begin{bmatrix} -2 & 0 \\ 0 & 2 \end{bmatrix}\begin{bmatrix} 0 \\ 1 \end{bmatrix} = \begin{bmatrix} 0 \\ 2 \end{bmatrix}$

$\underline{22.}$ $\begin{bmatrix} 0 & -3 \\ -3 & 0 \end{bmatrix}\begin{bmatrix} x \\ y \end{bmatrix} = \begin{bmatrix} -3y \\ -3x \end{bmatrix}$, the result is a magnification and a reflection in the line y = -x,

$\begin{bmatrix} 0 & -3 \\ -3 & 0 \end{bmatrix}\begin{bmatrix} 0 \\ 0 \end{bmatrix} = \begin{bmatrix} 0 \\ 0 \end{bmatrix}$, $\begin{bmatrix} 0 & -3 \\ -3 & 0 \end{bmatrix}\begin{bmatrix} 6 \\ 0 \end{bmatrix} = \begin{bmatrix} 0 \\ -18 \end{bmatrix}$,

$\begin{bmatrix} 0 & -3 \\ -3 & 0 \end{bmatrix}\begin{bmatrix} 3 \\ 4 \end{bmatrix} = \begin{bmatrix} -12 \\ -9 \end{bmatrix}$

$\underline{23.}$ $A = \begin{bmatrix} \cos 150° & -\sin 150° \\ \sin 150° & \cos 150° \end{bmatrix} = \begin{bmatrix} -\frac{\sqrt{3}}{2} & -\frac{1}{2} \\ \frac{1}{2} & -\frac{\sqrt{3}}{2} \end{bmatrix}$, $\begin{bmatrix} -\frac{\sqrt{3}}{2} & -\frac{1}{2} \\ \frac{1}{2} & -\frac{\sqrt{3}}{2} \end{bmatrix}\begin{bmatrix} 0 \\ 0 \end{bmatrix} = \begin{bmatrix} 0 \\ 0 \end{bmatrix}$,

$\begin{bmatrix} -\frac{\sqrt{3}}{2} & -\frac{1}{2} \\ \frac{1}{2} & -\frac{\sqrt{3}}{2} \end{bmatrix}\begin{bmatrix} 2 \\ 0 \end{bmatrix} = \begin{bmatrix} -\sqrt{3} \\ 1 \end{bmatrix}$, $\begin{bmatrix} -\frac{\sqrt{3}}{2} & -\frac{1}{2} \\ \frac{1}{2} & -\frac{\sqrt{3}}{2} \end{bmatrix}\begin{bmatrix} 0 \\ 2 \end{bmatrix} = \begin{bmatrix} -1 \\ -\sqrt{3} \end{bmatrix}$

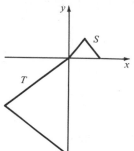

$\underline{24.}$ $A = \begin{bmatrix} \cos(-60°) & -\sin(-60°) \\ \sin(-60°) & \cos(-60°) \end{bmatrix} = \begin{bmatrix} \frac{1}{2} & \frac{\sqrt{3}}{2} \\ -\frac{\sqrt{3}}{2} & \frac{1}{2} \end{bmatrix}$,

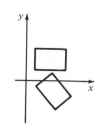

$$\begin{bmatrix} \frac{1}{2} & \frac{\sqrt{3}}{2} \\ -\frac{\sqrt{3}}{2} & \frac{1}{2} \end{bmatrix} \begin{bmatrix} 1 \\ 1 \end{bmatrix} = \begin{bmatrix} \frac{1 + \sqrt{3}}{2} \\ -\frac{\sqrt{3} + 1}{2} \end{bmatrix}, \quad \begin{bmatrix} \frac{1}{2} & \frac{\sqrt{3}}{2} \\ -\frac{\sqrt{3}}{2} & \frac{1}{2} \end{bmatrix} \begin{bmatrix} 1 \\ 3 \end{bmatrix} = \begin{bmatrix} \frac{1 + 3\sqrt{3}}{2} \\ -\frac{\sqrt{3} + 3}{2} \end{bmatrix}$$

$$\begin{bmatrix} \frac{1}{2} & \frac{\sqrt{3}}{2} \\ -\frac{\sqrt{3}}{2} & \frac{1}{2} \end{bmatrix} \begin{bmatrix} 4 \\ 1 \end{bmatrix} = \begin{bmatrix} 2 + \frac{\sqrt{3}}{2} \\ -2\sqrt{3} + \frac{1}{2} \end{bmatrix}, \quad \begin{bmatrix} \frac{1}{2} & \frac{\sqrt{3}}{2} \\ -\frac{\sqrt{3}}{2} & \frac{1}{2} \end{bmatrix} \begin{bmatrix} 4 \\ 3 \end{bmatrix} = \begin{bmatrix} 2 + \frac{3\sqrt{3}}{2} \\ -2\sqrt{3} + \frac{3}{2} \end{bmatrix}$$

Chapter 9

25. $\frac{5}{6}$, $\frac{8}{7}$, $\frac{11}{8}$, $\frac{14}{9}$

26. $\lim\limits_{n \to \infty} \dfrac{2n}{3n^2 + 1} = \lim\limits_{n \to \infty} \dfrac{\frac{2}{n}}{3 + \frac{1}{n^2}} = \dfrac{0}{3 + 0} = 0$

27. $A_1 = 1$, $A_2 = 1 + \frac{3}{2} = \frac{5}{2}$, $A_3 = \frac{5}{2} + \frac{9}{5} = \frac{43}{10}$, $A_4 = \frac{43}{10} + 2 = \frac{63}{10}$

28. $S = \dfrac{2}{1 - \frac{2}{3}} = \dfrac{2}{\frac{1}{3}} = 6.$ Also note: $\sum\limits_{i=1}^{\infty} 3\left(\frac{2}{3}\right)^i = \sum\limits_{i=1}^{\infty} 3\left(\frac{2}{3}\right)^{i-1} - 3 = 9 - 3 = 6$

29. $\dfrac{(n + 1)!}{(n - 2)!} = \dfrac{(n + 1)(n)(n - 1)(n - 2)!}{(n - 2)!} = (n + 1)(n)(n - 1) = n^3 - n.$

30. $(1 + x)^{\frac{1}{3}} = 1 + \frac{x}{3} + \dfrac{(\frac{1}{3})(-\frac{2}{3})x^2}{2!} + \dfrac{(\frac{1}{3})(-\frac{2}{3})(-\frac{5}{3})x^3}{3!} - \cdots = 1 + \frac{1}{3}x - \frac{1}{9}x^2 + \frac{5}{81}x^3 - \cdots$

31. $\sin 0.4 \doteq 0.4 - \dfrac{(0.4)^3}{3!} + \dfrac{(0.4)^5}{5!} \doteq 0.4 - .01066 + .000085 \doteq 0.389$

32. $[\cosh(-x) + \sinh(-x)][\cosh x + \sinh x] =$

$[\dfrac{e^x + e^{-x}}{2} + \dfrac{e^{-x} - e^x}{2}][\dfrac{e^x + e^{-x}}{2} + \dfrac{e^x - e^{-x}}{2}] = (e^{-x})(e^x) = e^0 = 1$

Comprehensive Test · Pages 345-355

Chapter 1

<u>1.</u> b   <u>2.</u> d   <u>3.</u> a   <u>4.</u> a   <u>5.</u> c   <u>6.</u> b   <u>7.</u> b   <u>8.</u> a

<u>9.</u> d   <u>10.</u> c

Chapter 2

<u>1.</u> a   <u>2.</u> c   <u>3.</u> b   <u>4.</u> c   <u>5.</u> b   <u>6.</u> b   <u>7.</u> c   <u>8.</u> c

<u>9.</u> c   <u>10.</u> d

Chapter 3

<u>1.</u> c   <u>2.</u> d   <u>3.</u> b   <u>4.</u> c   <u>5.</u> a   <u>6.</u> c   <u>7.</u> a   <u>8.</u> a

Chapter 4

<u>1.</u> b   <u>2.</u> b   <u>3.</u> d   <u>4.</u> a   <u>5.</u> b   <u>6.</u> c   <u>7.</u> b   <u>8.</u> b

Chapter 5

<u>1.</u> b   <u>2.</u> c   <u>3.</u> d   <u>4.</u> c   <u>5.</u> a   <u>6.</u> a   <u>7.</u> c   <u>8.</u> b

<u>9.</u> b   <u>10.</u> d

Chapter 6

<u>1.</u> a   <u>2.</u> b   <u>3.</u> c   <u>4.</u> a   <u>5.</u> c   <u>6.</u> b   <u>7.</u> c   <u>8.</u> a

Chapter 7

<u>1.</u> b   <u>2.</u> c   <u>3.</u> d   <u>4.</u> c   <u>5.</u> b   <u>6.</u> d   <u>7.</u> b   <u>8.</u> a

Chapter 8

<u>1.</u> c   <u>2.</u> a   <u>3.</u> b   <u>4.</u> b   <u>5.</u> d   <u>6.</u> a   <u>7.</u> d   <u>8.</u> b

Chapter 9

<u>1.</u> b   <u>2.</u> a   <u>3.</u> a   <u>4.</u> d   <u>5.</u> c   <u>6.</u> b   <u>7.</u> b   <u>8.</u> b

APPENDIX B. USING LOGARITHMS

1. Let $y = \log_{10} 10$.   Then $10^y = 10$, $10^y = 10^1$, and $y = 1$.

2. Let $y = \log_{10} 100$.   Then $10^y = 100$, $10^y = 10^2$, and $y = 2$.

3. Let $y = \log_{10} 1$.   Then $10^y = 1$, $10^y = 10^0$, and $y = 0$.

4. Let $y = \log_{10} \frac{1}{10}$.   Then $10^y = \frac{1}{10}$, $10^y = 10^{-1}$, and $y = -1$.

5. Let $y = \log_{10} \frac{1}{100}$.   Then $10^y = \frac{1}{100}$, $10^y = 10^{-2}$, and $y = -2$.

6. Let $y = \log_{10} \sqrt{10}$.   Then $10^y = \sqrt{10}$, $10^y = 10^{\frac{1}{2}}$, and $y = \frac{1}{2}$.

7. $\log_{10}(235)(172) = \log_{10} 235 + \log_{10} 172$.

8. $\log_{10}(3.1)(0.172) = \log_{10} 3.1 + \log_{10} 0.172$.

9. $\log_{10}(28)^{18} = 18 \log_{10} 28$.

10. $\log_{10} \sqrt[3]{84} = \frac{1}{3} \log_{10} 84$.

11. $\log_{10} \frac{81.7}{23.9} = \log_{10} 81.7 - \log_{10} 23.9$

12. $\log_{10} \frac{3.07}{0.0123} = \log_{10} 3.07 - \log_{10} 0.0123$

13. $\log_{10} \frac{(21.2)(7.6)^2}{1.15} = \log_{10} 21.2 + 2 \log_{10} 7.6 - \log_{10} 1.15$

14. $\log_{10} \frac{(\sqrt{31})(\sqrt[3]{17})}{\sqrt[5]{8}} = \frac{1}{2} \log_{10} 31 + \frac{1}{3} \log_{10} 17 - \frac{1}{5} \log_{10} 8$

15. $\log_{10} \frac{[(21.3)^2(\sqrt[3]{5})]^{\frac{1}{2}}}{(6.1)(8.7)^2} = \log_{10} 21.3 + \frac{1}{6} \log_{10} 5 - \log_{10} 6.1 - 2 \log_{10} 8.7$

16. $\log_{10} \frac{(\sqrt{5.1})(\sqrt[4]{3})}{\sqrt{(\sqrt[3]{21})(18)^5}} = \frac{1}{2} \log_{10} 5.1 + \frac{1}{4} \log_{10} 3 - \frac{1}{6} \log_{10} 21 - \frac{5}{2} \log_{10} 18$

17. $83 = 8.3 \times 10$, $\log_{10} 83 = \log_{10}(8.3 \times 10) = \log_{10} 8.3 + \log_{10} 10 = \log_{10} 8.3 + 1$

18. $1741 = 1.741 \times 1000$, $\log_{10} 1741 = \log_{10}(1.741 \times 1000) = \log_{10} 1.741 + \log_{10} 1000$

$= \log_{10} 1.741 + 3$

19. $0.675 = 6.75 \times \frac{1}{10}$, $\log_{10} 0.675 = \log_{10}(6.75 \times \frac{1}{10}) = \log_{10} 6.75 + \log_{10} \frac{1}{10} =$

$\log_{10} 6.75 - 1$

20. $0.0021 = 2.1 \times \frac{1}{1000}$, $\log_{10} 0.0021 = \log_{10}(2.1 \times \frac{1}{1000}) = \log_{10} 2.1 + \log_{10} \frac{1}{1000} =$

$\log_{10} 2.1 - 3$

21. $823.1 = 8.231 \times 100$, $\log_{10} 823.1 = \log_{10}(8.231 \times 100) = \log_{10} 8.231 + \log_{10} 100 =$

$\log_{10} 8.231 + 2$

22. $671.5 = 6.715 \times 100$, $\log_{10} 671.5 = \log_{10}(6.715 \times 100) = \log_{10} 6.715 + \log_{10} 100 =$

$\log_{10} 6.715 + 2$

Exercises B-2 · Page 375

1. $\log_{10} 9.23 \doteq 0.9652$ 2. $\log_{10} 7.81 \doteq 0.8927$

3. $\log_{10} 60 = \log_{10} 6.0 + 1 \doteq 1.7782$ 4. $\log_{10} 40 = \log_{10} 4.0 + 1 \doteq 1.6021$

5. $\log_{10} 0.0140 = \log_{10} 1.40 - 2 \doteq 8.1461 - 10$

6. $\log_{10} 0.00140 = \log_{10} 1.40 - 3 \doteq 7.1461 - 10$

7. $\log_{10} 720{,}000 = \log_{10} 7.20 + 5 \doteq 5.8573$

8. $\log_{10} 32{,}500 = \log_{10} 3.25 + 4 \doteq 4.5119$

9. $\log_{10} 2.176 \doteq 0.3365 + 0.6(0.3385 - 0.3365) = 0.3365 + 0.6(0.002) = 0.3377$

10. $\log_{10} 3.855 \doteq 0.5855 + 0.5(0.5866 - 0.5855) = 0.5855 + 0.5(0.0011) \doteq 0.5860$

11. $\log_{10} 0.01542 = \log_{10} 1.542 - 2 \doteq 0.1875 + 0.2(0.1903 - 0.1875) - 2 =$

$0.1875 + 0.2(0.0028) - 2 = 8.1881 - 10$

12. $\log_{10} 0.01257 = \log_{10} 1.257 - 2 \doteq 0.0969 + 0.7(0.1004 - 0.0969) - 2 =$

$0.0969 + 0.7(0.0035) - 2 \doteq 0.0969 + 0.0024 - 2 = 8.0993 - 10$

13. $x = \text{antilog}_{10} 1.4048$, $\log_{10} x = 0.4048 + 1$, $x \doteq 2.54 \times 10 = 25.4$

14. $x = \text{antilog}_{10} 2.9727$, $\log_{10} x = 0.9727 + 2$, $x \doteq 9.39 \times 10^2 = 939$

15. $x = \text{antilog}_{10}(9.6875 - 10)$, $\log_{10} x = 0.6875 - 1$, $x \doteq 4.87 \times 10^{-1} = 0.487$

16. $x = \text{antilog}_{10}(8.7903 - 10)$, $\log_{10} x = 0.7903 - 2$, $x \doteq 6.17 \times 10^{-2} = 0.0617$

17. $x = \text{antilog}_{10}(1.8756 - 3)$, $\log_{10} x = 0.8756 - 2$, $x \doteq 7.51 \times 10^{-2} = 0.0751$

18. $x = \text{antilog}_{10}(4.6776 - 5)$, $\log_{10} x = 0.6776 - 1$, $x \doteq 4.76 \times 10^{-1} = 0.476$

19. $x = \text{antilog}_{10} 0.6028$, $\log_{10} x = 0.6028$, $x \doteq 4.000 +$

$0.01\left(\dfrac{0.6028 - 0.6021}{0.6031 - 0.6021}\right) = 4.000 + 0.01\left(\dfrac{7}{10}\right) = 4.007$

20. $x = \text{antilog}_{10} 1.6700$, $\log_{10} x = 0.6700 + 1$, $x \doteq$

$\left[4.670 + 0.01\left(\dfrac{0.6700 - 0.6693}{0.6702 - 0.6693}\right)\right] \times 10 = \left[4.670 + 0.01\left(\dfrac{7}{9}\right)\right] \times 10 \doteq$

$4.678 \times 10 = 46.78$

21. $x = \text{antilog}_{10}(9.9225 - 10)$, $\log_{10} x = 0.9225 - 1$, $x \doteq$

$[8.360 + 0.01(\dfrac{0.9225 - 0.9222}{0.9227 - 0.9222})] \times 10^{-1} = [8.360 + 0.01(\dfrac{3}{5})] \times 10^{-1} =$

$8.366 \times 10^{-1} = 0.8366$

22. $x = \text{antilog}_{10}(8.8135 - 10)$, $\log_{10} x = 0.8135 - 2$, $x \doteq$

$[6.500 + 0.01(\dfrac{0.8135 - 0.8129}{0.8136 - 0.8129})] \times 10^{-2} = [6.500 + 0.01(\dfrac{6}{7})] \times 10^{-2} \doteq$

$6.509 \times 10^{-2} = 0.06509$

23. $x = \text{antilog}_{10} 5.6017$, $\log_{10} x = 0.6017 + 5$, $x \doteq$

$[3.990 + 0.01(\dfrac{0.6017 - 0.6010}{0.6021 - 0.6010})] \times 10^{5} = [3.990 + 0.01(\dfrac{7}{11})] \times 10^{5} \doteq$

$3.996 \times 10^{5} = 399{,}600$

24. $x = \text{antilog}_{10} 5.7237$, $\log_{10} x = 0.7237 + 5$, $x \doteq$

$[5.290 + 0.01(\dfrac{0.7237 - 0.7235}{0.7243 - 0.7235})] \times 10^{5} = [5.290 + 0.01(\dfrac{2}{8})] \times 10^{5} \doteq$

$5.293 \times 10^{5} = 529{,}300$

Exercises B-3 · Pages 377-378

1. $\log_{10} 2.67 + \log_{10} 8.11 + 3 = \log_{10} N$

2. $\log_{10} 8.21 + \log_{10} 1.37 - 2 = \log_{10} N$

3. $\log_{10} 2.56 - \log_{10} 7.26 + 1 = \log_{10} N$

4. $\log_{10} 8.73 - \log_{10} 2.89 + 1 = \log_{10} N$

5. $2 \log_{10} 6.13 + \log_{10} 8.1 - \log_{10} 5.06 + 2 = \log_{10} N$

6. $\log_{10} 2.17 + 2 \log_{10} 4.13 - \log_{10} 7.09 + 1 = \log_{10} N$

7. $\dfrac{1}{5}(\log_{10} 7.13 + 3 - 5) = \log_{10} N$

8. $\log_{10} 4 - \dfrac{1}{3}(\log_{10} 3.17 + 2 - 3) = \log_{10} N$

9. $\dfrac{1}{3}(\log_{10} 4.37 + \log_{10} 8.14 - \log_{10} 3.68 + 3) = \log_{10} N$

10. $\dfrac{1}{3}(\log_{10} 9.310 - \log_{10} 1.08 - 3 \log_{10} 6.24) = \log_{10} N$

11. $\dfrac{1}{3}(\log_{10} 7 + \log_{10} 7.826 - \log_{10} 4 - \log_{10} 3.142 + 2) = \log_{10} N$

12. $\log_{10} 3.142 + \dfrac{1}{3}(2 \log_{10} 2.891 - \log_{10} 5.3 - \log_{10} 1.02 + 1) = \log_{10} N$

13. $\dfrac{1}{7}[\dfrac{1}{4}(2 \log 3.1)] = \log_{10} N$ 14. $\dfrac{1}{4}[\dfrac{1}{7}(2 \log 3.1)] = \log_{10} N$

15. $\log_{10} 1.7 + 1 - \log_{10} 7.1 + \dfrac{1}{2}[3(\log_{10} 2.801 + 1) - \log_{10} 5.39 - 3 - \log_{10} 2.03] =$

$\log_{10} N$

16. $5(\log_{10} 9.2 - 1) + \log_{10} 7.032 + 3 + \log_{10} 1.367 - \log_{10} 3.172 - 2 -$
$\frac{1}{2}(\log_{10} 6.84 - 2) = \log_{10} N$

17. $0.4265 + 0.9090 + 3 = 4.3355$, antilog$_{10}$ $4.3355 \doteq$
$[2.16 + 0.01(\frac{0.3355 - 0.3345}{0.3365 - 0.3345})] \times 10^4 = [2.16 + 0.01(\frac{1}{2})] \times 10^4 =$
$2.165 \times 10^4 = 21{,}650$

18. $0.9143 + 0.1367 - 2 = 0.0510 - 1$, antilog$_{10}(0.0510 - 1) \doteq$
$[1.12 + 0.01(\frac{0.0510 - 0.0492}{0.0531 - 0.0492})] \times 10^{-1} = [1.12 + 0.01(\frac{18}{39})] \times 10^{-1} \doteq$
$1.125 \times 10^{-1} = 0.1125$

19. $0.4082 - 0.8609 + 1 = 0.5473$, antilog$_{10}$ $0.5473 \doteq$
$3.52 + 0.01(\frac{0.5473 - 0.5465}{0.5478 - 0.5465}) = 3.52 + 0.01(\frac{8}{13}) \doteq 3.526$

20. $0.9410 - 0.4609 + 1 = 1.4801$, antilog$_{10}$ $1.4801 \doteq$
$[3.02 + 0.01(\frac{0.4801 - 0.4800}{0.4814 - 0.4800})] \times 10 = [3.02 + 0.01(\frac{1}{14})] \times 10 \doteq 30.21$

21. $2(0.7875) + 0.9085 - 0.7042 + 2 = 3.7793$, antilog$_{10}$ $3.7793 \doteq$
$[6.01 + 0.01(\frac{0.7793 - 0.7789}{0.7796 - 0.7789})] \times 10^3 = [6.01 + 0.01(\frac{4}{7})] \times 10^3 \doteq 6016$

22. $0.3365 + 2(0.6160) - 0.8506 + 1 = 1.7179$, antilog$_{10}$ $1.7179 \doteq$
$[5.22 + 0.01(\frac{0.7179 - 0.7177}{0.7185 - 0.7177})] \times 10 = [5.22 + 0.01(\frac{2}{8})] \times 10 \doteq 52.22$

23. $\frac{1}{5}(0.8531 + 3 - 5) \doteq 0.7706 - 1$, antilog$_{10}(0.7706 - 1) \doteq$
$[5.89 + 0.01(\frac{0.7706 - 0.7701}{0.7709 - 0.7701})] \times 10^{-1} = [5.89 + 0.01(\frac{5}{8})] \times 10^{-1} \doteq 0.5896$

24. $0.6021 - \frac{1}{3}(0.5011 + 2 - 3) = 0.7684$, antilog$_{10}$ $0.7864 \doteq$
$5.86 + 0.01(\frac{0.7684 - 0.7679}{0.7686 - 0.7679}) = 5.86 + 0.01(\frac{5}{7}) \doteq 5.867$

25. $\frac{1}{3}(0.6405 + 0.9106 - 0.5658 + 3) \doteq 1.3284$, antilog$_{10}$ $1.3284 \doteq 2.130 \times 10 = 21.30$

26. $\frac{1}{3}([0.9689 - 0.0334 - 3(0.7952) + 3] - 3) = 0.5166 - 1$,
antilog$_{10}(0.5166 - 1) \doteq [3.28 + 0.01(\frac{0.5166 - 0.5159}{0.5172 - 0.5159})] \times 10^{-1} =$
$[3.28 + 0.01(\frac{7}{13})] \times 10^{-1} \doteq 0.3285$

27. $\frac{1}{3}[0.8451 + [0.8932 + 0.6(0.8938 - 0.8932)] - 0.6021 -$
$[0.4969 + 0.2(0.4983 - 0.4969)] + 2\} \doteq \frac{1}{3}\{0.8451 + 0.8936 - 0.6021 - 0.4972 + 2\} =$
0.8798, antilog$_{10}$ $0.8798 \doteq 7.58 + 0.01(\frac{0.8798 - 0.8797}{0.8802 - 0.8797}) = 7.58 + 0.01(\frac{1}{5}) = 7.582$

28. $[0.4969 + 0.2(0.4983 - 0.4969)] + \frac{1}{3}\{2[0.4609 + 0.1(0.4624 - 0.4609)] -$
$0.7243 - 0.0086 + 1\} = [0.4969 + 0.2(0.0014)] + \frac{1}{3}\{2[0.4609 + 0.1(0.0015)] -$
$0.7243 - 0.0086 + 1\} \doteq [0.4969 + 0.0003] + \frac{1}{3}\{2[0.4609 + 0.0003] -$
$0.7243 - 0.0087 + 1\} = 0.4972 + \frac{1}{3}\{2[0.4612] - 0.7330 + 1\} = 0.4972 +$
$\frac{1}{3}\{1.1894\} \doteq 0.4972 + 0.3965 = 0.8937$, antilog$_{10}$ $0.8937 \doteq 7.82 +$
$0.01(\frac{0.8937 - 0.8932}{0.8938 - 0.8932}) = 7.82 + 0.01(\frac{5}{6}) \doteq 7.828$

29. $\frac{1}{7}\{\frac{1}{4}[2(0.4914)]\} = \frac{1}{14}(0.4914) = 0.0351$, antilog$_{10}$ $0.0351 \doteq 1.08 +$
$0.01(\frac{0.0351 - 0.0334}{0.0374 - 0.0334}) = 1.08 + 0.01(\frac{17}{40}) \doteq 1.084$

30. $\frac{1}{4}\{\frac{1}{7}[2(0.4914)]\} = \frac{1}{14}(0.4914) = 0.0351$, antilog$_{10}$ $0.0351 \doteq 1.084$ (see Exercise 29)

31. $0.2304 + 1 - 0.8513 + \frac{1}{2}\{3[0.4472 + 0.1(0.4487 - 0.4472) + 1] - 0.7316 -$
$3 - 0.3075\} = 0.3791 + \frac{1}{2}\{3[1.4472 + 0.1(0.0015)] - 4.0391\} \doteq 0.3791 +$
$\frac{1}{2}\{3[1.4474] - 4.0391\} = 0.3791 + \frac{1}{2}\{4.3422 - 4.0391\} = 0.3791 + \frac{1}{2}\{0.3031\} \doteq$
$0.3791 + 0.1515 = 0.5306$, antilog$_{10}$ $0.5306 = 3.39 + 0.01(\frac{0.5306 - 0.5302}{0.5315 - 0.5302}) =$
$3.39 + 0.01(\frac{4}{13}) \doteq 3.393$

32. $5(0.9638 - 1) + [0.8470 + 0.2(0.8476 - 0.8470)] + 3 +$
$[0.1335 + 0.7(0.1367 - 0.1335)] - [0.5011 + 0.2(0.5024 - 0.5011)] - 2 -$
$\frac{1}{2}(0.8351 - 2) \doteq 4.8190 - 5 + [0.8470 + 0.0001] + 3 + [0.1335 + 0.0022] -$
$[0.5011 + 0.0003] - 2 - 0.4176 + 1 = 1.8828$, antilog$_{10}$ $1.8828 \doteq$
$[7.63 + 0.01(\frac{0.8828 - 0.8825}{0.8831 - 0.8825})] \times 10 = [7.63 + 0.01(\frac{3}{6})] \times 10 = 76.35$

Exercises B-4 · Page 379

1. $\log_{10} \sin 39°40' \doteq 9.8050 - 10$

2. $\log_{10} \tan 67°20' \doteq 10.3792 - 10 = 0.3792$

3. $\log_{10} \cos 13°17' \doteq 9.9884 - 10 + 0.7(-0.0003) \doteq 9.9884 - 10 - 0.0002 =$
   $9.9882 - 10$

4. $\log_{10} \cot 39°42' \doteq 10.0813 - 10 + 0.2(-0.0025) = 10.0813 - 10 - 0.0005 =$
   $10.0808 - 10 = 0.0808$

<u>5.</u> $\log_{10} \sec 59°23' \doteq -\log_{10} \cos 59°23' \doteq -[9.7076 - 10 + 0.3(-0.0021)] \doteq 0.2930$

<u>6.</u> $\log_{10} \csc 28°48' \doteq -\log_{10} \sin 28°48' \doteq -[9.6810 - 10 + 0.8(0.0023)] \doteq 0.3172$

<u>7.</u> $\log_{10} \sin 24' \doteq 7.7648 - 10 + 0.4(0.1760) = 7.8352 - 10$

<u>8.</u> $\log_{10} \cot 89°7' \doteq 8.2419 - 10 + 0.7(-0.0792) \doteq 8.1865 - 10$

<u>9.</u> $\log_{10} \csc 75°18' = -\log_{10} \sin 75°18' \doteq -[9.9853 - 10 + 0.8(0.0003)] \doteq 0.0145$

<u>10.</u> $\log_{10} \sec 70°3' = -\log_{10} \cos 70°3' \doteq -[9.5341 - 10 + 0.3(-0.0035)] \doteq 0.4670$

<u>11.</u> $y \doteq 31°10'$ <u>12.</u> $y \doteq 48°10'$ <u>13.</u> $y \doteq 58°20'$ <u>14.</u> $y \doteq 54°30'$

<u>15.</u> $y \doteq 40°40' + 10'(\frac{10}{11}) \doteq 40°49'$ <u>16.</u> $y \doteq 68°30' + 10'(\frac{3}{5}) = 68°36'$

<u>17.</u> $y \doteq 8°20' + 10'(\frac{42}{87}) \doteq 8°25'$ <u>18.</u> $y \doteq 46°20' + 10'(\frac{18}{26}) \doteq 46°27'$

<u>19.</u> $\text{L} \cos y = -0.1100 = 9.8900 - 10, \; y \doteq 39°0' + 10'(\frac{5}{10}) = 39°5'$

<u>20.</u> $\text{L} \sin y = -0.2370 = 9.7630 - 10, \; y \doteq 35°20' + 10'(\frac{8}{18}) \doteq 35°24'$

<u>Exercises B-5 · Page 381</u>

<u>1.</u> $\frac{a}{82.1} = \tan 28°32', \; \log_{10} a = \log_{10} 82.1 + \log_{10} \tan 28°32' =$

$1.9143 + 9.7354 - 10 = 1.6497, \; a \doteq 44.6;$

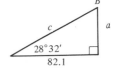

$\frac{c}{82.1} = \sec 28°32', \; \log_{10} c = \log_{10} 82.1 - \log_{10} \cos 28°32' \doteq$

$1.9143 - (9.9438 - 10) = 1.9705, \; c \doteq 93.4; \; B = 90 - 28°32' = 61°28'$

<u>2.</u> $\frac{3.02}{b} = \tan 10°12', \; \log_{10} b = \log_{10} 3.02 -$

$\log_{10} \tan 10°12' \doteq 0.4800 - [9.2551 - 10] = 1.2249,$

$b \doteq 16.78; \; \frac{3.02}{c} = \sin 10°12', \; \log_{10} c = \log_{10} 3.02 - \log_{10} \sin 10°12' \doteq$

$0.4800 - [9.2482 - 10] = 1.2318, \; c \doteq 17.05; \; B = 90° - 10°12' = 79°48'$

<u>3.</u> $\sin A = \frac{8.03}{17}, \; \log_{10} \sin A = \log_{10} 8.03 - \log 17 \doteq$

$10.9047 - 1.2304 - 10 = 9.6743 - 10, \; A \doteq 28°11';$

$B \doteq 90° - 28°11' = 61°49'; \; \frac{b}{17} = \cos A,$

$\log_{10} b = \log_{10} 17 + \log_{10} \cos 28°11' \doteq 1.2304 + 9.9452 - 10 = 1.1756, \; b \doteq 14.98$

<u>4.</u> $\sin A = \frac{7.202}{25.1}, \; \log_{10} \sin A = \log_{10} 7.202 - \log_{10} 25.1$

$\doteq 10.8574 - 1.3997 - 10 = 9.4577 - 10, \; A \doteq 16°40';$

$B \doteq 90° - 16°40' = 73°20'$; $\dfrac{b}{25.1} = \cos 16°40'$, $\log_{10} b = \log_{10} 25.1 +$

$\log_{10} \cos 16°40' \doteq 1.3997 + 9.9814 - 10 = 1.3811$, $b \doteq 24.05$

5.

$\tan A = \dfrac{5.03}{12.07}$, $\log_{10} \tan A = \log_{10} 5.03 - \log_{10} 12.07$

$\doteq 10.7016 - 1.0817 - 10 = 9.6199 - 10$, $A \doteq 22°37'$;

$B \doteq 90° - 22°37' = 67°23'$; $\dfrac{5.03}{c} = \sin A$, $\log_{10} c =$

$\log_{10} 5.03 - \log_{10} \sin A \doteq 0.7016 - (9.5850 - 10) = 1.1166$, $c \doteq 13.08$

6.

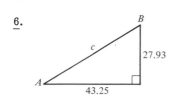

$\tan A = \dfrac{27.93}{43.25}$, $\log_{10} \tan A = \log_{10} 27.93 -$

$\log_{10} 43.25 \doteq 11.4461 - 1.6360 - 10 = 9.8101 - 10$,

$A \doteq 32°51$; $B \doteq 90° - 32°51' = 57°9'$; $\dfrac{27.93}{c} = \sin A$,

$\log_{10} c = \log_{10} 27.93 - \log_{10} \sin 32°51' \doteq$

$1.4461 - [9.7344 - 10] = 1.7117$, $c \doteq 51.49$

7.

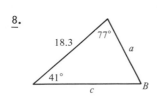

$A = 180° - 76° - 29° = 75°$; $\dfrac{5.02}{\sin 75°} = \dfrac{b}{\sin 29°}$,

$\log_{10} b = \log_{10} 5.02 + \log_{10} \sin 29° -$

$\log_{10} \sin 75° \doteq 0.7007 + (9.6856 - 10) -$

$(9.9849 - 10) = 0.4014$, $b \doteq 2.52$;

$\dfrac{5.02}{\sin 75°} = \dfrac{c}{\sin 76°}$, $\log_{10} c = \log_{10} 5.02 + \log_{10} \sin 76° - \log_{10} \sin 75° \doteq$

$0.7007 + (9.9869 - 10) - (9.9849 - 10) = 0.7027$, $c \doteq 5.043$

8.

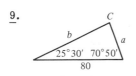

$B = 180° - 41° - 77° = 62°$, $\log_{10} a =$

$\log_{10} 18.3 + \log_{10} \sin 41° - \log_{10} \sin 62° \doteq$

$1.2625 + (9.8169 - 10) - (9.9459 - 10) = 1.1335$,

$a \doteq 13.60$, $\log_{10} c = \log_{10} 18.3 + \log_{10} \sin 77° -$

$\log_{10} \sin 62° \doteq 1.2625 + (9.9887 - 10) - (9.9459 - 10) = 1.3053$, $c \doteq 20.20$

9.

$C = 180° - 25°30' - 70°50' = 83°40'$; $\log_{10} a =$

$\log_{10} 80 + \log_{10} \sin 25°30' - \log_{10} 83°40' \doteq$

$1.9031 + (9.6340 - 10) - (9.9973 - 10) = 1.5398$,

$a \doteq 34.66$; $\log_{10} b = \log_{10} 80 + \log_{10} \sin 70°50' - \log_{10} \sin 83°40' = 1.9031 +$

$(9.9752 - 10) - (9.9973 - 10) = 1.8810$, $b \doteq 76.03$

10.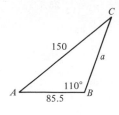

$\log_{10} \sin C = \log_{10} 85.5 + \log_{10} \sin 110° -$

$\log_{10} 150 \doteq 1.9320 + (9.9730 - 10) - 2.1761 =$

$9.7289 - 10,\ C \doteq 32°23';\ A \doteq 180° - 110° - 32°23' =$

$37°37';\ \log_{10} a = \log_{10} 150 + \log_{10} \sin 37°37' -$

$\log_{10} \sin 110° \doteq 2.1761 + (9.7856 - 10) - (9.9730 - 10) =$

$1.9887,\ a \doteq 97.43$

APPENDIX C. SPHERICAL TRIGONOMETRY

1. a. $34°$; b. 2040 naut. mi.   2. a. $43°$; 2580 naut. mi.

3. a. $63°$; b. 3780 naut. mi.   4. a. $66°$; 3960 naut. mi.

5. a. $15°$; b. 900 naut. mi.   6. a. $114°$; 6840 naut. mi.

1. $\cos c = \cos a \cos b = \cos 115°24' \cos 64°17' = -\cos 64°36' \cos 64°17' \doteq$

$-0.4289(0.4339) \doteq -0.1861$, $c \doteq 180° - 79°17' = 100°43'$; $\tan A =$

$\dfrac{\tan a}{\sin b} = \dfrac{\tan 115°24'}{\sin 64°17'} = -\dfrac{\tan 64°36'}{\sin 64°17'} \doteq -\dfrac{2.106}{0.9009} \doteq -2.338$, $A \doteq 180° - 66°51' =$

$113°9'$; $\tan B = \dfrac{\tan b}{\sin a} = \dfrac{\tan 64°17'}{\sin 115°24'} = \dfrac{\tan 64°17'}{\sin 64°36'} \doteq \dfrac{2.076}{0.9033} \doteq 2.298$, $B \doteq 66°29'$

2. $\cos a = \dfrac{\cos c}{\cos b} = \dfrac{\cos 92°10'}{\cos 97°56'} = \dfrac{\cos 87°50'}{\cos 82°4'} = \dfrac{0.0378}{0.1380} \doteq 0.2739$, $a \doteq 74°6'$;

$\sin B = \dfrac{\sin b}{\sin c} \doteq \dfrac{\sin 97°56'}{\sin 92°10'} = \dfrac{\sin 82°4'}{\sin 87°50'} \doteq \dfrac{0.9905}{0.9993} \doteq 0.9912$, we know from (5)

that $\tan B < 0$, $B \doteq 180° - 82°23' = 97°37'$; $\cos A = \dfrac{\tan b}{\tan c} = \dfrac{\tan 97°56'}{\tan 92°10'} =$

$\dfrac{\tan 82°4'}{\tan 87°50'} \doteq \dfrac{7.177}{26.43} \doteq 0.2715$, $A \doteq 74°15'$

3. $\tan b = \sin a \tan B = \sin 35°35' \tan 78°2' \doteq 0.5819(4.719) \doteq 2.746$, $b \doteq 70°$;

$\tan c = \dfrac{\tan a}{\cos B} = \dfrac{\tan 35°35'}{\cos 78°2'} \doteq \dfrac{0.7155}{0.2073} \doteq 3.452$, $c \doteq 73°51'$; $\cos A = \cos a \sin B$

$= \cos 35°35' \sin 78°2' \doteq 0.8133(0.9782) = 0.7956$, $A \doteq 37°17'$

4. $\tan a = \tan c \cos B = \tan 44°48' \cos 110°27' = -\tan 44°48' \cos 69°33' =$

$-0.9930(0.3494) \doteq -0.3470$, $a \doteq 180° - 19°8' = 160°52'$; $\sin b = \sin c \sin B =$

$\sin 44°48' \sin 110°27' = \sin 44°48' \sin 69°33' \doteq 0.7046(0.9370) \doteq 0.6602$,

from (1) we have $\cos b < 0$, $\therefore b \doteq 180° - 41°19' = 138°41'$; $\cot A =$

$\dfrac{\cos c}{\cot B} = \dfrac{\cos 44°48'}{\cot 110°27'} = -\dfrac{\cos 44°48'}{\cot 69°33'} \doteq -\dfrac{0.7096}{0.3729} \doteq -1.903$,

$A \doteq 180° - 27°43' = 152°17'$

5. $\sin c = \dfrac{\sin a}{\sin A} = \dfrac{\sin 37°40'}{\sin 37°40'} = 1$, $c = 90°$; $\sin b = \dfrac{\tan a}{\tan A} = 1$, $b = 90°$;

$\sin B = \dfrac{\cos A}{\cos a} = 1$, $B = 90°$

6. $\cos a = \dfrac{\cos A}{\sin B} = \dfrac{\cos 25°52'}{\sin 75°14'} \doteq \dfrac{0.8998}{0.9670} \doteq 0.9305$, $a \doteq 21°29'$; $\cos b =$

$\dfrac{\cos B}{\sin A} = \dfrac{\cos 75°14'}{\sin 25°52'} \doteq \dfrac{0.2549}{0.4363} \doteq 0.5842$, $b \doteq 54°15'$;

$\cos c = \cot A \cot B =$

$\cot 25°52' \cot 75°14' =$

7.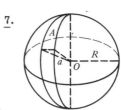

$2.063(0.2636) \doteq 0.5438$,

$c \doteq 57°3'$

8. From (2), we have $\sin a = \sin c \sin A = \sin 90° \sin 90° = 1$, $a = 90°$, and from (9), $\cos B = \cos b \sin A = \cos b \sin 90° = \cos b$, $B = b$, and b may take any value. Such a triangle could form one-half of any line.

9. $\sin c = \dfrac{\sin a}{\sin A} = \dfrac{\sin 37°18'}{\sin 73°22'} \doteq \dfrac{0.6060}{0.9582} \doteq 0.6324$, $c \doteq 39°14'$ or $140°46'$;

$\sin b = \dfrac{\tan a}{\tan A} = \dfrac{\tan 37°18'}{\tan 73°22'} \doteq \dfrac{0.7618}{3.347} \doteq 0.2276$, $b \doteq 13°9'$ or $166°51'$;

$\sin B = \dfrac{\cos A}{\cos a} = \dfrac{\cos 73°22'}{\cos 37°18'} \doteq \dfrac{0.2862}{0.7955} \doteq 0.3598$, $B \doteq 21°5'$ or $158°55'$;

when $c \doteq 39°14'$, we have from (1), by consideration of signs, that $b \doteq 13°9'$ and from (5) that $B \doteq 21°5'$.

10. $\sin c = \dfrac{\sin a}{\sin A} = \dfrac{\sin 150°}{\sin 120°} = \dfrac{\sin 30°}{\sin 60°} = \dfrac{\frac{1}{2}}{\frac{\sqrt{3}}{2}} = \dfrac{\sqrt{3}}{3} \doteq 0.5773$, $c \doteq 35°16'$ or $144°44'$;

$\sin b = \dfrac{\tan a}{\tan A} = \dfrac{\tan 150°}{\tan 120°} = \dfrac{-\frac{\sqrt{3}}{3}}{-\sqrt{3}} = \dfrac{1}{3} \doteq 0.3333$, $b \doteq 19°28'$ or $160°32'$;

$\sin B = \dfrac{\cos A}{\cos a} = \dfrac{\cos 120°}{\cos 150°} = \dfrac{-\frac{1}{2}}{-\frac{\sqrt{3}}{2}} = \dfrac{\sqrt{3}}{3} \doteq 0.5773$, $B \doteq 35°16'$ or $144°44'$;

when $c \doteq 35°16'$, we have from (1), by consideration of signs, that $b \doteq 160°32'$ and from (5) that $B \doteq 144°44'$.

11. $\sin c = \dfrac{\sin a}{\sin A} = \dfrac{\sin 10°27'}{\sin 50°50'} \doteq \dfrac{0.1814}{0.7753} \doteq 0.2340$, $c \doteq 13°32'$ or $166°28'$;

$\sin b = \dfrac{\tan a}{\tan A} = \dfrac{\tan 10°27'}{\tan 50°50'} \doteq \dfrac{0.1844}{1.228} \doteq 0.1502$, $b \doteq 8°38'$ or $171°22'$;

$\sin B = \dfrac{\cos A}{\cos a} = \dfrac{\cos 50°50'}{\cos 10°27'} \doteq \dfrac{0.6316}{0.9834} \doteq 0.6423$, $B \doteq 39°58'$ or $140°2'$;

when $c \doteq 13°32'$ we have from (1), by consideration of signs, that $b \doteq 8°38'$ and from (5) that $B \doteq 39°58'$.

<u>12.</u> $\sin A = \dfrac{\sin a}{\sin c} = \dfrac{\sin 90°}{\sin 90°} = \dfrac{1}{1} = 1$, $A = 90°$; $\cos B = \cos b \sin A = \cos b$,

$B = b$; any two values for b and B will suffice.

<u>13.</u> $\cos a \sin B = \cos a \dfrac{\sin b}{\sin c} = \dfrac{\cos c \sin b}{\cos b \sin c} = \dfrac{\tan b}{\tan c} = \cos A$ $(\cos b, \sin c \neq 0)$

<u>14.</u> $\cot A \cot B = \dfrac{\cos A \cos B}{\sin A \sin B} = \dfrac{\cos A \cos b \sin A}{\sin A \sin B} = \dfrac{\cos A \cos b}{\sin B} =$

$\dfrac{\cos a \sin B \cos b}{\sin B} = \cos a \cos b = \cos c$

Exercises C-3 · Page 387

<u>1.</u> $\sin (90° - A) = \tan (90° - c) \tan b$, $\cos A = \cot c \tan b$, $\tan b = \tan c \cos A$ (6),

$\sin (90° - A) = \cos (90° - B) \cos a$, $\cos A = \sin B \cos a$ (8)

<u>2.</u> $\sin (90° - c) = \tan (90° - A) \tan (90° - B)$, $\cos c = \cot A \cot B$ (10),

$\sin (90° - c) = \cos a \cos b$, $\cos c = \cos a \cos b$ (1)

<u>3.</u> $\sin (90° - B) = \tan (90° - c) \tan a$, $\cos B = \cot c \tan a$, $\tan a = \tan c \cos B$ (7),

$\sin (90° - B) = \cos (90° - A) \cos b$, $\cos B = \sin A \cos b$ (9)

Exercises C-4 · Page 388

<u>Note for Exs. 1-2:</u> Where the Law of Sines is employed, the principle that the angle opposite a greater side is greater, and conversely, must be used to determine in which quadrant an angle or side lies.

<u>1.</u> $\cos A = \dfrac{\cos a - \cos b \cos c}{\sin b \sin c} = \dfrac{\cos 80°40' - \cos 40°40' \cos 121°12'}{\sin 40°40' \sin 121°12'} =$

$\dfrac{\cos 80°40' + \cos 40°40' \cos 58°48'}{\sin 40°40' \sin 58°48'} \doteq \dfrac{0.1622 + 0.7585(0.5180)}{0.6517(0.8554)} \doteq \dfrac{0.5551}{0.5575} \doteq$

0.9957, $A \doteq 5°20'$; $\sin B = \dfrac{\sin b \sin A}{\sin a} \doteq \dfrac{\sin 40°40' \sin 5°20'}{\sin 80°40'} \doteq$

$\dfrac{0.6517(0.0929)}{0.9868} \doteq \dfrac{0.0605}{0.9868} \doteq 0.0613$, $B \doteq 3°31'$; $\sin C = \dfrac{\sin c \sin A}{\sin a} \doteq$

$\dfrac{\sin 121°12' \sin 5°20'}{\sin 80°40'} = \dfrac{\sin 58°48' \sin 5°20'}{\sin 80°40'} \doteq \dfrac{0.8554(0.0929)}{0.9868} \doteq \dfrac{0.0795}{0.9868} \doteq$

0.0806, $C \doteq 180° - 4°37' = 175°23'$

<u>2.</u> $\cos c = \cos b \cos a + \sin b \sin a \cos C = \cos 82°21' \cos 108°34' +$

$\sin 82°21' \sin 108°34' \cos 28°22' = -\cos 82°21' \cos 71°26' +$

$\sin 82°21' \sin 71°26' \cos 28°22' \doteq -0.1331(0.3184) + 0.9911(0.9479)(0.8799) \doteq$

$-0.0424 + 0.8266 \doteq 0.7842$, $c \doteq 38°21'$; $\sin A = \dfrac{\sin a \sin C}{\sin c} \doteq$

$\dfrac{\sin 71°26' \sin 28°22'}{\sin 38°21'} \doteq \dfrac{0.9479(0.4751)}{0.6204} \doteq \dfrac{0.4503}{0.6204} \doteq 0.7258$, $A \doteq 180° - 46°32' =$

$133°28'$; $\sin B = \dfrac{\sin b \sin C}{\sin c} \doteq \dfrac{\sin 82°21' \sin 28°22'}{\sin 38°21'} \doteq \dfrac{0.9911(0.4751)}{0.6204} \doteq \dfrac{0.4709}{0.6204} \doteq$

0.7590, $B \doteq 49°23'$

3. $\dfrac{\sin a}{\sin A} = \dfrac{\sin b}{\sin B}$, $\sin a \sin B = \sin A \sin b$, $\sin a \sin B - \sin A \sin b = 0 = \sin A \sin b -$

$\sin a \sin B$, $\sin a \sin A + \sin a \sin B - \sin A \sin b - \sin B \sin b = \sin a \sin A +$

$\sin A \sin b - \sin a \sin B - \sin B \sin b$, $\sin a(\sin A + \sin B) - \sin b(\sin A + \sin B) =$

$\sin a(\sin A - \sin B) + \sin b(\sin A - \sin B)$, $(\sin a - \sin b)(\sin A + \sin B) =$

$(\sin a + \sin b)(\sin A - \sin B)$, $\dfrac{\sin a - \sin b}{\sin a + \sin b} = \dfrac{\sin A - \sin B}{\sin A + \sin B}$, provided no

denominator $= 0$.

4. $\dfrac{\tan \frac{1}{2}(a - b)}{\tan \frac{1}{2}(a + b)} = \dfrac{\tan \frac{1}{2}(A - B)}{\tan \frac{1}{2}(A + B)}$, $\tan \frac{1}{2}(a - b) \tan \frac{1}{2}(A + B) =$

$\tan \frac{1}{2}(a + b) \tan \frac{1}{2}(A - B)$, $\left(\dfrac{\sin a - \sin b}{\cos a + \cos b}\right)\left(\dfrac{\sin A + \sin B}{\cos A + \cos B}\right) =$

$\left(\dfrac{\sin a + \sin b}{\cos a + \cos b}\right)\left(\dfrac{\sin A - \sin B}{\cos A + \cos B}\right)$, see Exs. 20, 21, section 5-3,

$(\sin a - \sin b)(\sin A + \sin B) = (\sin a + \sin b)(\sin A - \sin B)$,

$\dfrac{\sin a - \sin b}{\sin a + \sin b} = \dfrac{\sin A - \sin B}{\sin A + \sin B}$, provided no denominator $= 0$; a reversal of the

steps gives the proof.

Exercises C-5 · Page 389

1. $\cos a = \dfrac{\cos A + \cos B \cos C}{\sin B \sin C} = \dfrac{\cos 38°40' + \cos 72°50' \cos 124°32'}{\sin 72°50' \sin 124°32'} =$

$\dfrac{\cos 38°40' - \cos 72°50' \cos 55°28'}{\sin 72°50' \sin 55°28'} \doteq \dfrac{0.7808 - 0.2952(0.5669)}{0.9555(0.8238)} \doteq \dfrac{0.6135}{0.7871} \doteq$

0.7794, $a \doteq 38°48'$, $\cos b = \dfrac{\cos B + \cos A \cos C}{\sin A \sin C} =$

$\dfrac{\cos 72°50' - \cos 38°40' \cos 55°28'}{\sin 38°40' \sin 55°28'} \doteq \dfrac{0.2952 - 0.7808(0.5669)}{0.6248(0.8238)} \doteq -\dfrac{0.1474}{0.5147} \doteq$

-0.2864, $b \doteq 180° - 73°21' = 106°39'$,

$\cos c = \dfrac{\cos C + \cos B \cos A}{\sin B \sin A} = \dfrac{-\cos 55°28' + \cos 72°50' \cos 38°40'}{\sin 72°50' \sin 38°40'} \doteq$

$\dfrac{-0.5669 + 0.2952(0.7808)}{0.9555(0.6248)} \doteq -\dfrac{0.3364}{0.5970} \doteq -0.5635$, $c \doteq 180° - 55°42' = 124°18'$

2. $\dfrac{\sin a'}{\sin A'} = \dfrac{\sin b'}{\sin B'} = \dfrac{\sin c'}{\sin C'},\ \dfrac{\sin (180° - A)}{\sin (180° - a)} = \dfrac{\sin (180° - B)}{\sin (180° - b)} = \dfrac{\sin (180° - C)}{\sin (180° - c)},$

$\dfrac{\sin A}{\sin a} = \dfrac{\sin B}{\sin b} = \dfrac{\sin C}{\sin c}$; the Law of Sines might be said to be self-dual because it

has exactly the same form as its dual.

APPENDIX D. GRAPHS OF PURE WAVES

1. A = 3, B = 4, $c_1 = 0$, $c_2 = 0$,

  P = 3 cos 0 + 4 sin 0 = 3,

  Q = 3 sin 0 - 4 cos 0 = -4,

  $C = \pm\sqrt{3^2 + (-4)^2} = \pm 5$, (u, v) =

  $(\cos c_3,\ \sin c_3) = (\dfrac{3}{\pm 5},\ \dfrac{-4}{\pm 5})$;

  letting C = -5, $\cos c_3 = -\dfrac{3}{5}$,

  $c_3 \doteq \pi - 0.59\dfrac{\pi}{2} = 1.41\dfrac{\pi}{2} \doteq 2.21$,

  y = -5 cos (x + 2.21)

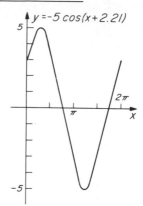

amplitude: 5, period: 2π

phase shift: 2.21

2. A = -4, B = 3, $c_1 = 0$, $c_2 = 0$, P = -4 cos 0 + 3 sin 0 = -4, Q = -4 sin 0 -

  3 cos 0 = -3, $C = \pm\sqrt{(-4)^2 + (-3)^2} = \pm 5$, (u, v) = $(\cos c_3,\ \sin c_3) = (\dfrac{-4}{\pm 5},\ \dfrac{-3}{\pm 5})$;

  letting C = -5, $\cos c_3 = \dfrac{4}{5}$, $c_3 \doteq 0.41\dfrac{\pi}{2} \doteq 0.64$, y = -5 cos (x + 0.64)

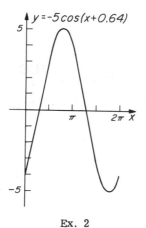

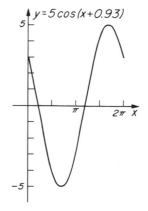

Ex. 2

amplitude: 5, period: 2π

phase shift: 0.64

Ex. 3

amplitude: 5, period: 2π

phase shift: 0.93

3. $A = 3$, $B = -4$, $c_1 = 0$, $c_2 = 0$, $P = 3 \cos 0 - 4 \sin 0 = 3$, $Q = 3 \sin 0 + 4 \cos 0 =$

4, $C = \pm\sqrt{3^2 + 4^2} = \pm 5$, $(u, v) = (\cos c_3, \sin c_3) = (\frac{3}{\pm 5}, \frac{4}{\pm 5})$; letting $C = 5$,

$\cos c_3 = \frac{3}{5}$, $c_3 \doteq 0.59 \frac{\pi}{2} \doteq 0.93$, $y = 5 \cos (x + 0.93)$

4. $A = 4$, $B = 3$, $c_1 = 0$, $c_2 = 0$, $P = 4 \cos 0 + 3 \sin 0 = 4$, $Q = 4 \sin 0 - 3 \cos 0 =$

-3, $C = \pm\sqrt{4^2 + (-3)^2} = \pm 5$, $(u, v) = (\cos c_3, \sin c_3) = (\frac{4}{\pm 5}, \frac{-3}{\pm 5})$; letting $C = -5$,

$\cos c_3 = -\frac{4}{5}$, $c_3 \doteq 3.14 - 0.41 (\frac{3.14}{2}) = 2.50$, $y = -5 \cos (x + 2.50)$

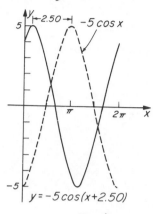

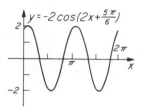

Ex. 4

amplitude: 5, period: 2π

phase shift: 2.50

Ex. 5

amplitude: 2, period: π

phase shift: $\frac{5\pi}{12}$

5. $A = \sqrt{3}$, $B = 1$, $c_1 = 0$, $c_2 = 0$, $P = \sqrt{3} \cos 0 + 1 \sin 0 = \sqrt{3}$, $Q = \sqrt{3} \sin 0 -$

$1 \cos 0 = -1$, $C = \pm\sqrt{(\sqrt{3})^2 + (-1)^2} = \pm 2$, $(u, v) = (\cos c_3, \sin c_3) = (\frac{\sqrt{3}}{\pm 2}, \frac{-1}{\pm 2})$;

letting $C = -2$, $\cos c_3 = -\frac{\sqrt{3}}{2}$, $c_3 = \frac{5\pi}{6}$, $y = -2 \cos (2x + \frac{2\pi}{3})$.

6. $A = 1$, $B = \sqrt{3}$, $c_1 = 0$, $c_2 = 0$, $P = 1 \cos 0 + \sqrt{3} \sin 0 = 1$, $Q = 1 \sin 0 -$

$\sqrt{3} \cos 0 = -\sqrt{3}$, $C = \pm\sqrt{1^2 + (-\sqrt{3})^2} = \pm 2$, $(u, v) = (\cos c_3, \sin c_3) = (\frac{1}{\pm 2}, \frac{-\sqrt{3}}{\pm 2})$;

letting $C = -2$, $\cos c_3 = -\frac{1}{2}$, $c_3 = \frac{2\pi}{3}$, $y = -2 \cos (2x + \frac{2\pi}{3})$

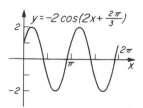

Ex. 6

amplitude: 2, period: π

phase shift: $\frac{\pi}{3}$

APPENDIX A. SETS, RELATIONS, AND FUNCTIONS

(Solutions to exercises in Instructor's Guide pages 46-49)

A-1

<u>1</u>. {1, 2, 3, 4, 5, 6}   <u>2</u>. {1, 4}   <u>3</u>. {1, 2, 3, 4, 5, 6, 7}   <u>4</u>. {4}

<u>5</u>. {1, 3, 4, 5, 7}   <u>6</u>. {1, 2, 3, 4, 5}   <u>7</u>. A-2   <u>8</u>. D   <u>9</u>. A-5

<u>10</u>. D   <u>11</u>. M-2   <u>12</u>. True   <u>13</u>. False   <u>14</u>. True   <u>15</u>. False

<u>16</u>. False   <u>17</u>. True   <u>18</u>. False   <u>19</u>. False   <u>20</u>. False   <u>21</u>. False

<u>22</u>. <u>a</u>. $\{10, 7^2, \frac{16}{2}\}$ <u>b</u>. $\{\sqrt[3]{-8}, 10, 7^2, \frac{16}{2}\}$

    <u>c</u>. $\{\sqrt[3]{-8}, 10, 3.14, 7^2, 2\frac{1}{4}, \frac{16}{2}\}$ <u>d</u>. $\{-\sqrt{3}\}$

<u>23</u>. <u>a</u>. $\{\sqrt{25}, \sqrt[3]{27}\}$ <u>b</u>. $\{-3^2, \sqrt{25}, \sqrt[3]{27}\}$

    <u>c</u>. $\{1.1, -3^2, \sqrt{25}, \sqrt[3]{27}, \frac{5}{3}\}$ <u>d</u>. $\{\sqrt{7}\}$

A-2

<u>1</u>. $12x = 12$; $x = 1$; $\{1\}$ <u>2</u>. $-\frac{1}{2}x = 2$; $x = -4$; $\{-4\}$

<u>3</u>. $3x + 9 = 7x - 7$; $-4x = -16$; $x = 4$; $\{4\}$

<u>4</u>. $11 - 2x - 8 = 28 - 14x$; $12x + 3 = 28$; $12x = 25$; $x = \frac{25}{12}$; $\{\frac{25}{12}\}$

<u>5</u>. $4(5x - 2) + 3(2x + 4) = 12(2x + 1)$; $20x - 8 + 6x + 12 = 24x + 12$; $26x - 24x = 8$;

    $2x = 8$; $x = 4$; $\{4\}$

<u>6</u>. $x^2 + 3x + 2 = x^2 + x + 2x + 2$; $x^2 + 3x + 2 = x^2 + 3x + 2$; R

<u>7</u>. $x^2 - x = 0$; $x(x - 1) = 0$; $x = 0$ or $x = 1$; $\{0, 1\}$

<u>8</u>. $x(x + 4) = 0$; $x = 0$ or $x + 4 = 0$; $x = 0$ or $x = -4$; $\{0, -4\}$

<u>9</u>. $10x^2 - 8x = 0$; $x(10x - 8) = 0$; $x = 0$ or $10x - 8 = 0$; $x = 0$ or $10x = 8$; $x = 0$

    or $x = \frac{4}{5}$; $\{0, \frac{4}{5}\}$

<u>10</u>. $x^2 - 2x - 15 = 0$; $(x - 5)(x + 3) = 0$; $x - 5 = 0$ or $x + 3 = 0$; $x = 5$ or $x = -3$;

    $\{5, -3\}$

<u>11</u>. $x^2 - 2x - 15 = 0$; $\{5, -3\}$ (see ex. 10)

<u>12</u>. $x^2 = 5$; $x = \pm\sqrt{5}$; $\{-\sqrt{5}, \sqrt{5}\}$

<u>13</u>. $x = \dfrac{-8 \pm \sqrt{64 - 4}}{2} = \dfrac{-8 \pm \sqrt{60}}{2} = -4 \pm \sqrt{15}$; $\{-4 + \sqrt{15}, -4 - \sqrt{15}\}$

14. $(x + 3) - (x - 3) = 6$; $6 = 6$; $\{x \in R: x \neq \pm 3\}$

15. $x - 2 = \pm 7$; $x = 7 + 2 = 9$ or $x = -7 + 2 = -5$; $\{9, -5\}$

16. $x - 5 = \pm\sqrt{10}$; $x = 5 + \sqrt{10}$; $x = 5 - \sqrt{10}$; $\{5 + \sqrt{10}, 5 - \sqrt{10}\}$

A-3

1. $-x \leq -3$; $x \geq 3$; $\{x: x \geq 3\}$

2. $3x - 6 \leq x + 8$; $2x \leq 14$; $x \leq 7$; $\{x: x \leq 7\}$

3. $6x - 18 < -2x - 10$; $8x < 8$; $x < 1$; $\{x: x < 1\}$

4. $2x^2 - 8x \geq 2x^2 - 5x + 1$; $-8x \geq -5x + 1$; $-3x \geq 1$; $x \leq -\frac{1}{3}$; $\{x: x \leq -\frac{1}{3}\}$

5. $x + 5 > 14$; $x > 9$; $\{x: x > 9\}$

6.

7.

8.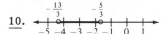

9.

10.

11.

12.

13. R (by def. of abs. value)

14. R^+ (by def. of abs. value)

15. R^+ **16.** R^- **17.** R

18. $\emptyset$ (the square of any real number is non-negative)   **19.** R **20.** R

A-4

1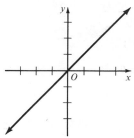

Function

2.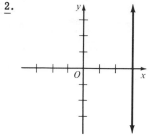

Not a function

3.

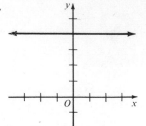

Function

4.

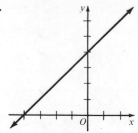

Function

5.

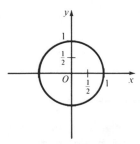

Not a function

6.

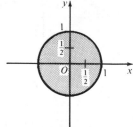

Not a function

7.

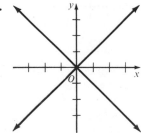

Not a function

8.

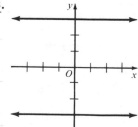

Not a function

9. {1}   10. R 11. {y: y ≥ 0}   12. R 13. {y: y ≥ 0}

14. {y: y ≥ 0}   15. {y: y ≠ 0}   16. R

17.

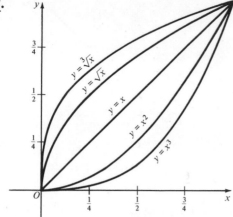

18. $\{y: -6 \le y \le 6\}$

19. $\{y: 0 \le y \le 36\}$

20. $\{y: -8 \le y \le 8\}$

21. $\{y: 2 \le y \le 4\}$

22. $\{y: 0 \le y \le 2\}$

23. $\{y: 0 < y \le \frac{1}{2}\}$

24. $\{y: -1 \le y \le 4\}$

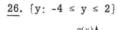

25. $\{y: 0 \le y \le 4\}$

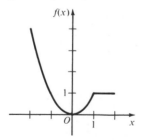

26. $\{y: -4 \le y \le 2\}$

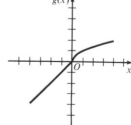

27. $\{y: -8 \leq y \leq 9\}$

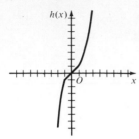

<u>A-5</u>

1. $D_f = \{1, 2, 3, 4\}$ $\qquad\qquad$ $R_f = \{1, 2, 3, 4\}$

$\quad$ $D_g = \{1, 2, 3, 4\}$ $\qquad\qquad$ $R_g = \{0, 2, 4, 6\}$

$\quad$ $f + g = \{(1, 4), (2, 5), (3, 6), (4, 7)\}$

$\quad$ $D_{f+g} = \{1, 2, 3, 4\}$ $\qquad\qquad$ $R_{f+g} = \{4, 5, 6, 7\}$

$\quad$ $f \cdot g = \{(1, 0), (2, 6), (3, 8), (4, 6)\}$

$\quad$ $D_{f \cdot g} = \{1, 2, 3, 4\}$ $\qquad\qquad$ $R_{f \cdot g} = \{0, 6, 8\}$

$\quad$ $\dfrac{f}{g} = \{(2, \frac{3}{2}), (3, \frac{1}{2}), (4, \frac{1}{6})\}$

$\quad$ $D_{\frac{f}{g}} = \{2, 3, 4\}$ $\qquad\qquad$ $R_{\frac{f}{g}} = \{\frac{3}{2}, \frac{1}{2}, \frac{1}{6}\}$

2. $D_f = \{-2, -1, 0, 1, 2\}$ $\qquad\qquad$ $R_f = \{0, 2, 4\}$

$\quad$ $D_g = \{-2, -1, 0, 1, 2\}$ $\qquad\qquad$ $R_g = \{0, 1, 2\}$

$\quad$ $f + g = \{(-2, 6), (-1, 3), (0, 0), (1, 3), (2, 6)\}$

$\quad$ $D_{f+g} = \{-2, -1, 0, 1, 2\}$ $\qquad\qquad$ $R_{f+g} = \{0, 3, 6\}$

$\quad$ $f \cdot g = \{(-2, 8), (-1, 2), (0, 0), (1, 2), (2, 8)\}$

$\quad$ $D_{f \cdot g} = \{-2, -1, 0, 1, 2\}$ $\qquad\qquad$ $R_{f \cdot g} = \{0, 2, 8\}$

$\quad$ $\dfrac{f}{g} = \{(-2, 2), (-1, 2), (1, 2), (2, 2)\}$

$\quad$ $D_{\frac{f}{g}} = \{-2, -1, 1, 2\}$ $\qquad\qquad$ $R_{\frac{f}{g}} = \{2\}$

3. $(f + g)(x) = f(x) + g(x) = 6x + 1 + 3x = 9x + 1$

$\quad$ $(f \cdot g)(x) = f(x) \cdot g(x) = (6x + 1)(3x) = 18x^2 + 3x$

$\quad$ $\dfrac{f}{g}(x) = \dfrac{f(x)}{g(x)} = \dfrac{6x + 1}{3x} = 2 + \dfrac{1}{3x}$

$\quad$ $f \circ g(x) = f(g(x)) = f(3x) = 6(3x) + 1 = 18x + 1$

$\underline{4}$. $(f + g)(x) = f(x) + g(x) = x^2 - 1 + (x - 1)^2 = 2x^2 - 2x$

$(f \cdot g)(x) = f(x) \cdot g(x) = (x^2 - 1)(x - 1)^2 = (x - 1)^3 (x + 1) = x^4 - 2x^3 + 2x - 1$

$\dfrac{f}{g}(x) = \dfrac{f(x)}{g(x)} = \dfrac{x^2 - 1}{(x - 1)^2} = \dfrac{x + 1}{x - 1}$

$(f \circ g)(x) = f(g(x)) = f((x - 1)^2) = (x - 1)^4 - 1 = x^4 - 4x^3 + 6x^2 - 4x$

$\underline{5}$. $(f + g)(x) = f(x) + g(x) = \dfrac{x + 4}{x - 2} + (x - 2) = \dfrac{x^2 - 3x + 8}{x - 2}$

$(f \cdot g)(x) = f(x) \cdot g(x) = \dfrac{x + 4}{x - 2} \cdot (x - 2) = x + 4 \ (x \neq 2)$

$\dfrac{f}{g}(x) = \dfrac{f(x)}{g(x)} = \dfrac{\dfrac{x + 4}{x - 2}}{x - 2} = \dfrac{x + 4}{(x - 2)^2}$

$f \circ g(x) = f(g(x)) = f(x - 2) = \dfrac{(x - 2) + 4}{(x - 2) - 2} = \dfrac{x + 2}{x - 4}$

$\underline{6}$. $R_f = \{y: -8 \le y \le 8\}$

$R_g = \{y: -7 \le y \le 9\}$

$R_h = \{y: -16 \le y \le 16\}$

$\underline{7}$. $R_f = \{y: 0 \le y \le 3\}$

$R_g = \{y: -4 \le y \le -1\}$

$R_h = \{y: 0 \le y \le 9\}$

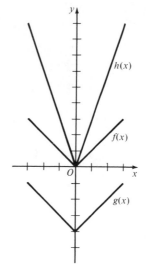

<u>8.</u> $R_f = \{y: \frac{1}{4} \le y \le 1\}$

$R_g = \{y: 2\frac{1}{4} \le y \le 3\}$

$R_h = \{y: -6 \le y \le -4\frac{1}{2}\}$

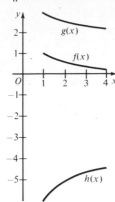

<u>9.</u> $R_f = \{y: \frac{1}{4} \le y \le 1\}$

$R_g = \{y: -1 \le y \le -\frac{1}{4}\}$

$R_h = \{y: 0 \le y \le \frac{3}{4}\}$

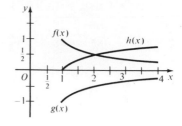

<u>10.</u> $D_{f \circ g} = \{x: 5 - x \ge 0\} = \{x: x \le 5\}$

$f \circ g(x) = f(5 - x) = \sqrt{5 - x}$

$D_{g \circ f} = \{x: x \ge 0\}; \; g \circ f = (\sqrt{x}) = 5 - \sqrt{x}$

<u>11.</u> $D_{f \circ g} = R \qquad f \circ g(x) = f(x^2) = -x^2$

$D_{g \circ f} = R \qquad g \circ f(x) = g(-x) = (-x)^2 = x^2$

<u>12.</u> $g(x) + 1 = 4x - 2; \; g(x) = 4x - 3$